Innovative Nanomaterial Properties and Applications in Chemistry, Physics, Medicine, or Environment

Innovative Nanomaterial Properties and Applications in Chemistry, Physics, Medicine, or Environment

Editor

Thomas Dippong

Basel • Beijing • Wuhan • Barcelona • Belgrade • Novi Sad • Cluj • Manchester

Editor
Thomas Dippong
Chemistry and Biology Department
Technical University of Cluj Napoca
Baia Mare
Romania

Editorial Office
MDPI
St. Alban-Anlage 66
4052 Basel, Switzerland

This is a reprint of articles from the Special Issue published online in the open access journal *Nanomaterials* (ISSN 2079-4991) (available at: www.mdpi.com/journal/nanomaterials/special_issues/D86Q2K7624).

For citation purposes, cite each article independently as indicated on the article page online and as indicated below:

Lastname, A.A.; Lastname, B.B. Article Title. *Journal Name* **Year**, *Volume Number*, Page Range.

ISBN 978-3-7258-0134-3 (Hbk)
ISBN 978-3-7258-0133-6 (PDF)
doi.org/10.3390/books978-3-7258-0133-6

© 2024 by the authors. Articles in this book are Open Access and distributed under the Creative Commons Attribution (CC BY) license. The book as a whole is distributed by MDPI under the terms and conditions of the Creative Commons Attribution-NonCommercial-NoDerivs (CC BY-NC-ND) license.

Contents

About the Editor vii

Thomas Dippong, Erika Andrea Levei, Iosif Grigore Deac, Mihaela Diana Lazar and Oana Cadar
Influence of SiO_2 Embedding on the Structure, Morphology, Thermal, and Magnetic Properties of $Co_{0.4}Zn_{0.4}Ni_{0.2}Fe_2O_4$ Particles
Reprinted from: *Nanomaterials* **2023**, *13*, 527, doi:10.3390/nano13030527 1

Thomas Dippong, Erika Andrea Levei, Ioan Petean, Iosif Grigore Deac and Oana Cadar
A Strategy for Tuning the Structure, Morphology, and Magnetic Properties of $MnFe_2O_4/SiO_2$ Ceramic Nanocomposites via Mono-, Di-, and Trivalent Metal Ion Doping and Annealing
Reprinted from: *Nanomaterials* **2023**, *13*, 2129, doi:10.3390/nano13142129 18

Thomas Dippong, Dana Toloman, Mihaela Diana Lazar and Ioan Petean
Effects of Lanthanum Substitution and Annealing on Structural, Morphologic, and Photocatalytic Properties of Nickel Ferrite
Reprinted from: *Nanomaterials* **2023**, *13*, 3096, doi:10.3390/nano13243096 37

Roman Atanasov, Dorin Ailenei, Rares Bortnic, Razvan Hirian, Gabriela Souca and Adam Szatmari et al.
Magnetic Properties and Magnetocaloric Effect of Polycrystalline and Nano-Manganites $Pr_{0.65}Sr_{(0.35-x)}Ca_xMnO_3$ ($x \leq 0.3$)
Reprinted from: *Nanomaterials* **2023**, *13*, 1373, doi:10.3390/nano13081373 54

Xiaoqian Ai, Shun Yan, Chao Lin, Kehong Lu, Yujie Chen and Ligang Ma
Facile Fabrication of Highly Active CeO_2@ZnO Nanoheterojunction Photocatalysts
Reprinted from: *Nanomaterials* **2023**, *13*, 1371, doi:10.3390/nano13081371 72

Ximena Jaramillo-Fierro and Ricardo León
Effect of Doping TiO_2 NPs with Lanthanides (La, Ce and Eu) on the Adsorption and Photodegradation of Cyanide—A Comparative Study
Reprinted from: *Nanomaterials* **2023**, *13*, 1068, doi:10.3390/nano13061068 86

Elena Segura-Sanchis, Rocío García-Aboal, Roberto Fenollosa, Fernando Ramiro-Manzano and Pedro Atienzar
Scanning Photocurrent Microscopy in Single Crystal Multidimensional Hybrid Lead Bromide Perovskites
Reprinted from: *Nanomaterials* **2023**, *13*, 2570, doi:10.3390/nano13182570 119

Chulsoo Kim, Byungyou Hong and Wonseok Choi
Surface-Enhanced Raman Spectroscopy (SERS) Investigation of a 3D Plasmonic Architecture Utilizing Ag Nanoparticles- Embedded Functionalized Carbon Nanowall
Reprinted from: *Nanomaterials* **2023**, *13*, 2617, doi:10.3390/nano13192617 132

Aisha Hamidu, William G. Pitt and Ghaleb A. Husseini
Recent Breakthroughs in Using Quantum Dots for Cancer Imaging and Drug Delivery Purposes
Reprinted from: *Nanomaterials* **2023**, *13*, 2566, doi:10.3390/nano13182566 142

Melanie Fritz, Susanne Körsten, Xiaochen Chen, Guifang Yang, Yuancai Lv and Minghua Liu et al.
Time-Dependent Size and Shape Evolution of Gold and Europium Nanoparticles from a Bioproducing Microorganism, a Cyanobacterium: A Digitally Supported High-Resolution Image Analysis
Reprinted from: *Nanomaterials* **2022**, *13*, 130, doi:10.3390/nano13010130 177

Xing Zhou and Yuliang Mao
The Adsorption Effect of Methane Gas Molecules on Monolayer PbSe with and without Vacancy Defects: A First-Principles Study
Reprinted from: *Nanomaterials* **2023**, *13*, 1566, doi:10.3390/nano13091566 192

About the Editor

Thomas Dippong

Thomas Dippong (an associate professor at the Technical University of Cluj-Napoca) is a chemical engineer with a Ph.D. and habilitation in chemistry. His current research activities are related to the characterization of nanoparticles for various applications as part of an ongoing research project in partnership with the Technical University of Cluj-Napoca within the field of ferrite embedded in silica matrix. He is an expert in analytical chemistry, organic and inorganic chemistry, thermal treatment, instrumental analysis, and the synthesis of nanomaterials. Dr. Dippong has published 146 peer-reviewed publications (90 papers in high-ranked scientific ISI-Thomson journals (51 Q1, 15 Q2, and 24 Q3) and 56 in other national and international journals), with 1950 citations and an h-index of 37 (WoS). He has given 35 lectures at international conferences (the 14th ICTAC in Brazil, the 5th CEEC-TAC in Roma, the JTACC in Budapest, the ESTAC in Brasov, etc.). He has also published two books with international publishing houses and fifteen books with national publishing houses. In the last three consecutive years (2021–2023), Dr. Thomas Dippong was included in the prestigious list of the world's top 2% researchers. He has been a contract manager for many projects and is currently an active member of four projects. Dr. Dippong has reviewed 470 scientific articles for 95 ISI-Thomson journals. He has been a guest editor for six Special Issues published by four prestigious Q1 ISI-Thomson journals.

Article

Influence of SiO$_2$ Embedding on the Structure, Morphology, Thermal, and Magnetic Properties of Co$_{0.4}$Zn$_{0.4}$Ni$_{0.2}$Fe$_2$O$_4$ Particles

Thomas Dippong [1], Erika Andrea Levei [2], Iosif Grigore Deac [3], Mihaela Diana Lazar [4] and Oana Cadar [2,*]

1. Faculty of Science, Technical University of Cluj-Napoca, 76 Victoriei Street, 430122 Baia Mare, Romania
2. INCDO-INOE 2000, Research Institute for Analytical Instrumentation, 67 Donath Street, 400293 Cluj-Napoca, Romania
3. Faculty of Physics, Babes-Bolyai University, 1 Kogalniceanu Street, 400084 Cluj-Napoca, Romania
4. National Institute for Research and Development of Isotopic and Molecular Technologies, 67-103 Donath Street, 400293 Cluj-Napoca, Romania
* Correspondence: oana.cadar@icia.ro

Citation: Dippong, T.; Levei, E.A.; Deac, I.G.; Lazar, M.D.; Cadar, O. Influence of SiO$_2$ Embedding on the Structure, Morphology, Thermal, and Magnetic Properties of Co$_{0.4}$Zn$_{0.4}$Ni$_{0.2}$Fe$_2$O$_4$ Particles. *Nanomaterials* **2023**, *13*, 527. https://doi.org/10.3390/nano13030527

Academic Editor: Alessandro Lascialfari

Received: 20 December 2022
Revised: 23 January 2023
Accepted: 26 January 2023
Published: 28 January 2023

Copyright: © 2023 by the authors. Licensee MDPI, Basel, Switzerland. This article is an open access article distributed under the terms and conditions of the Creative Commons Attribution (CC BY) license (https://creativecommons.org/licenses/by/4.0/).

Abstract: (Co$_{0.4}$Zn$_{0.4}$Ni$_{0.2}$Fe$_2$O$_4$)$_\alpha$(SiO$_2$)$_{(100-\alpha)}$ samples obtained by embedding Co$_{0.4}$Zn$_{0.4}$Ni$_{0.2}$Fe$_2$O$_4$ nanoparticles in SiO$_2$ in various proportions were synthesized by sol-gel process and characterized using thermal analysis, Fourier-transform infrared spectroscopy, X-ray diffraction, transmission electron microscopy, inductively coupled plasma optical emission spectrometry, and magnetic measurements. Poorly crystalline Co–Zn–Ni ferrite at low annealing temperatures (500 °C) and highly crystalline Co–Zn–Ni ferrite together with traces of crystalline Fe$_2$SiO$_4$ (800 °C) and SiO$_2$ (tridymite and cristobalite) (1200 °C) were obtained. At 1200 °C, large spherical particles with size increasing with the ferrite content (36–120 nm) were obtained. Specific surface area increased with the SiO$_2$ content and decreased with the annealing temperature above 500 °C. Magnetic properties were enhanced with the increase in ferrite content and annealing temperature.

Keywords: Co$_{0.4}$Zn$_{0.4}$Ni$_{0.2}$Fe$_2$O$_4$; silica matrix; crystalline phase; annealing temperature; magnetic behavior

1. Introduction

Spinel ferrite nanoparticles are widely studied due to their outstanding electrical and magnetic properties, high thermal and chemical stability, and applicability in different areas such as electronic, microwave, and communication devices, information storage systems, ferrofluid technology, solid oxide fuel cell, gas sensors, magnetocaloric refrigeration, and medical diagnosis [1–4]. The physicochemical properties of ferrites are determined by the preparation method, heat treatment, and chemical composition, as well as the type, stoichiometric ratio, and distribution of cations [5–9]. The particle morphology and surface coating may also influence the magnetic behavior of ferrites having the same compositions [10].

The synthesis route is a key factor in high-purity spinel ferrite nanoparticles' preparation [5,9,11]. The most common ways to synthesize nanostructured ferrites are sol-gel, solid-phase, hydrothermal, coprecipitation, auto combustion, sonochemical, microwave refluxing, etc. [7,10,11]. The solid-state synthesis produces nanoparticles with high yields and well-controllable grain size [3]. At the same time, the conventional ceramic method produces particles in the micrometer range that tend to agglomerate due to slow reaction kinetics [11]. Generally, wet chemical synthesis methods such as hydrothermal, sol-gel, and auto combustion are used to produce high-purity crystalline ferrite nanoparticles at low annealing temperatures [2,3,11]. The sol-gel process allows the easy, low-cost production of ferrite nanocomposites with controlled structure and properties [12]. Moreover, the sol-gel process may produce nanocomposite materials comprising highly dispersed magnetic ferrite nanoparticles [13–16]. This method consists of incorporating metal nitrates in tetraethyl orthosilicate (TEOS), polycondensation of the SiO_2 network, thermal-assisted formation of metal-carboxylate precursors in the reaction between the metal ions and diol, and thermal decomposition of carboxylate precursors into a hybrid oxidic system [14–16].

In the soft ferrites family, $ZnFe_2O_4$ is a spinel, with tetrahedral (A) sites occupied by Zn^{2+} ions and octahedral (B) sites by Fe^{3+} ions; $NiFe_2O_4$ is an inverse spinel, with Ni^{2+} ions occupying the octahedral (B) sites and Fe^{3+} ions equally distributed between octahedral (B) and tetrahedral (A) sites; whereas $CoFe_2O_4$ is a mixed spinel with high inversion degree [6]. The high electrical resistivity of $NiFe_2O_4$ is due to the lack of electron hopping, as Ni exists only in the divalent form [6]. By partial substitution of Ni^{2+} with Zn^{2+} in $NiFe_2O_4$, the Zn^{2+} ions will occupy the tetrahedral (A) sites, forcing the Fe^{3+} ions to occupy both octahedral (B) and tetrahedral (A) sites. This arrangement of the cations will increase the saturation magnetization (M_S) of Ni–Zn ferrite, compared to that of $NiFe_2O_4$ and $ZnFe_2O_4$ [6,10,17]. Hence, by varying the substitution degree of the divalent cation, the magnetic behavior of Ni-Zn ferrites can be enhanced, making them fit for a broad range of applications [2,6,17,18]. Besides the high M_S, Ni–Zn ferrites also present large electrical resistivity, narrow dielectric loss, low coercivity, good mechanical hardness, high magnetic permeability, and high operating frequency, which make them potential candidates for transformer cores, microwave devices, noise filters, recording heads, magnetic fluids, chokes, coils, etc., [2–4,7].

Ni–Zn ferrite has an inverse spinel structure with Fe^{3+} ions occupying both tetrahedral (A) and octahedral (B) sites, Ni^{2+} ions preferably located in octahedral (B) sites, and Zn^{2+} ions in tetrahedral (A) sites [3]. Adding Co^{2+} to Ni–Zn ferrite induces magnetic anisotropy and reduces the permeability due to the preferential orientation of the Co^{2+} ions' magnetic moment along a particular direction [6,7]. The low dielectric loss, low magnetic loss, high saturation magnetizations, and high resistivity of mixed Co–Zn–Ni ferrites make them widely used as capacitors, filters, magnetic antennas, and absorbing materials [5,19,20].

The SiO_2 embedding is used to control the particle size, reduce the particle agglomeration, and enhance the ferrites' magnetic properties and biocompatibility, as SiO_2 is biologically inert and diminishes the inflammatory risk [12,14]. One of the most used network-forming agents in sol-gel synthesis is TEOS, as it has a short gelation time, produces strong networks with moderate reactivity, and allows the embedding of both organic and inorganic molecules [14–16]. Previous studies demonstrated that transitional metal ferrites embedded in the SiO_2 matrix display high magnetocrystalline anisotropy, unique magnetic structure, and high correlation between the coercivity, crystallite sizes, and annealing temperature [14–16]. Also, the partial substitution of Zn^{2+} ions by Co^{2+} ions in Zn–Ni ferrite was expected to enhance the magnetic properties of the nanoparticles.

This study investigates the relationship between the $Co_{0.4}Zn_{0.4}Ni_{0.2}Fe_2O_4$ content embedded in the SiO_2 matrix and the crystallite size, specific surface area, porosity particle size, thermal behavior, and magnetic properties (saturation magnetization–M_S, remanent magnetization–M_R, coercive field–H_c, magnetic anisotropy–K), and the $Co_{0.4}Zn_{0.4}Ni_{0.2}Fe_2O_4$ content in the SiO_2 matrix, at different annealing temperatures.

2. Materials and Methods

All chemical reagents were of analytical grade and were purchased from Merck (Darmstadt, Germany). $(Co_{0.4}Zn_{0.4}Ni_{0.2}Fe_2O_4)_\alpha(SiO_2)_{(100-\alpha)}$ samples were produced by sol-gel process dissolving $Co(NO_3)_2 \cdot 6H_2O$, $Zn(NO_3)_2 \cdot 6H_2O$, $Ni(NO_3)_2 \cdot 6H_2O$, and $Fe(NO_3)_3 \cdot 9H_2O$ in 1,3-propanediol (1,3PD), in a molar ratio of 0.4:0.4:0.2:2:8. Afterwards, TEOS dissolved in ethanol was added to the nitrate-1,3PD mixture, using 0:2 (α = 0%), 0.5:1.5 (α = 25%), 1:1 (α = 50%), 1.5:0.5 (α = 75%), and 2:0 (α = 100%) NO_3:TEOS molar ratio. Diluted nitric acid was slowly added till the reaction mixture reached pH = 2 and then, the mixture was thoroughly stirred for 1 h. The obtained samples were dried at 40 °C for 5 h and 300 °C for 5 h, powdered in an agate mortar and annealed for 5 h at 500, 800, and 1200 °C, respectively, using an LT9 muffle furnace (Nabertherm, Lilienthal, Germany).

The thermal behavior of samples was studied by thermogravimetry (TG) and differential thermal analysis (DTA) by using a Q600 SDT (TA Instruments, New Castle, DE, USA) thermal analyzer, in air, up to 1200 °C, with a 10 °C/min⁻ heating rate using an SDT Q600 thermogravimeter and alumina standards. A D8 Advance (Bruker, Karlsruhe, Germany) diffractometer equipped with a LynxEye linear detector was used for the investigation of crystalline phases, using the CuKα radiation (λ = 1.54060 Å) in the 2θ range 10–80°. The FT-IR spectra of the samples were recorded using a Spectrum BX II (Perkin Elmer, Waltham, MA, USA) Fourier-transform infrared (FT-IR) spectrometer, while the composition of Ni-Zn-Co ferrites was confirmed by Perkin Elmer Optima 5300 DV (Norwalk, CT, USA) inductively coupled plasma optical emission spectrometry (ICP-OES) after aqua regia digestion using a Speedwave Xpert (Berghof, Germany) microwave digestion system. N_2 adsorption-desorption isotherms were recorded at −196 °C by a Sorptomatic 1990 (Thermo Fisher Scientific, Waltham, MA, USA) instrument, which was used for calculation of the specific surface area (SSA) using the Brunauer–Emmett–Teller (BET) model. The particle morphology was investigated using a transmission electron microscope (TEM, HD-2700, Hitachi, Tokyo, Japan) and a digital image recording system on samples deposited on carbon-coated copper grids. The average particle size was estimated from TEM measurements using the UTHSCSA ImageTool image software for over 100 nanoparticles in each sample. The hysteresis loops were recorded in magnetic fields between −2 to 2 T, at room temperature, and magnetization versus applied field was measured on samples embedded in an epoxy matrix by a 7400 vibrating-sample magnetometer (VSM, Lake Shore Cryotronics, Westerville, OH, USA). The magnetic measurement uncertainty was 10%.

3. Results and Discussion

The TG (Figure 1a) and DTA curves (Figure 1b indicates the maximum of the exothermic and endothermic effects, respectively) of sample α = 0% dried at 40 °C show two weak endothermic effects at 64 and 173 °C attributed to the loss of water from TEOS and an intense exothermic effect at 300 °C ascribed to 1,3PD decomposition [12]. These two processes result in a mass loss of 63.8% [12,15,16].

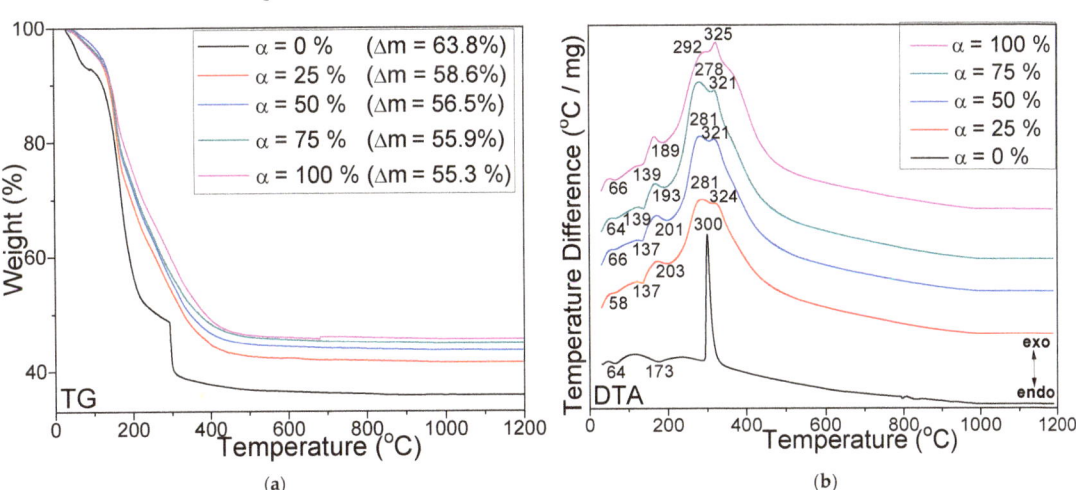

Figure 1. TG (**a**) and DTA (**b**) curves of $(Co_{0.4}Zn_{0.4}Ni_{0.2}Fe_2O_4)_\alpha(SiO_2)_{100-\alpha}$ samples dried at 40 °C.

For samples α = 25, 50, 75, and 100%, the total mass loss slightly decreases (58.6–55.3%) with the $Co_{0.4}Zn_{0.4}Ni_{0.2}Fe_2O_4$ content embedded in the SiO_2 matrix. The formation of Co, Zn, and Ni malonates is indicated by the endothermic effect at 137–139 °C, whereas the formation of Fe malonate is indicated by the exothermic effect at 189–203 °C [12,14–16]. The exothermic effects at 278–292 °C are attributed to Co, Zn, and Ni malonates' decomposition, while those at 321–325 °C are attributed to Fe malonates' decomposition [12,14–16]. The temperature corresponding to the formation of divalent metal (Co, Ni, Zn) malonates slightly increases, whereas that of trivalent metal (Fe) malonates slightly decreases [12,14–16]. The transformations of the SiO_2 matrix during the thermal process make it challenging to delimitate the effects ascribed to malonate precursors' formation and decomposition [12,14–16].

Except for sample $\alpha = 0\%$, the FT-IR spectra of samples dried at 40 °C (Figure 2a) show a band at around 1380 cm^{-1}, characteristic of nitrates. This band is missing for the samples heated at 300 °C, confirming the metal malonates' formation and nitrates' decomposition up to 300 °C [12,15,16]. For samples dried at 40 °C, the band at 1590–1620 cm^{-1} is specific to O–H vibrations in 1,3PD and adsorbed molecular water, and the bands at around 2950 and 2870 cm^{-1} are specific to stretching vibration of C-H in the methylene groups of 1,3PD. This band does not appear in the FT-IR spectra of samples annealed at high temperatures, indicating the precursor's decomposition [15,16,21].

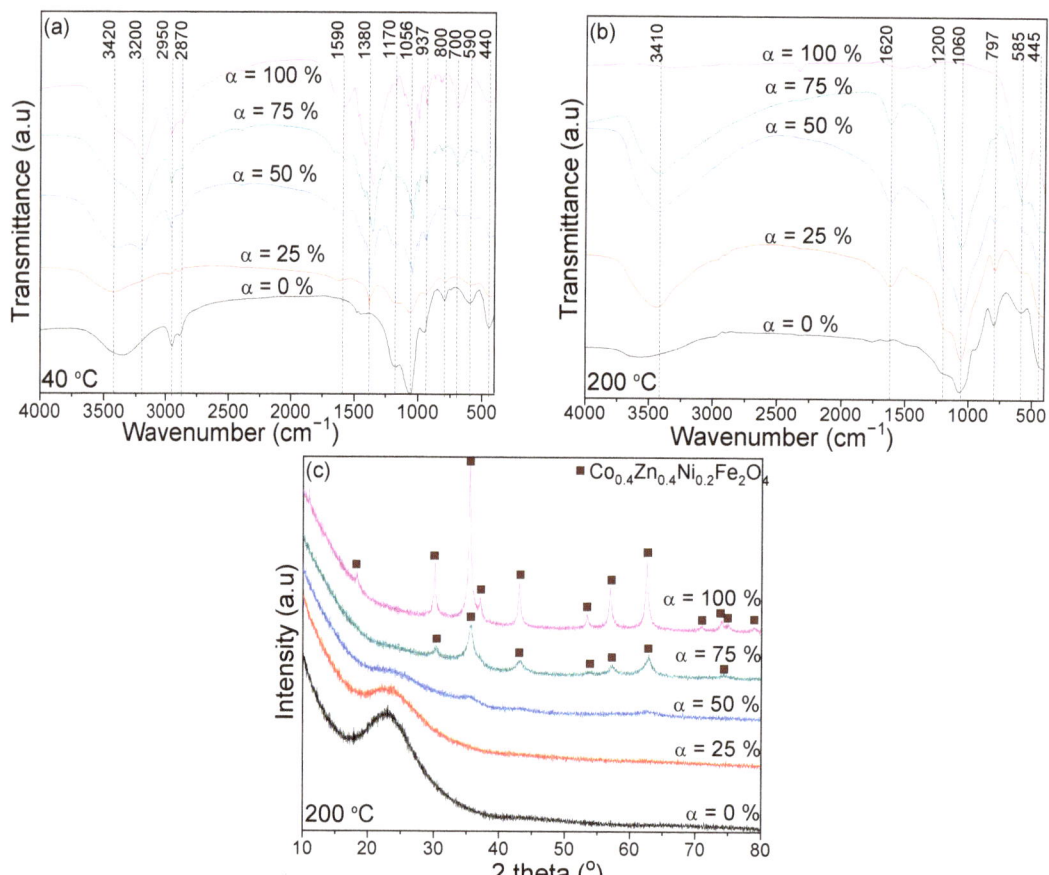

Figure 2. FT-IR spectra of $(Co_{0.4}Zn_{0.4}Ni_{0.2}Fe_2O_4)_\alpha$ $(SiO_2)_{100-\alpha}$ samples heated at 40 °C (**a**) and 300 °C (**b**) and XRD patterns of $(Co_{0.4}Zn_{0.4}Ni_{0.2}Fe_2O_4)_\alpha$ $(SiO_2)_{100-\alpha}$ samples at 300 °C (**c**).

In the FT-IR spectra of sample $\alpha = 25$–100% heated at 300 °C (Figure 2b), the band at around 1620 cm^{-1} characteristic to C=O of COO– groups' vibration indicates the formation of malonate-metal complexes [15,16,21]. For samples $\alpha = 25$–100%, the bands at 585–590 and 440–445 cm^{-1} indicate the presence of M-O bonds in tetrahedral (A) and octahedral (B) sites, respectively [15,16,21]. For samples $\alpha = 0$–75% heated both at 40 and 300 °C, the characteristic bands for Si–O (440–445 cm^{-1}) and Si–O–Si (1170–1200, 1056–1060, 792–829, 797–800, 585–590 cm^{-1}) vibration suggests the formation of the SiO$_2$ matrix [12,15,21]. The bands at 3420–3410 cm^{-1} are attributed to the vibration of O-H and hydrogen bonds in 1,3PD (40 °C) and metal malonates (300 °C) [15,16].

For sample α = 100% annealed at 300 °C (Figure 2c), the presence of well-defined, single-phase Co$_{0.4}$Zn$_{0.4}$Ni$_{0.2}$Fe$_2$O$_4$ containing low-crystallized CoFe$_2$O$_4$ (JCPDS card no. 22–1086), ZnFe$_2$O$_4$ (JCPDS card no. 70–6491), and NiFe$_2$O$_4$ (JCPDS card no. 74–2081) is observed. The peaks pointed out at 2θ values of 18.29, 30.08, 35.43, 37.05, 43.05, 53.05, 56.97, 62.58, 70.95, 74.01, 75.01, 78.97° (CoFe$_2$O$_4$); 18.25, 30.01, 35.35, 36.98, 42.96, 53.30, 56.81, 62.39, 70.77, 73.79, 74.80, 78.74° (ZnFe$_2$O$_4$); and 18.42, 30.29, 35.68, 37.33, 43.37, 53.82, 57.37, 63.01, 71.51, 74.57, 75.58, 79.59° (NiFe$_2$O$_4$), corresponding to Miller indices of (111), (220), (311), (222), (400), (422), (511), (440), (620), (533), (622), and (444) confirms the formation of single-phase spinel-like structure (space group *Fd3m*). The degree of crystallinity was calculated using the highest intensity peak of spinel ferrite (311) [8,21,22]. In sample α = 75% annealed at 300 °C, the presence of Co$_{0.4}$Zn$_{0.4}$Ni$_{0.2}$Fe$_2$O$_4$ is also observed, although the degree of crystallinity is lower than in sample α = 100%. In samples with low ferrite content α = 0, 25, and 50%, the crystalline ferrite is not remarked; the intensity of halo between 16 and 30° (2θ) ascribed to the amorphous SiO$_2$ matrix increases with the SiO$_2$ content.

Figure 3 displays the FT-IR spectra (left) and XRD patterns (right) of samples annealed at 500, 800, and 1200 °C. For samples α = 0–75%, the occurrence of SiO$_2$ matrix is supported by the symmetric and asymmetric stretching vibrations of SiO$_4$ tetrahedron (794–796 cm^{-1}), the Si–O–Si stretching vibrations (1067–1093 cm^{-1}), the shoulder at 1220–1250 cm^{-1} and the Si–O bond vibration (458–482 cm^{-1}) [12,15,16]. These bands are lacking for samples α = 100%, the bands at 578–584 cm^{-1} being ascribed to Zn–O, Co–O and Ni–O vibrations, and at around 400 cm^{-1}, they are ascribed to Fe–O bonds' vibration [12,15,16]. For samples with high SiO$_2$ content (α = 25%) annealed at 1200 °C, the vibration band at 620 cm^{-1} is attributed to Si–O–Si cyclic structures [12,15,16].

The XRD patterns of samples α = 100% annealed at 500, 800, and 1200 °C (Figure 3) display no impurities or unreacted Fe, Ni, Co, and Zn oxides, the broadening of diffraction peaks being ascribed to ultrafine Co$_{0.4}$Zn$_{0.4}$Ni$_{0.2}$Fe$_2$O$_4$ particles [12,15,16]. The intensity of diffraction peaks matching to ferrites increases at high annealing temperatures indicating a high degree of crystallinity, high crystal nucleation (owing to the small growth rate and homogenous distribution), and large crystallites (owing to the coalescence process) [6,8,14–16,22]. The degree of crystallinity (DC) was determined as the ratio between the area under all diffraction peaks and the total area under the amorphous halo and diffraction peaks [14–16]. The intensity of the main diffraction peak of cubic spinel ferrite at (311) plane was considered as a measure of the degree of crystallinity [14–16].

By annealing at 500 °C, the samples α = 25, 50 and 75% display single-phase Co$_{0.4}$Zn$_{0.4}$Ni$_{0.2}$Fe$_2$O$_4$, but less crystallized than the sample α = 100%, the degree of crystallinity increases with the ferrite content embedded in the SiO$_2$ matrix. With sample α = 75% annealed at 800 °C, single phase Co$_{0.4}$Zn$_{0.4}$Ni$_{0.2}$Fe$_2$O$_4$ is observed, while for sample α = 50%, the presence of trace Fe$_2$SiO$_4$ (JCPDS card no. 70–1861) with peaks at 2θ values of 25.04, 29.30, 31.62, 34.97, 35.91, 37.29, 41.2, 43.67, 51.39, 54.67, 55.73, 57.42, 61.17, 65.46, 68.38, 72.11, and 75.43° associated with Miller indices of (111), (002) (130), (131), (112), (200), (220), (132), (222), (061), (133), (043), (062), (162), (322), (080), and (303) is also remarked.

The formation of Fe$_2$SiO$_4$ appears due to the incomplete reduction of Fe^{3+} to Fe^{2+}, which further reacts with the SiO$_2$ matrix, forming Fe$_2$SiO$_4$ [16].

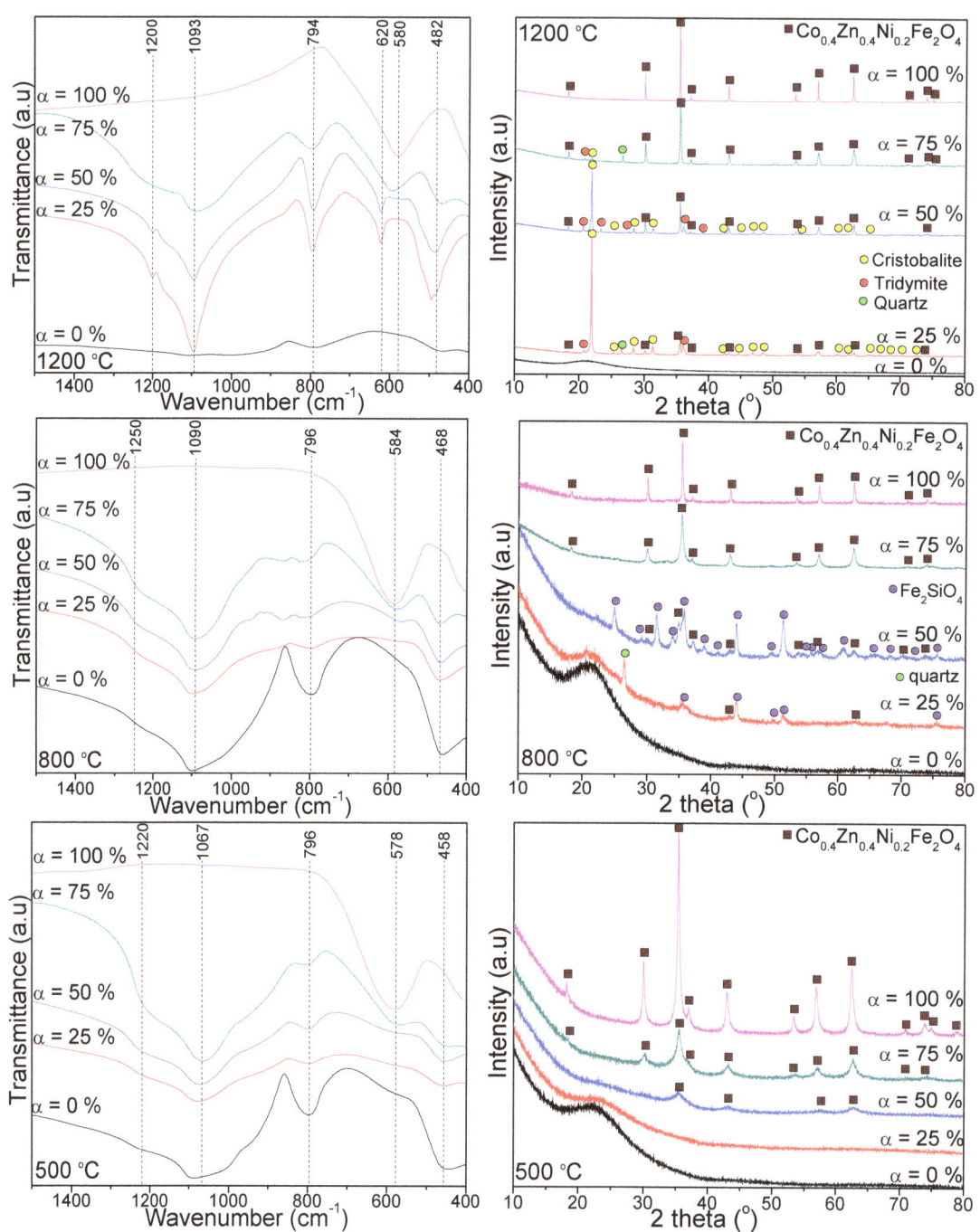

Figure 3. FT-IR spectra (**left**) and XRD patterns (**right**) of $(Co_{0.4}Zn_{0.4}Ni_{0.2}Fe_2O_4)_\alpha$ $(SiO_2)_{100-\alpha}$ samples annealed at 500, 800, and 1200 °C.

For sample α = 25% annealed at 800 °C, the main phase of $Co_{0.4}Zn_{0.4}Ni_{0.2}Fe_2O_4$ is accompanied by Fe_2SiO_4 and quartz (JCPDS card no. 79–1910) at 2θ value of 26.6° corresponding to Miller indices of (011), whereas for sample α = 75% annealed at 1200 °C, beside the main phase of $Co_{0.4}Zn_{0.4}Ni_{0.2}Fe_2O_4$, crystalline phases of SiO_2 (quartz and tridymite (JCPDS card no. 042–1401)) with peaks at 2θ values of 20.76, 23.38, 27.49, 36.17, and 37.72 corresponding to Miller indices of (220), (222), (420), (040), and (240) are remarked. Unexpectedly, the samples α = 25 and 50% annealed at 1200 °C display cristobalite (JCPDS card no. 39–1425) with peaks at 2θ values of 21.98, 25.32, 28.43, 31.46, 42.66, 44.84, 47.06, 48.61, 56.22, 60.30, 62.01, 65.65, 66.81, 68.67, 69.42, 69.79, 70.54, and 72.69 associated with the Miller indices of (101), (110), (111), (102), (321), (202), (113), (212), (104), (311), (302), (204), (223), (214), (321), (303), (105), and (313) as the main crystalline phase attended by $Co_{0.4}Zn_{0.4}Ni_{0.2}Fe_2O_4$ and tridymite. Additionally, for sample α = 25%, quartz is also present. The intensity of diffraction peaks belonging to the cristobalite increases with the SiO_2 content.

At all annealing temperatures, for samples with no ferrite content (α = 0%), no crystalline phases are observed and the halo between 16 and 30° (2θ) matches the amorphous SiO_2 matrix. A possible explanation for the absence of crystalline phases corresponding to the SiO_2 could be the difficult diffusion of oxygen within the pores of the silica matrix [14–16]. Usually, the amorphous phase content is proportional to the area under the diffraction halo, but not all the amorphous phases produce diffraction halos due to the lack of a significant local order. The area under the amorphous halo and the total area of the diffraction peaks were used to explore the evolution of amorphous and crystalline phases [14–16]. Similar behavior is observed for the sample with high SiO_2 matrix content (α = 25%) annealed at 500 °C. Consequently, low annealing temperature and high SiO_2 content led to highly amorphous content. However, not all the amorphous phases imply diffraction halos owing to the absence of a large local order [14,21].

The average crystallite size (D_{XRD}) was calculated using the Scherrer formula (Equation (1)):

$$D_{XRD} = \frac{0.9 \cdot \lambda}{\beta \cdot \cos\theta} \quad (1)$$

where λ is the wavelength of CuK_α radiation (1.5406 Å), β is the broadening of full width at half-maximum intensity (FWHM), and θ is the Bragg angle (°) [12–16].

The average crystallite size increases with the annealing temperature and $Co_{0.4}Zn_{0.4}Ni_{0.2}Fe_2O_4$ content by the grain growth blocking effect of the SiO_2 matrix (see Table 1) [14–16]. The samples with low ferrite content comprise both amorphous and crystalline phases. The changes in crystallite size may be associated with the influence of the SiO_2 matrix on the grain growth and lattice strains, in such a way that SiO_2 content increase, while the annealing temperature reduces the grain growth [23–25]. The largest crystallite size was obtained for non-embedded $Co_{0.4}Zn_{0.4}Ni_{0.2}Fe_2O_4$ (α = 100%), following the assumption that the SiO_2 matrix contributes to the reduction of crystallite size. One plausible explanation could be the improvement of the crystal-nuclei coalescence process, which occurs at high annealing temperatures (1200 °C). In addition, the annealing temperature reduces lattice strains and defects [14–16].

The lattice constant (a) was calculated using Bragg's law with Nelson–Riley function according to Equation (2) [14–16]:

$$a = \frac{\lambda\sqrt{h^2 + k^2 + l^2}}{2 \cdot \sin\theta} \quad (2)$$

where λ is the wavelength of CuK_α radiation (1.5406 Å) [15,16].

Table 1. Particle size (D_{TEM}), crystallites size (D_{XRD}), crystallinity degree (DC), lattice constant (a), hopping length (L_A and L_B), specific surface area (SSA), and Co/Zn/Ni/Fe molar ratio of $(Co_{0.4}Zn_{0.4}Ni_{0.2}Fe_2O_4)_\alpha(SiO_2)_{100-\alpha}$ samples.

Parameter	Temp. (°C)	α				
		0	25	50	75	100
D_{TEM} (nm)	1200	-	34 ± 2	50 ± 3	78 ± 4	120 ± 6
D_{XRD} (nm)	500	-	15 ± 1	17 ± 1	22 ± 1	30 ± 2
	800	-	26 ± 2	32 ± 2	42 ± 3	51 ± 3
	1200	-	33 ± 2	49 ± 3	75 ± 4	120 ± 5
DC (%)	500	51 ± 3	59 ± 3	65 ± 3	69 ± 4	76 ± 5
	800	61 ± 3	66 ± 3	68 ± 3	73 ± 4	79 ± 4
	1200	73 ± 4	81 ± 4	86 ± 4	91 ± 5	94 ± 5
a (Å)	500	-	8.40 ± 0.02	8.41 ± 0.02	8.43 ± 0.02	8.45 ± 0.02
	800	-	8.38 ± 0.02	8.39 ± 0.02	8.40 ± 0.02	8.43 ± 0.02
	1200	-	8.36 ± 0.02	8.37 ± 0.02	8.38 ± 0.02	8.40 ± 0.02
SSA (m²/g)	200	270 ± 14	220 ± 10	170 ± 9	24 ± 1	19 ± 1
	500	260 ± 14	270 ± 14	260 ± 14	120 ± 6	26 ± 2
	800	≤0.5	≤0.5	≤0.5	≤0.5	≤0.5
	1200	≤0.5	≤0.5	≤0.5	≤0.5	≤0.5
Co/Zn/Ni/Fe molar ratio	500	-	0.37/0.38/0.19/1.97	0.38/0.39/0.18/1.97	0.39/0.38/0.17/1.98	0.41/0.38/0.19/2.02
	800	-	0.31/0.29/0.18/2.20	0.32/0.30/0.17/2.33	0.39/0.38/0.18/2.03	0.41/0.38/0.19/2.02
	1200	-	0.39/0.38/0.18/1.99	0.39/0.41/0.19/1.99	0.39/0.40/0.19/2.01	0.41/0.41/0.20/2.00

The lattice constant (a) increases with the $Co_{0.4}Zn_{0.4}Ni_{0.2}Fe_2O_4$ content embedded in the SiO_2 matrix and decreases with the annealing temperature (Table 1). The high surface energy and tension, surface dipole interactions, and the cation distribution inside the nanocrystallite do not produce the lattice shrinking [14–16,26]. The tetrahedral (A) site (0.52 Å) displays a smaller radius than the octahedral (B) site (0.81 Å) [2], while the ionic radius of Co^{2+} (0.75), Zn^{2+} (0.74,) and Ni^{2+} (0.69 Å) are larger than the ionic radius of Fe^{3+} (0.64 Å) [15]. By increasing the number of Fe^{3+} ions in the octahedral (B) sites, the system changes from inverse spinel to normal spinel structure [14–16]. Consequently, adding Co^{2+}, Zn^{2+}, and Ni^{2+} ions lead to a strained lattice.

The elemental composition is confirmed by the Co/Zn/Ni/Fe molar ratio using the MW/ICP-OES analysis (Table 1). The best experimental and theoretical data correlation is observed for the samples annealed at 1200 °C. For samples α = 25 and 50% annealed at 800 °C, the higher Fe content confirms the presence of Fe_2SiO_4 as a secondary phase observed by XRD.

The shape of N_2 adsorption-desorption isotherms of $(Co_{0.4}Zn_{0.4}Ni_{0.2}Fe_2O_4)_\alpha(SiO_2)_{100-\alpha}$ (α = 0, 25, 50, 75, and 100%) samples annealed at 300 and 500 °C (Figure 4a,b) is preserved, confirming the stability of the porous structure up to 500 °C. The isotherms of samples annealed at 800 and 1200 °C could not be recorded, indicating the breakdown of porous structure at temperatures above 500 °C. The isotherm for SiO_2 (α = 0%) is of type IV, and for samples α = 25, 50, and 75% is of type I [27].

The SSA decreases with the increase of α, by the increase of D_{XRD} (Table 1). For SiO_2 (α = 0%) and ferrite (α = 100%) samples, the SSA does not depend on the annealing temperature, while for samples α = 25, 50, and 75% annealed at 500 °C, an increase of SSA value was observed. A possible explanation could be the better organization and crystallization of samples annealed at 500 °C than at 300 °C. The pore size distribution (Figure 4c,d) shows that all samples contain different-sized pores up to 550 Å. The pores are generally under 100 Å in samples annealed at 500 °C and up to 200 Å in samples annealed at 300 °C, respectively. These results follow the variation of SSA described above.

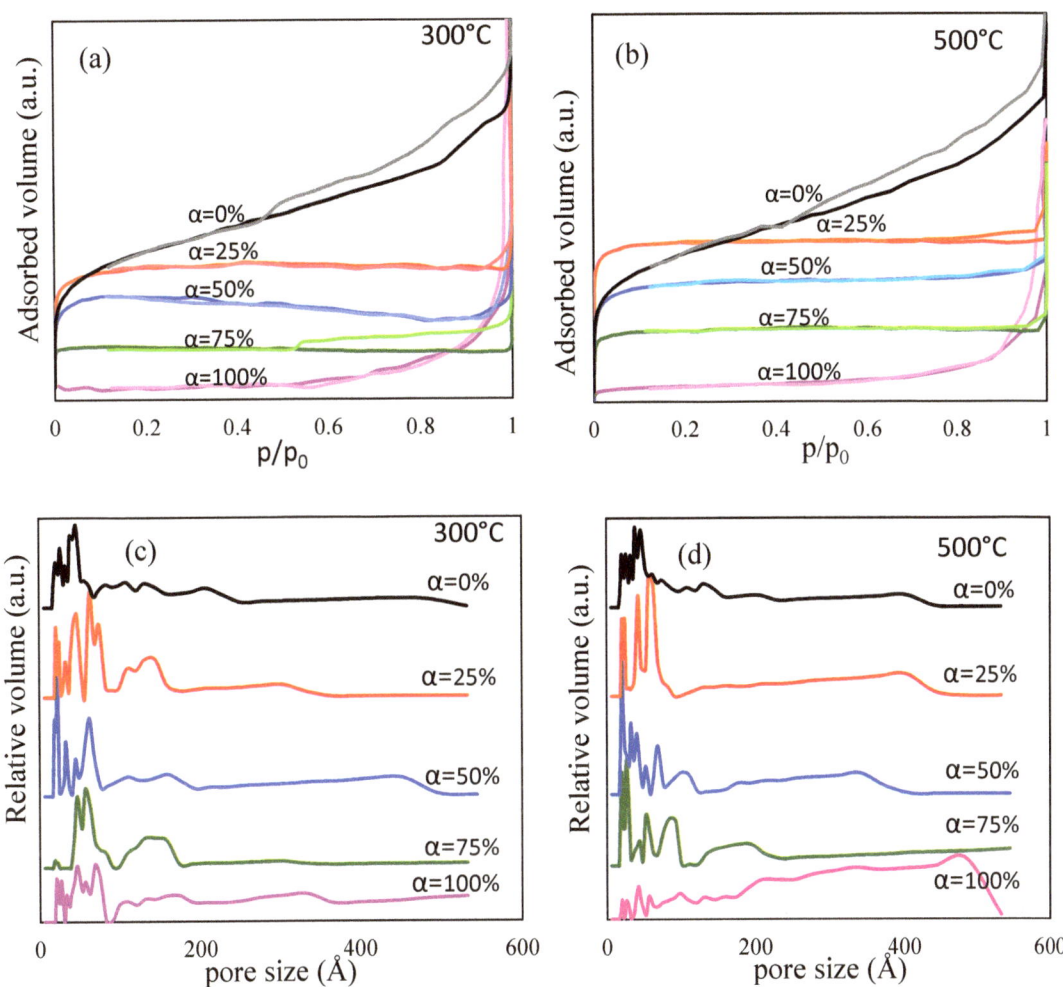

Figure 4. N$_2$ adsorption-desorption isotherms of (Co$_{0.4}$Zn$_{0.4}$Ni$_{0.2}$Fe$_2$O$_4$)$_\alpha$(SiO$_2$)$_{100-\alpha}$ (α = 0, 25, 50, 75, and 100%) samples annealed at 300 °C (**a**) and 500 °C (**b**) pore size distribution at 300 °C (**c**) and 500 °C (**d**).

The TEM image of the SiO$_2$ matrix (α = 0%) consists of a dark area, without any possibility of identifying the matrix network, whereas those of samples α = 25–100% annealed at 500 and 800 °C are blurry, with low contrast, due to the small size poorly crystalline Co$_{0.4}$Zn$_{0.4}$Ni$_{0.2}$Fe$_2$O$_4$ particles (~1 nm). For samples α = 25–100% annealed at 1200 °C (Figure 5), the higher ferrite content embedded in the SiO$_2$ matrix results in large spherical particles. The increase in particle size from 34 to 122 nm (Table 1, Figure 6) with the ferrite content could be the outcome of different reaction kinetics, variation of the particle growth rate, or crystalline clusters formation [14–16,28]. The different particle arrangement could be due to the solid bodies formed by well-faceted grains, while the particle agglomeration could be a consequence of small particle size, inter-particle interactions, interfacial surface tensions, and strong intermolecular friction produced during the conversion of thermal energy into internal heat energy [14–16]. The porous surface formed by the gases generated during the thermal decomposition also favors the particle's agglomeration [14–16].

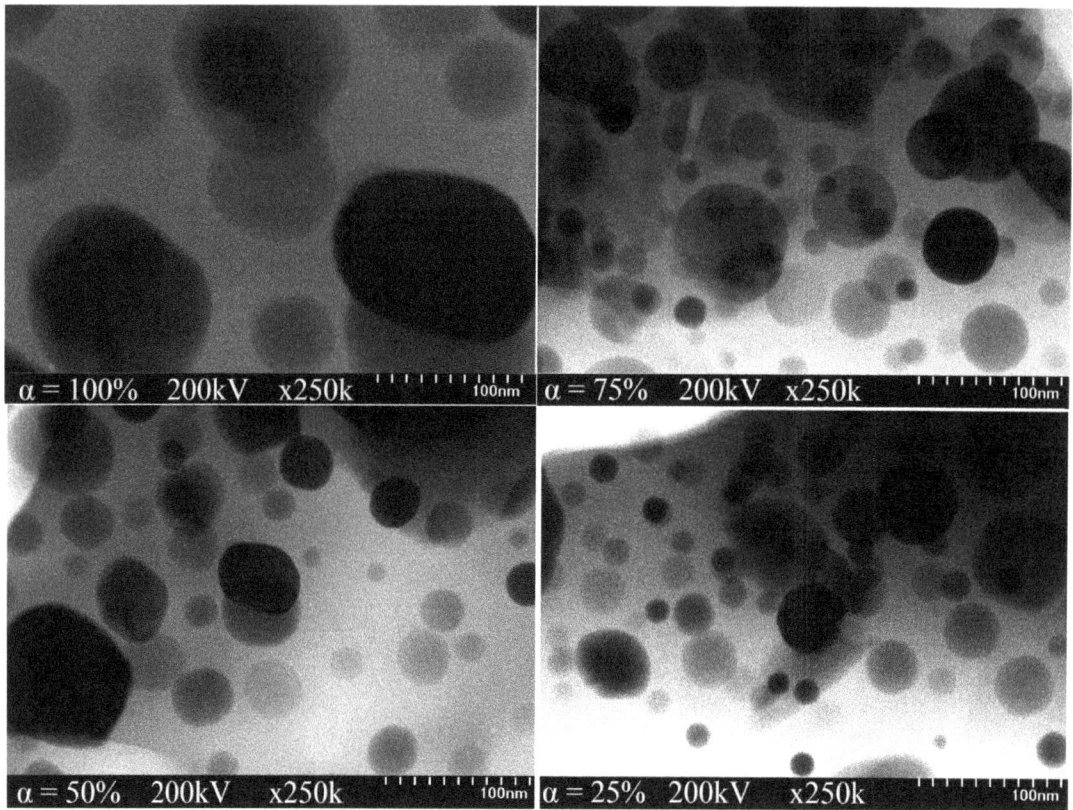

Figure 5. TEM images of $(Co_{0.4}Zn_{0.4}Ni_{0.2}Fe_2O_4)_\alpha(SiO_2)_{100-\alpha}$ samples (α = 25–100%) annealed at 1200 °C.

The average crystallite sizes are consistent with the particle sizes determined from TEM, the differences being attributed to the interference in the diffraction patterns introduced by the amorphous SiO_2 and large-size nanoparticles [12–16]. The particle size determined via TEM is generally larger than the crystallite size estimated by XRD, considering that a particle typically consists of several crystallites. The crystallite size can be calculated by analyzing the broadening of diffraction peaks without considering the effects of other factors that contribute to the diffraction peak width (i.e., instrumental contribution, temperature, microstrain, etc.) [29]. Moreover, even if they are few in number, the large nanoparticles significantly contribute to the diffraction patterns since they comprise a large fraction of atoms. The interference of the amorphous SiO_2 with particle size lower than that of the embedded ferrite crystallites should also be considered [14–16].

The SiO_2 matrix (α = 0%) displays a diamagnetic behavior (Figure 7), while samples (α = 25–100%) show a typical ferromagnetic behavior (Figure 8) both at 800 and 1200 °C.

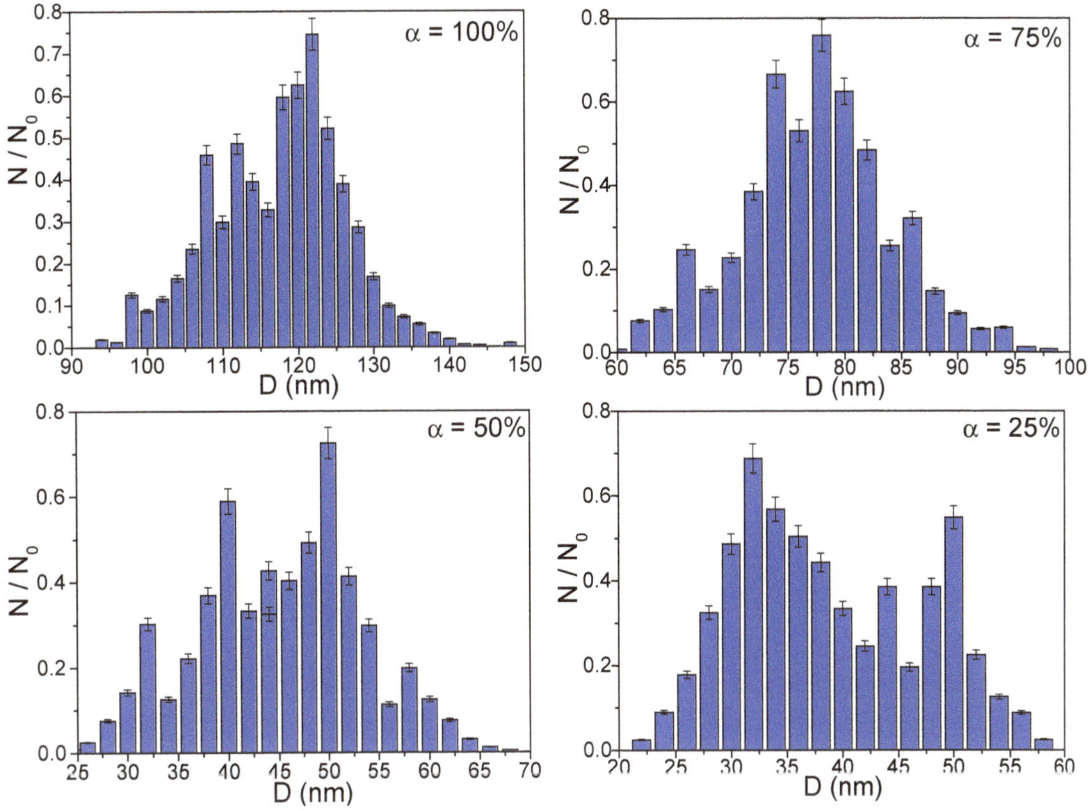

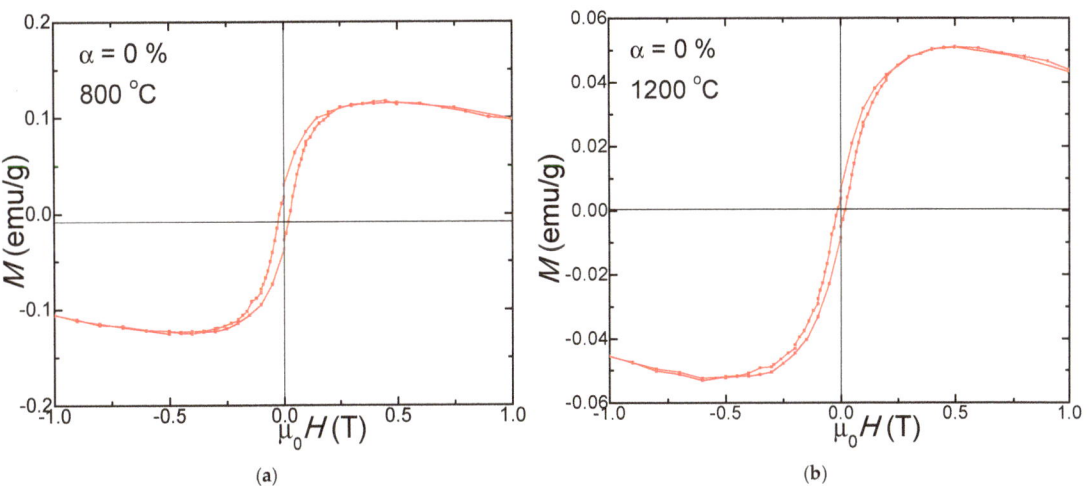

Figure 6. Particle size distributions for $(Co_{0.4}Zn_{0.4}Ni_{0.2}Fe_2O_4)_\alpha(SiO_2)_{100-\alpha}$ samples (α = 25–100%) annealed at 1200 °C.

Figure 7. Magnetic hysteresis loops of SiO_2 matrix (α = 0%) annealed at 800 °C (**a**) and 1200 °C (**b**).

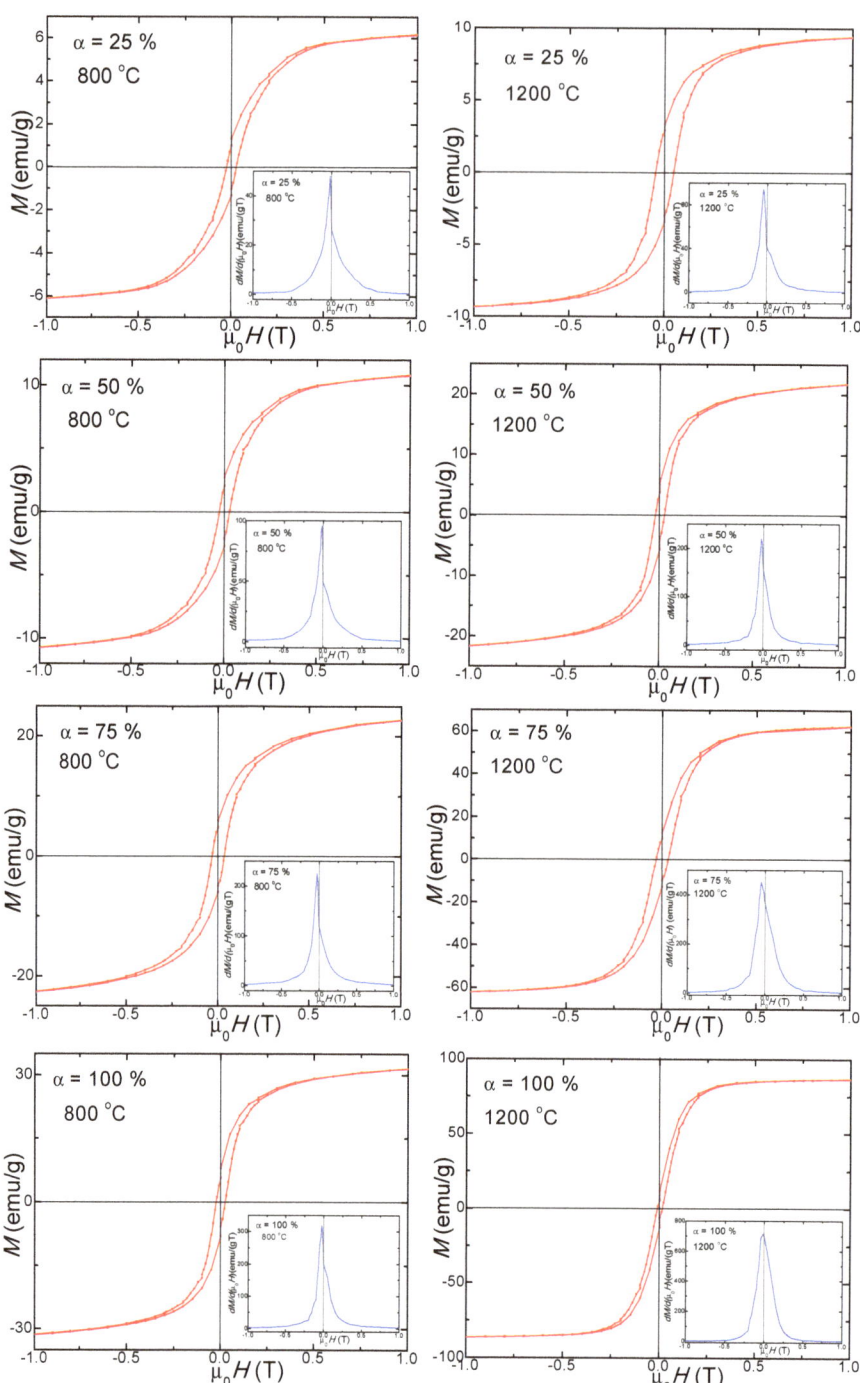

Figure 8. Magnetic hysteresis loops and magnetization derivative of $(Co_{0.4}Zn_{0.4}Ni_{0.2}Fe_2O_4)_\alpha$ $(SiO_2)_{100-\alpha}$ samples (α = 25–100%) annealed at 800 and 1200 °C.

The samples show diamagnetic behavior in the presence of some accidental low concentration of ferromagnetic impurities, which likely come from the manipulation of the samples (from cutters, tweezers, spatula, etc.). The influence of these ferromagnetic impurities on the ferrites containing samples is negligible since the magnetization of the samples reaches 10 emu/g, while the ferromagnetic impurities can provide around 0.1 emu/g.

Generally, the magnetic properties of nano-ferrites are determined by the structure, crystal defects, porosity, particle size, and K [6]. The annealing temperature and SiO_2 content are critical for producing single-phase nano-ferrites with enhanced magnetic properties. The M_S increases with the $Co_{0.4}Zn_{0.4}Ni_{0.2}Fe_2O_4$ content and in samples with the same composition with the annealing temperature [2,6,17]. The highest value of M_S (90.1 emu/g) was obtained for non-embedded $Co_{0.4}Zn_{0.4}Ni_{0.2}Fe_2O_4$ ($\alpha = 100\%$) annealed at 1200 °C. The increase of M_S from 10.0 emu/g ($\alpha = 25\%$) to 90.1 emu/g ($\alpha = 100\%$) with the ferrite content is a consequence of the non-magnetic SiO_2 matrix that has a magnetic dilution effect [12–15].

The nanoparticle size is contingent on the annealing temperature, which controls the formation of assemblies of weakly interacting particles through reduced magnetostatic energy and weakly bonded interfaces [14–16,30]. According to Neel's theory of ferrimagnetism, the magnetic properties of ferrites depend on the distribution of magnetic ions between the tetrahedral (A) and octahedral (B) sites [2,3]. The presence of impurity atoms or oxygen vacancies results in the break of the superexchange spin coupling between magnetic ions inducing a supplementary surface spin disorder [2]. Thus, magnetization is influenced by the crystalline structure, defects, and cationic distribution. The net magnetization of these spinel structures is given by the difference between the magnetic moments of tetrahedral (A) and octahedral (B) sites [5]. Magnetization is also affected by the migration of the magnetic ions between tetrahedral (A) and octahedral (B) sites and by the magnetic moment alignment from A and B sites [5]. When the Co^{2+} ions occupy the octahedral (B) sites, the magnetic moment ($3\mu_B$) will be higher in this position than in tetrahedral (A) sites which are occupied by Ni^{2+} and Zn^{2+} ions ($2\mu_B$) [3].

The higher ferrite content in samples annealed at 1200 °C results in an M_R increase from 3.3 emu/g ($\alpha = 25\%$) to 13.7 emu/g ($\alpha = 100\%$). The remanence ratio $S = M_R/M_S$ indicates how square is the hysteresis loop. A theoretical value of this parameter $S < 0.5$ indicates samples consisting of an assembly of single domain, non-interacting particles [31–33]. As can be deduced from Table 2, this ratio is lower than 0.5 for all samples.

Table 2. Saturation magnetization (M_S), remanent magnetization (M_R), coercivity (H_c) and anisotropy constant (K) of $(Co_{0.4}Zn_{0.4}Ni_{0.2}Fe_2O_4)_\alpha (SiO_2)_{100-\alpha}$ samples annealed at 800 and 1200 °C.

Sample $(Co_{0.4}Zn_{0.4}Ni_{0.2}Fe_2O_4)_\alpha (SiO_2)_{100-\alpha}$	Temperature (°C)	M_S (emu/g)	M_R (emu/g)	S	H_c (Oe)	$K \cdot 10^5$ (erg/cm^3)
$\alpha = 25\%$	800	7 ± 0.4	1.2 ± 0.1	0.17 ± 0.02	260 ± 13	119 ± 6
	1200	10 ± 1	3.3 ± 0.2	0.33 ± 0.03	490 ± 25	305 ± 15
$\alpha = 50\%$	800	14 ± 1	2.6 ± 0.1	0.19 ± 0.02	270 ± 14	229 ± 11
	1200	29 ± 2	5.3 ± 0.3	0.18 ± 0.02	320 ± 16	577 ± 29
$\alpha = 75\%$	800	23 ± 1	5.7 ± 0.3	0.25 ± 0.03	300 ± 15	437 ± 22
	1200	62 ± 4	9.8 ± 0.5	0.16 ± 0.02	220 ± 11	857 ± 43
$\alpha = 100\%$	800	38 ± 2	8.7 ± 0.4	0.23 ± 0.02	320 ± 16	754 ± 38
	1200	90 ± 6	14 ± 1	0.15 ± 0.02	170 ± 10	934 ± 47

The H_c increases from 260 Oe ($\alpha = 25\%$) to 320 Oe ($\alpha = 100\%$) with ferrite content embedded in the matrix in samples annealed at 800 °C and decreases from 485 Oe ($\alpha = 25\%$) to 165 Oe ($\alpha = 100\%$) in samples annealed at 1200 °C. The samples $\alpha = 50, 75,$ and 100% have small H_c and high M_S values due to the large particles with high magnetic coupling [31,32]. Generally, H_c depends on crystallite sizes, magnetocrystalline anisotropy, domain walls and M_S [3]. In the case of small single magnetic domain particles, they change the magnetization by spin rotation [34]. At high H_c values, the thermal energy cannot induce the magnetization fluctuations for change of the magnetization [3]. Consequently, besides the particles' shape and the density of disordered surface spins, H_c has an important contribution to the magnetic order inside a monodomain particle. For large particles, the density of surface spins is low and results in higher magnetization [26].

The magnetocrystalline anisotropy constant (K) was calculated using Equation (3):

$$K = \frac{\mu_0 \cdot M_S \cdot H_c}{2} \quad (3)$$

where M_S is the saturation magnetization, μ_0 is vacuum permeability ($\mu_0 = 1.256 \times 10^{-6}$ N/A^2), and H_c is the coercivity field (T) [35].

The K values (Table 2) increase with the $Co_{0.4}Zn_{0.4}Ni_{0.2}Fe_2O_4$ content embedded in the silica matrix, with the highest K observed for the non-embedded $Co_{0.4}Zn_{0.4}Ni_{0.2}Fe_2O_4$ ($\alpha = 100\%$). The strain on the ferrite nanoparticle surface induced by the SiO_2 matrix hinders the rotation of the magnetic moments from the particle-matrix interface [32]. As can be seen from Table 2, all the main magnetic parameters were strongly affected by the SiO_2 embedding. M_S, M_R, and K depreciate substantially with the increase of SiO_2 content due to the non-magnetic nature of the matrix. The H_c value of the samples annealed at 1200 °C has a different trend; it is enhanced, suggesting the pinning of the magnetic moments from the surface of the particles, which is induced by the strain of the SiO_2 layer.

The chemical composition, crystallographic structure, particle size, atomic packing density, and internal defects highly influence the ferrites' magnetic properties [20]. Previous studies reported a decrease in the magnetocrystalline anisotropy of cubic ferrites when they are doped with Zn^{2+} ions [36]. Ni-Co ferrites have large K due to the Co^{2+} ions preference for the octahedral (B) sites [19].

The magnetization derivative curves (dM/d($\mu_0 H$)) vs. applied magnetic field are shown in the insets of Figure 8. The presence of a single peak indicates a single magnetic phase for samples $\alpha = 100\%$, considering that in a pure magnetic sample, the peak occurs at the nominal coercive field, suggesting crystalline samples with a single magnetic phase [14,16,31]. The sharp peaks indicate high magnetic purity, whereas the broad peaks suggest wide particle size distributions. The magnetization derivative for samples $\alpha = 25-75\%$ annealed at 800 and 1200 °C indicate that two magnetic phases (from the triple Co–Zn–Ni ferrite) are magnetically coupled inside of the particle along their magnetic moments; the crystalline phases of Fe_2SiO_4 (a typical paramagnet at room temperature [34,37]), cristobalite and tridymite identified by XRD do not display magnetic properties, the hard magnetic phase being dominant (since it has a larger H_c). If not instrumental, the peaks' asymmetry reveals the presence of two magnetic phases inside the particles, one forming from a solid solution of two ferrites and another being the third ferrite.

In our previous studies on Co–Ni, Co–Zn and Zn–Ni ferrites, only a single magnetic phase was obtained [14–16]. For samples annealed at 1200 °C, the hysteresis loops are, generally, broader, and the (dM/d($\mu_0 H$)) vs. $\mu_0 H$ curves are narrower and sharper than in samples annealed at 800 °C. The peak heights and their horizontal shifts are related to the strength of the magnetic phases [14,26,30]. For samples annealed at 800 °C, the broader peaks suggest a large particle size distribution associated with a large H_c.

The structure and the magnetic properties of $(Co_{0.4}Zn_{0.4}Ni_{0.2}Fe_2O_4)_\alpha(SiO_2)_{100-\alpha}$ are highly influenced by the SiO_2 content and the annealing temperature. Our future studies intend to identify the metallic ion which does not couple with the other two metallic ions of the mixed ferrite leading to a second magnetic phase. Thus, by $Co_{0.4}Zn_{0.4}Ni_{0.2}Fe_2O_4$ nanoparticles, the magnetic behavior can be easily tuned. Furthermore, combining the best magnetic properties and morphological configuration may be of interest for several technical applications.

4. Conclusions

The influence of SiO_2 embedding on the structure, morphology, thermal, and magnetic properties of $Co_{0.4}Zn_{0.4}Ni_{0.2}Fe_2O_4$ particles obtained by the sol-gel process was investigated. Fe, Co, Zn, and Ni malonates formed in two stages, as indicated on the DTA curve. At 1200 °C, the XRD and FT-IR results supported the formation of a single-phase spinel structure and SiO_2 matrix. Highly crystalline single-phase ferrite starting from 300 °C for non-embedded $Co_{0.4}Zn_{0.4}Ni_{0.2}Fe_2O_4$ (α = 100%) and an amorphous halo without any crystalline phases for SiO_2 (α = 0%) sample were remarked. For samples α = 25–75%, the single crystal phase Co–Ni–Zn ferrite at 500 °C was accompanied by Fe_2SiO_4 and quartz at 800 °C, while at 1200 °C the major cristobalite phase was accompanied by Co–Ni–Zn ferrite, tridymite, and quartz. The crystallite size increased with ferrite content in the SiO_2 matrix, namely: 14.5–29.6 nm (500 °C), 26.3–52.4 nm (800 °C), and 33.3–118 nm (1200 °C), respectively. TEM images confirmed that the particles are in the nanometer range. The SSA gradually increased with the SiO_2 content and decreased with the annealing temperature above 500 °C. The main magnetic parameters increased with the $Co_{0.4}Zn_{0.4}Ni_{0.2}Fe_2O_4$ content: M_S from 7.3 emu/g to 90.1 emu/g, M_R from 1.2 to 13.7 emu/g, and K from $0.119 \cdot 10^{-3}$ to $0.934 \cdot 10^{-3}$ erg/cm^3 (at 1200 °C). H_c increased with $Co_{0.4}Zn_{0.4}Ni_{0.2}Fe_2O_4$ content from 260 to 320 (at 800 °C) and decreased from 485 to 165 Oe (at 1200 °C). The M_S and M_R increased with the annealing temperature. As expected, the non-embedded $Co_{0.4}Zn_{0.4}Ni_{0.2}Fe_2O_4$ (α = 100%) was ferromagnetic with high M_S, while the SiO_2 matrix (α = 0%) was diamagnetic with a small ferromagnetic fraction. $Co_{0.4}Zn_{0.4}Ni_{0.2}Fe_2O_4$ non-embedded into the SiO_2 matrix displays the behavior of a single magnetic phase, while the $Co_{0.4}Zn_{0.4}Ni_{0.2}Fe_2O_4$ embedded in the SiO_2 matrix shows two magnetic phases, the solid solution of two ferrites, and the third ferrite. Moreover, when the ferrite is embedded in the SiO_2 matrix, the particle sizes decreased and the main magnetic parameters depreciated.

Author Contributions: Conceptualization, T.D.; methodology, T.D., O.C. and E.A.L.; formal analysis, T.D., O.C., I.G.D., M.D.L. and E.A.L.; investigation, T.D., O.C., I.G.D., M.D.L. and E.A.L.; resources, T.D., O.C. and E.A.L.; data curation, T.D.; writing—original draft preparation, T.D., O.C., I.G.D. and E.A.L.; writing—review and editing, T.D., O.C. and E.A.L.; visualization, T.D.; supervision, T.D. All authors have read and agreed to the published version of the manuscript.

Funding: The APC was funded by the Technical University of Cluj-Napoca.

Institutional Review Board Statement: Not applicable.

Informed Consent Statement: Not applicable.

Data Availability Statement: Data are available on request from the corresponding author.

Conflicts of Interest: The authors declare no conflict of interest.

References

1. Vatsalya, V.L.S.; Sundari, G.S.; Sridhar, C.S.L.N.; Lakshmi, C.S. Evidence of superparamagnetism in nano phased copper doped nickel zinc ferrites synthesized by hydrothermal method. *Optik* **2021**, *247*, 167874. [CrossRef]
2. Mallapur, B.M.M.; Chougule, K. Synthesis, characterization and magnetic properties of nanocrystalline Ni–Zn–Co ferrites. *Mater. Lett.* **2010**, *64*, 231–234. [CrossRef]
3. Mohapatra, P.P.; Singh, H.K.; Kiran, M.S.R.N.; Dobbidi, P. Co substituted Ni–Zn ferrites with tunable dielectric and magnetic response for high-frequency applications. *Ceram. Int.* **2022**, *48*, 29217–29228. [CrossRef]
4. Ghodake, J.S.; Shinde, T.J.; Patil, R.P.; Patil, S.B.; Suryavanshi, S.S. Initial permeability of Zn–Ni–Co ferrite. *J. Magn. Magn. Mater.* **2015**, *378*, 436–439. [CrossRef]
5. Chakrabarty, S.; Bandyopadhyay, S.; Pal, M.; Dutta, A. Sol-gel derived cobalt containing Ni–Zn ferrite nanoparticles: Dielectric relaxation and enhanced magnetic property study. *Mat. Chem. Phys.* **2021**, *259*, 124193. [CrossRef]
6. Dalal, M.; Mallick, A.; Mahapatra, A.S.; Mitra, A.; Das, A.; Das, D.; Chakrabarti, P.K. Effect of cation distribution on the magnetic and hyperfine behavior of nanocrystalline Co doped Ni–Zn ferrite ($Ni_{0.4}Zn_{0.4}Co_{0.2}Fe_2O_4$). *Mater. Res.* **2016**, *76*, 389–401.
7. Sherstyuk, D.P.; Starikov, A.Y.; Zhivulin, V.E.; Zherebstov, D.A.; Gudkova, S.A.; Perov, N.S.; Alekhina, Y.A.; Astapovich, K.A.; Vinnik, D.A.; Trukhanov, A.V. Effect of Co content on magnetic features and SPIN states in Ni–Zn spinel ferrites. *Ceram. Int.* **2021**, *47*, 12163–12169. [CrossRef]

8. Raju, K.; Venkataiah, G.; Yoon, D.H. Effect of Zn substitution on the structural and magnetic properties of Ni–Co ferrites. *Ceram. Int.* **2014**, *40*, 9337–9344. [CrossRef]
9. Kumar, R.; Barman, P.B.; Singh, R.R. An innovative direct non-aqueous method for the development of Co doped Ni-Zn ferrite nanoparticles. *Mater. Today Commun.* **2021**, *27*, 102238. [CrossRef]
10. Vinnik, D.A.; Sherstyuk, D.P.; Zhivulin, V.E.; Zhivulin, D.E.; Starikov, A.Y.; Gudkova, S.A.; Zherebtsov, D.A.; Pankratov, D.A.; Alekhina, Y.A.; Perov, N.S.; et al. Impact of the Zn–Co content on structural and magnetic characteristics of the Ni spinel ferrites. *Ceram. Int.* **2022**, *48*, 18124–18133. [CrossRef]
11. Kaur, M.; Jain, P.; Singh, M. Studies on structural and magnetic properties of ternary cobalt magnesium zinc (CMZ) $Co_{0.6-x}Mg_xZn_{0.4}Fe_2O_4$ (x = 0.0, 0.2, 0.4, 0.6) ferrite nanoparticles. *Mat. Chem. Phys.* **2015**, *162*, 332–339. [CrossRef]
12. Stefanescu, M.; Stoia, M.; Dippong, T.; Stefanescu, O.; Barvinschi, P. Preparation of $Co_xFe_{3-x}O_4$ oxidic system starting from metal nitrates and propanediol. *Acta Chim. Slov.* **2009**, *56*, 379–385.
13. Shobana, M.K.; Sankar, S. Structural, thermal and magnetic properties of $Ni_{1-x}Mn_xFe_2O_4$ nanoferrites. *J. Magn. Mater.* **2009**, *321*, 2125–2128. [CrossRef]
14. Dippong, T.; Toloman, D.; Levei, E.A.; Cadar, O.; Mesaros, A. A possible formation mechanism and photocatalytic properties of $CoFe_2O_4$/PVA-SiO_2 nanocomposites. *Thermochim. Acta* **2018**, *666*, 103–115.
15. Dippong, T.; Cadar, O.; Levei, E.A.; Deac, I.G.; Borodi, G.; Barbu-Tudoran, L. Influence of polyol structure and molecular weight on the shape and properties of Ni0.5Co0.5Fe2O4 nanoparticles obtained by sol-gel synthesis. *Ceram. Int.* **2019**, *45*, 7458–7467. [CrossRef]
16. Dippong, T.; Levei, E.A.; Cadar, O. Formation, structure and magnetic properties of $MFe_2O_4@SiO_2$ (M = Co, Mn, Zn, Ni, Cu) nanocomposites. *Materials* **2021**, *14*, 1139. [CrossRef]
17. Ramesh, S.; Sekhar, B.C.; Rao, P.S.V.S.; Rao, B.P. Microstructural and magnetic behavior of mixed Ni–Zn–Co and Ni–Zn–Mn ferrites. *Ceram. Int.* **2014**, *40*, 8729–8735. [CrossRef]
18. Sarkar, D.; Bhattacharya, A.; Nandy, A.; Das, S. Enhanced broadband microwave reflection loss of carbon nanotube ensheathed Ni–Zn–Co-ferrite magnetic nanoparticles. *Mater. Lett.* **2014**, *120*, 259–262. [CrossRef]
19. Stergiou, C. Magnetic, dielectric and microwave adsorption properties of rare earth doped Ni-Co and Ni-Co-Zn spinel ferrite. *J. Magn. Magn. Mater.* **2017**, *426*, 629–635. [CrossRef]
20. Guo, H.S.; Zhang, L.; Yan, Y.L.; Zhang, J.; Wang, J.; Wang, S.Y.; Li, L.Z.; Wu, X.H. Effect of lanthanum substitution on structural, magnetic and electric properties of Ni–Zn–Co ferrites for radio frequency and microwave devices. *Ceram. Int.* **2022**, *48*, 22516–22522. [CrossRef]
21. Wang, Y.; Gao, X.; Wu, X.; Zhang, W.; Luo, C.; Liu, P. Facile design of 3D hierarchical $NiFe_2O_4$/N-CN/ZnO composites as a high-performance electromagnetic wave absorber. *Chem. Eng. J.* **2019**, *375*, 121942. [CrossRef]
22. Huili, H.; Grindi, B.; Viau, G.; Tahar, L.B. Influence of the stoichiometry and grain morphology on the magnetic properties of Co substituted Ni–Zn nanoferrites. *Ceram. Int.* **2016**, *42*, 17594–17604. [CrossRef]
23. Anupama, M.K.; Rudraswamy, B.; Dhananjaya, N. Investigation on impedance response and dielectric relaxation of Ni-Zn ferrites prepared by self-combustion technique. *J. Alloys Compd.* **2017**, *706*, 554–561. [CrossRef]
24. Amir, M.; Gungunes, H.; Baykal, A.; Almessiere, M.A.; Sozeri, H.; Ercan, I.; Sertkol, M.; Asiri, S.; Manikandan, A. Effect of annealing temperature on magnetic and Mossbauer properties of $ZnFe_2O_4$ nanoparticles by sol-gel approach. *Supercond. Nov. Magn.* **2018**, *31*, 3347–3356. [CrossRef]
25. Asiri, S.; Sertkol, M.; Güngüneş, H.; Amir, M.; Manikandan, A.; Ercan, I.; Baykal, A. the temperature effect on magnetic properties of $NiFe_2O_4$ nanoparticles. *J. Inorg. Organomet. Polym. Mater.* **2018**, *28*, 1587–1597. [CrossRef]
26. Ansari, S.M.; Suryawanshi, S.R.; More, M.A.; Sen, D.; Kolekar, Y.D.; Ramana, C.V. Field emission properties of nanostructured cobalt ferrite ($CoFe_2O_4$) synthesized by low-temperature chemical method. *Chem. Phys. Lett.* **2018**, *701*, 151–156. [CrossRef]
27. Thommes, M.; Kaneko, K.; Neimark, A.V.; Olivier, J.P.; Rodriguez-Reinoso, F.; Rouquerol, J.; Sing, K.S.W. Physisorption of gases, with special reference to the evaluation of surface area and pore size distribution (IUPAC Technical Report). *Pure Appl. Chem.* **2015**, *87*, 1051–1069. [CrossRef]
28. Bakhshi, H.; Vahdati, N.; Sedghi, A.; Mozharivskyj, Y. Comparison of the effect of nickel and cobalt cations addition on structural and magnetic properties of manganese-zinc ferrite nanoparticles. *J. Magn. Magn. Mater.* **2019**, *474*, 56–62. [CrossRef]
29. Pussi, K.; Gallo, J.; Ohara, K.; Carbo-Argibay, E.; Kolenko, Y.V.; Berbiellini, B.; Bansil, A.; Kamali, S. Structure of Manganese Oxide Nanoparticles Extracted via Pair Distribution Functions. *Condens. Mater.* **2020**, *5*, 19. [CrossRef]
30. Rao, K.S.; Choudary, G.S.V.R.K.; Rao, K.H.; Sujatha, C. Structural and magnetic properties of ultrafine $CoFe_2O_4$ nanoparticles. *Proc. Mat. Sci.* **2015**, *10*, 19–27. [CrossRef]
31. Ghosh, M.P.; Mukherjee, S. Microstructural, magnetic and hyperfine characterizations of Cu-doped cobalt ferrite nanoparticles. *J. Am. Chem. Soc.* **2019**, *102*, 7509–7520. [CrossRef]
32. Cullity, B.D.; Graham, C.D. *Introduction to Magnetic Materials*; John Wiley & Sons, Inc.: Hoboken, NJ, USA, 2009; 359p.
33. Dippong, T.; Levei, E.A.; Diamandescu, L.; Bibicu, I.; Leostean, C.; Borodi, G. Structural and magnetic properties of $Co_xFe_{3-x}O_4$ versus Co/Fe molar ratio. *J. Magn. Magn. Mater.* **2015**, *394*, 111–116. [CrossRef]
34. Santoro, R.P.; Newnham, R.E.; Nomura, S. Magnetic properties of Mn_2SiO_4 and Fe_2SiO_4. *J. Phys. Chem. Solids.* **1966**, 655–666. [CrossRef]

35. Sontu, U.B.; Yelasani, V.; Musugu, V.R.R. Structural, electrical and magnetic characteristics of nickel substituted cobalt ferrite nano particles, synthesized by self combustion method. *J. Magn. Magn. Mater.* **2015**, *374*, 376–380. [CrossRef]
36. Amer, M.A.; Tawfik, A.; Mostafa, A.G.; El-Shora, A.F.; Zaki, S.M. Spectral studies of Co substituted Ni–Zn ferrites. *J. Magn. Magn. Mater.* **2011**, *323*, 1445–1452. [CrossRef]
37. Eibschütz, M.; Ganiel, U. Mössbauer studies of Fe^{2+} in paramagnetic fayalite (Fe_2SiO_4). *Solid State Commun.* **1967**, *5*, 267–270. [CrossRef]

Disclaimer/Publisher's Note: The statements, opinions and data contained in all publications are solely those of the individual author(s) and contributor(s) and not of MDPI and/or the editor(s). MDPI and/or the editor(s) disclaim responsibility for any injury to people or property resulting from any ideas, methods, instructions or products referred to in the content.

Article

A Strategy for Tuning the Structure, Morphology, and Magnetic Properties of MnFe₂O₄/SiO₂ Ceramic Nanocomposites via Mono-, Di-, and Trivalent Metal Ion Doping and Annealing

Thomas Dippong [1], Erika Andrea Levei [2], Ioan Petean [3], Iosif Grigore Deac [4] and Oana Cadar [2,*]

1. Faculty of Science, Technical University of Cluj-Napoca, 76 Victoriei Street, 430122 Baia Mare, Romania
2. INCDO-INOE 2000, Research Institute for Analytical Instrumentation, 67 Donath Street, 400293 Cluj-Napoca, Romania
3. Faculty of Chemistry and Chemical Engineering, Babes-Bolyai University, 11 Arany Janos Street, 400028 Cluj-Napoca, Romania
4. Faculty of Physics, Babes-Bolyai University, 1 Kogalniceanu Street, 400084 Cluj-Napoca, Romania
* Correspondence: oana.cadar@icia.ro

Citation: Dippong, T.; Levei, E.A.; Petean, I.; Deac, I.G.; Cadar, O. A Strategy for Tuning the Structure, Morphology, and Magnetic Properties of MnFe₂O₄/SiO₂ Ceramic Nanocomposites via Mono-, Di-, and Trivalent Metal Ion Doping and Annealing. *Nanomaterials* 2023, 13, 2129. https://doi.org/10.3390/nano13142129

Academic Editor: Csaba Balázsi

Received: 4 July 2023
Revised: 19 July 2023
Accepted: 20 July 2023
Published: 22 July 2023

Copyright: © 2023 by the authors. Licensee MDPI, Basel, Switzerland. This article is an open access article distributed under the terms and conditions of the Creative Commons Attribution (CC BY) license (https://creativecommons.org/licenses/by/4.0/).

Abstract: This work presents the effect of monovalent (Ag^+, Na^+), divalent (Ca^{2+}, Cd^{2+}), and trivalent (La^{3+}) metal ion doping and annealing temperature (500, 800, and 1200 °C) on the structure, morphology, and magnetic properties of $MnFe_2O_4$/SiO_2 ceramic nanocomposites synthesized via sol–gel method. Fourier-transform infrared spectroscopy confirms the embedding of undoped and doped $MnFe_2O_4$ nanoparticles in the SiO_2 matrix at all annealing temperatures. In all cases, the X-ray diffraction (XRD) confirms the formation of $MnFe_2O_4$. In the case of undoped, di-, and trivalent metal-ion-doped gels annealed at 1200 °C, three crystalline phases (cristobalite, quartz, and tridymite) belonging to the SiO_2 matrix are observed. Doping with mono- and trivalent ions enhances the nanocomposite's structure by forming single-phase $MnFe_2O_4$ at low annealing temperatures (500 and 800 °C), while doping with divalent ions and high annealing temperature (1200 °C) results in additional crystalline phases. Atomic force microscopy (AFM) reveals spherical ferrite particles coated by an amorphous layer. The AFM images showed spherical particles formed due to the thermal treatment. The structural parameters calculated by XRD (crystallite size, crystallinity, lattice constant, unit cell volume, hopping length, density, and porosity) and AFM (particle size, powder surface area, and thickness of coating layer), as well as the magnetic parameters (saturation magnetization, remanent magnetization, coercivity, and anisotropy constant), are contingent on the doping ion and annealing temperature. By doping, the saturation magnetization and magnetocrystalline anisotropy decrease for gels annealed at 800 °C, but increase for gels annealed at 1200 °C, while the remanent magnetization and coercivity decrease by doping at both annealing temperatures (800 and 1200 °C).

Keywords: manganese ferrite; silica matrix; doping; annealing; magnetic behavior

1. Introduction

Nanoparticles display enhanced properties relative to microparticles and bulk materials, allowing their use in various applications [1]. Spinel ferrite (MFe_2O_4, M=Mn, Ni, Co, Cu, etc.) nanoparticles are developing as a family of versatile materials with controllable particle size and shape, tunable dielectric, catalytic, and magnetic properties, as well as easy and convenient synthesis processes [1–6]. Of these, manganese ferrite, $MnFe_2O_4$, displays a face-centered cubic structure with two types of cation lattice sites: tetrahedral (A) formed by four O^{2-} ions, and octahedral (B) sites composed of six O^{2-} ions. The percentage of Fe^{3+} ions occupying the A sites dictates the inversion degree. Thus, in normal spinel structure the A sites are occupied by Fe^{3+} ions, while in inverse spinel structure the A sites are occupied by Mn^{2+} ions [1]. The inversion degree highly influences the magnetic properties of $MnFe_2O_4$ nanoparticles. $MnFe_2O_4$ has attracted significant interest due to its controllable grain size, superparamagnetic nature, low coercivity (H_C), high magnetic permeability, moderate saturation magnetization (M_S), good chemical stability, high catalytic performance, capacity to be guided by an external magnetic field, surface tailoring possibility, good biocompatibility, and high crystal symmetry [1,5,7]. $MnFe_2O_4$ is also a non-toxic, non-corrosive, environmentally friendly, high thermal, and shock-resistant material often used for application in medicine, electronics, as well as in the paint and coating industry [1–5,8].

Doping with various cations enhances the ferrites' magnetic, optical, and electrical properties. The doped spinel ferrites have various benefits, i.e., they are less expensive, easy to produce, have good stability, and have different magnetic properties compared to undoped ferrites [9]. Cation distribution is significantly affected by the doping ion radius, charge, lattice energy, and crystal field stabilization energy in A and B sites [10]. In recent years, a large interest has been granted to doped ferrites due to their numerous technical applications, including magnetically controlled anticancer medication delivery, color imaging, and gas-sensitive and catalytic materials [9–11]. Metal ion doping generates oxygen vacancies and reactive oxygen species that enhance the catalytic performance [9]. In Zn^{2+}-doped $MnFe_2O_4$, Zn has a strong tendency to occupy A sites enhancing the magnetic properties [10,12]. Previous studies reported that the Mn–Zn ferrites decompose by annealing, leading to impure phases and, consequently, the decrease in magnetic and dielectric properties [10,13]. The high stability and outstanding electrical and thermal conductivity of Ag make it a dopant that improves the catalytic activity of $MnFe_2O_4$, allowing the degradation of refractory organic pollutants [6]. Ag–$MnFe_2O_4$ composites also display superparamagnetic and remarkably antibacterial activity [14]. A previous study on Ca–$MnFe_2O_4$ nanoparticles coated with citrate obtained by the sol-gel method revealed that high Ca content improves their capacity to be used as a hyperthermia agent without compromising their cytocompatibility or cellular internalization [15]. Recently, the structural tuning of $MnFe_2O_4$ by doping with rare earth ions has drawn attention as a novel technique to enhance its physical characteristics. Of these, the non-magnetic La^{3+} ion stands out due to its larger ionic radius compared to that of Fe^{3+} and Mn^{2+} ions that change the structural and magnetic properties of $MnFe_2O_4$ by the higher electron–hole pair recombination in the ferrites, supporting the shift of the electronic states [7,16]. Moreover, La^{3+} doping promotes a higher production of oxygen vacancies and photocatalytic degradation [7,16].

Given the diversity of experimental techniques (e.g., sol–gel, hydrothermal, thermal decomposition, colloid emulsion, and laser pyrolysis) used to obtain $MnFe_2O_4$ ferrites, the sol–gel route offers a flexible approach due to its low cost, low reaction temperature, simplicity, and good control of chemical composition, structural, physical–chemical, and magnetic properties [2–5]. The prolonged exposure to synthesis conditions, as well as the time of thermal processing, was found to influence the structure of the nanomaterials [17]. Solvo/hydrothermal synthesis is an environmentally friendly approach to producing small and uniformly distributed nanostructures. It also allows the easy doping and coating of the particles to generate composite materials [18]. Microwave-assisted solvothermal approach allows the fine control of process parameters, high productivity, exceptional phase purity, good reproducibility, and short reaction times concomitantly obtaining small particles with uniform particle morphology and high crystalline particles [19]. Baublytė et al. [20] showed a correlation between precursor concentration, particle size, and crystallinity.

The embedding of $MnFe_2O_4$ in mesoporous SiO_2 plays an important role in enhancing the stability in water, improving biocompatibility, and diminishing the degradation of $MnFe_2O_4$ nanoparticles. The SiO_2 coating also prevents agglomeration by controlling the dipolar attraction between the magnetic nanoparticles. Moreover, the silanol groups from the surface of mesoporous SiO_2 promote the binding of biomolecules, directing targeted ligands and drug loading on the nanocarrier surface [2–5]. Our previously reported sol–gel synthesis method allows the obtaining of homogeneous pure or mixed ferrite nanoparticles and their incorporation in inorganic or organic matrices, requires reduced time and energy and has a short gelation time. The obtaining of $MnFe_2O_4$ embedded in the SiO_2 matrix sol–gel method consist in the mixing of reactants with tetraethylorthosilicate (TEOS) and the formation of strong networks with moderate reactivity that permit the incorporation of various inorganic and organic molecules [2–5]. The simple variation in synthesis conditions such as pH, time and annealing temperature allows a high degree of control over the nucleation and particle growth [1–6]. The easily controllable magnetic, electrical, dielectric, and optical properties of $MnFe_2O_4/SiO_2$ nanocomposites recommend their use in various technological and scientific systems, such as magnetic devices, catalysis, and sensors [3,5,21,22]. The $MnFe_2O_4/SiO_2$ nanocomposites display remarkable electrical and magnetic properties, high chemical and thermal stability, improved microwave absorption performance owing to the strong eddy current loss, excellent attenuation characteristic, better impedance matching, and multiple Debye relaxation processes [23]. The magnetic $MnFe_2O_4/SiO_2$ nanocomposites are widely studied due to their potential applications in different areas such as electronic, microwave, and communication devices, information storage systems, ferrofluid technology, gas sensors, magnetocaloric refrigeration, and for photocatalytic activity [1,24–27]. Moreover, the $MnFe_2O_4$ and SiO_2 integrated into a single entity (nanocomposites particle) is of particular interest in magnetic fluid hyperthermia due to $MnFe_2O_4$ superior magnetization and biocompatibility of SiO_2 [28] and drug delivery applications by providing the advantages of mesoporous silica surface (e.g., drug loading and surface functionalization) and the magnetic nature of $MnFe_2O_4$ nanoparticles (e.g., magnetic controllability and targeted drug delivery) [29].

The architecture adopted by ferrites depends on the metal ion(s) size, charge and concentration, crystal field effects and electrostatic contribution to the lattice energy, while the particle size increase and the volume-to-surface ratio decreases with annealing temperature. Moreover, due to its high degree of magnetization compared to other nanoferrites, $MnFe_2O_4$ has become important for various biomedical applications. Additionally, tailoring $MnFe_2O_4$ by doping with various ions could enhance its magnetic, optical, and electrical properties. Thus, producing homogenous doped $MnFe_2O_4$ nanoparticles with tailored magnetic properties and crystalline structures is challenging, but it is important to discover novel approaches to increase their potential for existing and new conceivable applications. In this regard, this study was conducted to assess the changes in structure, morphology, surface, and magnetic properties of $MnFe_2O_4$ doped with monovalent (Ag^+, $Ag_{0.1}Mn_{0.95}Fe_2O_4$; Na^+, $Na_{0.1}Mn_{0.95}Fe_2O_4$), divalent (Ca^{2+}, $Ca_{0.1}Mn_{0.9}Fe_2O_4$; Cd^{2+}, $Cd_{0.1}Mn_{0.9}Fe_2O_4$), and trivalent (La^{3+}, $La_{0.1}MnFe_{1.9}O_4$) metal ions embedded in a SiO_2 matrix synthesized through a modified sol–gel method, followed by annealing at 500, 800, and 1200 °C.

2. Materials and Methods

2.1. Reagents

All chemicals were used as received without further purification and purchased from different commercial sources as follows: manganese nitrate tetrahydrate ($Mn(NO_3)_2 \cdot 4H_2O$, Merck, Darmstadt, Germany), ferric nitrate nonahydrate ($Fe(NO_3)_3 \cdot 9H_2O$, 98%, Merck, Darmstadt, Germany), silver nitrate ($AgNO_3$, 99%, Carlo Erba, Milan, Italy), sodium nitrate ($NaNO_3$, 99%, Merck, Darmstadt, Germany), calcium nitrate tetrahydrate ($Ca(NO_3)_2 \cdot 4H_2O$, 99%, Carlo Erba, Milan, Italy), cadmium nitrate tetrahydrate ($Cd(NO_3)_2 \cdot 4H_2O$, 99%, Carlo Erba, Milan, Italy), lanthanum nitrate hexahydrate ($La(NO_3)_3 \cdot 6H_2O$, 98%, Carlo Erba, Milan, Italy) 1,3 propanediol (1,3–PD, 99%, Merck, Darmstadt, Germany), TEOS (99%, Merck), and ethanol (96%, Merck, Darmstadt, Germany).

2.2. Synthesis

$MnFe_2O_4$, $Ag_{0.1}Mn_{0.95}Fe_2O_4$, $Na_{0.1}Mn_{0.95}Fe_2O_4$, $Ca_{0.1}Mn_{0.9}Fe_2O_4$, $Cd_{0.1}Mn_{0.9}Fe_2O_4$, and $La_{0.1}MnFe_{1.9}O_4$ embedded in SiO_2 gels containing 50 wt.% ferrite and 50 wt.% SiO_2 were prepared through a modified sol–gel route using different M/Co/Fe (M = Ag, Na, Ca, Cd, La) molar ratios, namely 0/1/2 ($MnFe_2O_4$), 1/9.5/20 ($Ag_{0.1}Mn_{0.95}Fe_2O_4$, $Na_{0.1}Co_{0.95}Fe_2O_4$),

1/9/20 ($Cd_{0.1}Co_{0.9}Fe_2O_4$, $Ca_{0.1}Co_{0.9}Fe_2O_4$), and 1/10/19 ($La_{0.1}CoFe_{1.9}O_4$). The key advantages of the sol-gel method are versatility, simplicity, effectiveness, achievement of high purity products, narrow particle size distribution, and uniform nanostructure at low temperatures. The main disadvantages refer to the presence of amorphous phases at low annealing temperatures and secondary crystalline phases at high annealing temperatures, respectively [22]. Briefly, the sol–gel method used here involves the following steps: the reactants (metal nitrates and polyols) are mixed with TEOS at ambient temperature, the sol is exposed to ambient temperature until the gelation of the SiO_2 network, followed by the thermal-assisted formation of carboxylate precursors and their decomposition to a multicomponent system (mixed oxide). Generally, to obtain spinel ferrites via the sol–gel method, nitrate salts are preferred as precursors, as they are a convenient source of aqueous metal ions and act as low-temperature oxidizing agents for the synthesis [22,30]. Accordingly, the initial sols were prepared by mixing the metal nitrates with 1,3–PD, TEOS and ethanol using a NO_3^-/1,3–PD/TEOS molar ratio of 1/1/1. The resulting sols were vigorously stirred over a 1 h and kept at ambient air until gelation occurred. The obtained gels consisting of a homogenous mixture of TEOS, 1,3–PD and metal nitrates were ground thoroughly using an agate mortar pestle and heated at 40 °C for 5 h and 200 °C for 5 h, respectively, in an UFE 400 universal oven (Memmert, Schwabach, Germany). Finally, the powder samples were annealed at different temperatures (500, 800 and 1200 °C) for 5 h at 10 °C/min using a LT9 (Nabertherm, Lilienthal, Germany) muffle furnace, at ambient temperature.

2.3. Characterization

X-ray diffraction patterns were recorded on a D8 Advance (Bruker, Karlsruhe, Germany) diffractometer equipped with an X-ray tube (CuKα radiation, λ = 1.54060 Å, 40 kV and 35 mA) and 1–dimensional LynxEye detector; data collection was carried out in the 2θ range of 15–80°, with a step size of 0.015° and counting time of 1s/step. The Fourier-transform infrared (FT–IR) spectra in the range of 400–4000 cm^{-1} were recorded in transmittance mode with a resolution of 2 cm^{-1} and 8 scans on KBr pellets containing 1% sample using a Perkin–Elmer Spectrum BX II (Perkin Elmer, Waltham, MA, USA) spectrometer equipped with DTGS detectors. For atomic force microscopy (AFM), the thermally treated powders were dispersed in ultrapure water by stirring to enable the finest particles to release. Bigger particles sediment on the bottom of the vials, while the finest ferrite particles remained dispersed due to Brownian motion. They were transferred onto a glass slide via vertical adsorption, and the formed thin films were dried at room temperature and investigated using a JSPM 4210 (JEOL, Tokyo, Japan) microscope. The AFM was operated in alternative current mode with NSC 15 (Mikromasch, Sofia, Bulgaria) cantilevers with a nominal force constant of 40 N/m frequency of 325 kHz. At least three macroscopic areas of 1 μm^2 of the thin films were scanned to obtain the topographic images. The thin ferrite films' height, surface roughness (Rq), particle diameter, and surface area were obtained by analyzing the images with WinSPM System Data Processing Version 2.0 by JEOL. The magnetic measurements were carried out at room temperature using a cryogen free vibrating-sample magnetometer (VSM), CFSM—12 T (Cryogenic Ltd., London, UK). The hysteresis loops were recorded with a maximum field of 2 T, while the magnetization was measured in a high magnetic field up to 7 T. The powder samples were embedded into epoxy resin to avoid the motion of particles.

3. Results and Discussion

3.1. FT–IR Analysis

The functional groups, molecular geometry and inter-molecular interactions remarked in the FT–IR spectra of undoped and doped $MnFe_2O_4$ thermally treated at 40, 200, 500, 800, and 1200 °C are presented in Figure 1.

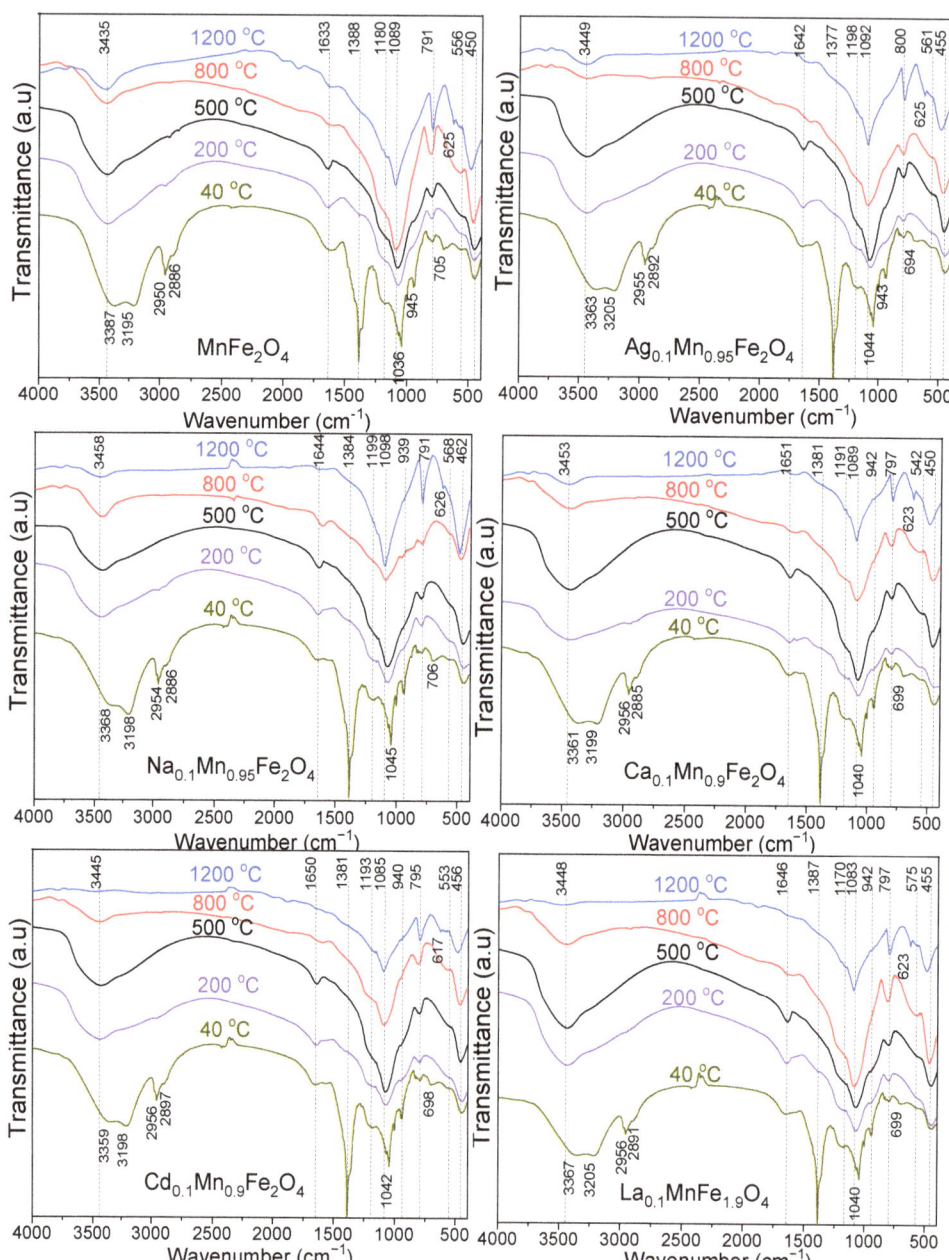

Figure 1. FT–IR spectra of $MnFe_2O_4$, $Ag_{0.1}Mn_{0.95}Fe_2O_4$, $Na_{0.1}Mn_{0.95}Fe_2O_4$, $Ca_{0.1}Mn_{0.9}Fe_2O_4$, $Cd_{0.1}Mn_{0.9}Fe_2O_4$, and $La_{0.1}MnFe_{1.9}O_4$ gels thermally treated at 40, 200, 500, 800, and 1200 °C.

In all cases, the FT-IR spectra show the representative bands of the SiO_2 matrix, namely: O–H bond vibration in the Si–OH group (3435–3458 cm^{-1}), H–O–H bond bending vibration (1633–1651 cm^{-1}), Si–O–Si bonds stretching vibration (1083–1098 cm^{-1}), Si–O chains vibration in SiO_4 tetrahedron (791–800 cm^{-1}), Si–O bond vibration (450–462 cm^{-1})

and Si–O–Si cyclic structures vibration (542–575 cm^{-1}) [2–5,16]. The deformation vibration of Si–OH resulted during the hydrolysis of –Si(OCH$_2$CH$_3$)$_4$ groups of TEOS in gels dried at 40 °C shown by the shoulder at 939–945 cm^{-1} disappears at higher temperatures. The water is present on the surface and in the volume of SiO$_2$ particles in physically and chemically (bounded to the surface and molecularly dispersed) bound forms [31,32]. The physiosorbed water does not interact strongly with the particle's surface and it can be easily removed at low temperatures, whereas the chemisorbed water is removed at higher temperatures [32]. Removing water from most metal oxide nanoparticles may be incomplete regardless of the used temperature and lead to coarsening and phase transformation [32].

For gels dried at 40 °C, the intense band at 1633 cm^{-1} is attributed to –O–H bond vibrations in the diol, physically and chemically bound water molecules [2–5,16,32], while for gels thermally treated at 200 °C this band shifts to 1642 cm^{-1} and is attributed to the C=O bond vibration in the carboxylate precursors and the chemically absorbed water molecules [31]. Increasing the thermal treatment temperature, this band progressively decreases following the decomposition of carboxylate precursors and loss of chemically absorbed water, until it disappears at 1200 °C [16]. The presence of this band at high temperatures could be explained by the high hygroscopicity of the nanoparticles and the presence of polar hydroxyl groups, both on the silica surface and in its volume [32]. The dissociation of chemically adsorbed water on the particle surface with the formation of hydroxyl groups stabilizes the surface, reduces the water mobility, and increases the nanoparticle's stability [32].

In the case of gels dried at 40 °C, the intense band at 1633 cm^{-1} is attributed to the overlapping of C–O and H–O deformation vibrations [2–5,16]. This band shifts to 1642 cm^{-1} in the case of gels thermally treated at 200 °C and is attributed to the C=O bond vibration in the carboxylate precursors; by further increasing the thermal treatment temperature, it progressively decreases till it disappears at 1200 °C [33]. In gels thermally treated at 40 °C, the broad bands at 3359–3387 cm^{-1} and 3195–3205 cm^{-1} are attributed to O–H stretching in precursors and to intermolecular hydrogen bonds [16,34]. The intense band around 1377–1388 cm^{-1} present only in gels dried at 40 °C is characteristic of N–O bond's asymmetric vibration in nitrates. The absence of this band at high temperatures suggests that the reaction between nitrates and 1,3–PD with the formation of metal–carboxylate precursors has already occurred [16]. The asymmetric and symmetric bands at 2950–2956 cm^{-1} and 2885–2897 cm^{-1} present only in gels at 40 °C, are characteristic of C–H bond vibration in the methylene groups of 1,3–PD and carboxylates precursors and disappear at higher temperatures when the precursors decompose [16,34].

The vibration of M–O bonds in A sites is indicated by the band at 542–575 cm^{-1} and in B sites by the band at 450–462 cm^{-1} [16]. The bands at 617–626 cm^{-1} (1200 °C) and 694–706 cm^{-1} (200 °C) are attributed to O–Fe–O and Fe–OH bond vibration [33,34]. The band shift is ascribed to the modification in M–O bond length at the A and B sites due to the introduction of large size rare earth La^{3+} ion [7].

3.2. Structural Analysis

The XRD patterns of gels annealed at 500, 800 and 1200 °C are presented in Figure 2. The diffraction peaks of MnFe$_2$O$_4$ gels match the reflection planes of (111), (220), (311), (222), (400), (422), (511), (440), (531), (620), (533), (622) and (444) confirming the presence of pure, low–crystallized MnFe$_2$O$_4$ (JCPDS #00–010–0319) phase with a cubic spinel structure (space group $Fd3m$) [16]. At 500 °C, single-phase poorly crystallized MnFe$_2$O$_4$ is remarked, while at 800 and 1200 °C, the better crystallized MnFe$_2$O$_4$ is accompanied by α–Fe$_2$O$_3$ (JCPDS #00–033–0664) and cristobalite (JCPDS #00–074–9378) [5,16]. XRD patterns of all doped ferrites annealed at 500 °C present single-phase, low-crystallized MnFe$_2$O$_4$. Contrary to undoped MnFe$_2$O$_4$, for MnFe$_2$O$_4$ doped with monovalent (Ag$^+$ and Na$^+$) ions annealed at 800 °C, the presence of single-phase crystalline MnFe$_2$O$_4$ is remarked.

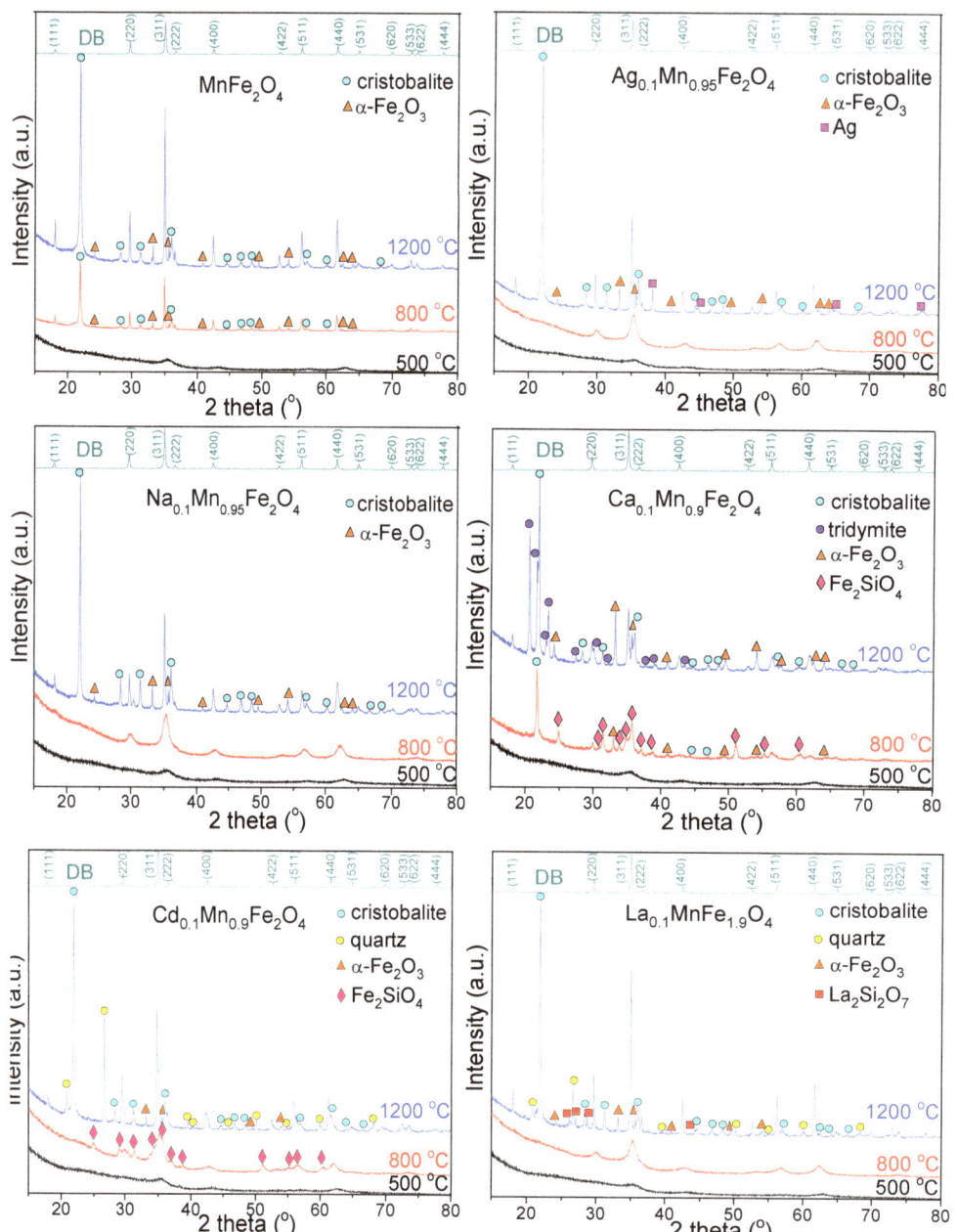

Figure 2. XRD patterns of $MnFe_2O_4$, $Ag_{0.1}Mn_{0.95}Fe_2O_4$, $Na_{0.1}Mn_{0.95}Fe_2O_4$, $Ca_{0.1}Mn_{0.9}Fe_2O_4$, $Cd_{0.1}Mn_{0.9}Fe_2O_4$, and $La_{0.1}MnFe_{1.9}O_4$ gels annealed at 500, 800, and 1200 °C.

For $Ag_{0.1}Mn_{0.95}Fe_2O_4$ annealed at 1200 °C, besides the $MnFe_2O_4$ crystalline phase, cristobalite and metallic Ag (JCPDS #00–033–0664) are also formed, indicating the presence of unreacted Ag in the SiO_2 matrix. The diffraction patterns of $MnFe_2O_4$ doped with divalent (Ca^{2+} and Cd^{2+}) ions annealed at 800 °C display various secondary phases.

For $Cd_{0.1}Mn_{0.9}Fe_2O_4$, the $MnFe_2O_4$ crystalline phase is accompanied by Fe_2SiO_4 (JCPDS #00–071–1400), while for $Ca_{0.1}Mn_{0.9}Fe_2O_4$, the barely crystalline $MnFe_2O_4$ is attended by Fe_2SiO_4, α–Fe_2O_3, and cristobalite. At 1200 °C, the $MnFe_2O_4$ is accompanied by the crystalline phases belonging to SiO_2 matrix (cristobalite and tridymite (JCPDS #00–074–8988)) and α–Fe_2O_3 for $Ca_{0.1}Mn_{0.9}Fe_2O_4$, and cristobalite, quartz (JCPDS #00–079–1910), and α–Fe_2O_3 (in a smaller amount than in the case of $Ca_{0.1}Mn_{0.9}Fe_2O_4$) for $Cd_{0.1}Mn_{0.9}Fe_2O_4$. For $MnFe_2O_4$ doped with trivalent metal ions ($La_{0.1}MnFe_{1.9}O_4$), single-crystalline-phase $MnFe_2O_4$ at 800 °C, and additional secondary phases (cristobalite, quartz, α–Fe_2O_3, and $La_2Si_2O_7$ (JCPDS #00–081–0461)) are observed at 1200 °C. Generally, the micron-sized SiO_2 is crystalline, while the amorphous SiO_2 refers to particle sizes up to 100 nm. Thus, in the case of $MnFe_2O_4$ doped with divalent (Ca^{2+}, Cd^{2+}) and trivalent (La^{3+}) ions annealed at 1200 °C, three polymorph crystalline phases of the SiO_2 matrix (cristobalite, quartz and tridymite) are formed, while in the case of doping with divalent (Ca^{2+} and Cd^{2+}) ions annealed at 800 °C, the presence of crystalline Fe_2SiO_4 is remarked.

The structural parameters, i.e., crystallite size (D_{XRD}), degree of crystallinity (DC), lattice constant (a), unit cell volume (V), distance between magnetic ions—hopping length in A (d_A) and B (d_B) sites, physical density (d_p), X-ray density (d_{XRD}), and porosity (P) of gels annealed at 500, 800, and 1200 °C determined by XRD are displayed in Table 1.

Table 1. Structural parameters of $MnFe_2O_4$, $Ag_{0.1}Mn_{0.95}Fe_2O_4$, $Na_{0.1}Mn_{0.95}Fe_2O_4$, $Ca_{0.1}Mn_{0.9}Fe_2O_4$, $Cd_{0.1}Mn_{0.9}Fe_2O_4$, and $La_{0.1}MnFe_{1.9}O_4$ gels annealed at 500, 800, and 1200 °C.

Parameter	Temp (°C)	$MnFe_2O_4$	$Ag_{0.1}Mn_{0.95}Fe_2O_4$	$Na_{0.1}Mn_{0.95}Fe_2O_4$	$Ca_{0.1}Mn_{0.9}Fe_2O_4$	$Cd_{0.1}Mn_{0.9}Fe_2O_4$	$La_{0.1}MnFe_{1.9}O_4$	Error
D_{XRD} (nm)	500	14.2	10.2	14.0	11.8	12.0	11.1	±1.3
	800	16.7	14.7	15.9	15.3	16.3	16.2	±1.6
	1200	66.3	55.4	40.1	58.0	56.5	50.0	±5.5
DC (%)	500	61.5	48.5	48.9	61.2	60.8	59.9	±5.0
	800	70.2	62.0	69.4	66.9	63.9	68.0	±6.6
	1200	90.1	88.8	86.3	85.5	89.5	88.6	±8.7
a (Å)	500	8.445	8.414	8.427	8.409	8.418	8.441	±0.01
	800	8.485	8.457	8.462	8.443	8.467	8.478	±0.01
	1200	8.544	8.504	8.510	8.491	8.533	8.517	±0.01
V (Å^3)	500	602.3	595.7	598.4	594.6	596.5	601.4	±0.01
	800	610.9	604.6	605.9	601.9	607.0	609.4	±0.01
	1200	623.7	615.0	616.3	612.2	621.3	617.8	±0.01
d_A (Å)	500	3.657	3.643	3.649	3.641	3.645	3.655	±0.01
	800	3.674	3.662	3.664	3.656	3.666	3.671	±0.01
	1200	3.700	3.682	3.685	3.677	3.695	3.688	±0.01
d_B (Å)	500	2.986	2.975	2.979	2.973	2.976	2.984	±0.01
	800	2.999	2.990	2.992	2.985	2.994	2.997	±0.01
	1200	3.021	3.007	3.009	3.002	3.017	3.011	±0.01
d_p (g/cm^3)	500	4.133	4.375	4.187	4.298	4.340	4.388	±0.01
	800	4.255	4.554	4.299	4.420	4.471	4.472	±0.01
	1200	4.334	4.633	4.474	4.549	4.555	4.577	±0.01
d_{XRD} (g/cm^3)	500	5.087	5.322	5.110	5.119	5.264	5.278	±0.01
	800	5.015	5.244	5.047	5.057	5.173	5.209	±0.01
	1200	4.912	5.155	4.962	4.972	5.054	5.138	±0.01
P (%)	500	18.7	17.8	18.0	16.0	17.5	16.9	± 1.6
	800	15.1	13.2	14.8	12.6	13.6	14.1	± 1.2
	1200	11.8	10.1	9.83	8.51	9.87	10.9	± 1.0

The effect of doping with various ions results in lower D_{XRD}, i.e., 14.2–10.2 nm (500 °C), 16.7–14.7 nm (800 °C), and 66.3–40.1 nm (1200 °C) compared to undoped $MnFe_2O_4$. By increasing the temperature, the diffraction peaks become narrower and sharper indicating that the crystallite size increases and the surface area decreases. The D_{XRD} increases with the annealing temperature due to the crystallite agglomeration without recrystallization, leading to the formation of a single crystal instead a polycrystalline structure at high temperatures (1200 °C) [2–5]. The DC was calculated as the ratio between the area of the crystalline contribution and the total area. The lattice constant of undoped and doped $MnFe_2O_4$ gels increases with the annealing temperature, but the replacement of Mn^{2+} ion by metal ions leads to a decrease in lattice constant and the formation of a compositionally homogeneous solid solution. Moreover, the variation in the lattice constant causes internal stress and defeats further grain growth during the annealing process [2–5,16,33,35]. When a large-sized La^{3+} ion (1.6061 Å) replaces the small-sized Fe^{3+} ion (0.645 Å) initiates higher asymmetry in the lattice structure [6]. The increase in molecular weight of ferrites is more significant than the increase in volume (V, Table 1), but the molecular weight is more affected by the increase in unit cell volume [2–5]. The distance between magnetic ions (d, hopping length) in A and B sites of gels annealed at 800 and 1200 °C is higher for undoped $MnFe_2O_4$ than doped $MnFe_2O_4$ (Table 1). The lower value of physical density (d_p, Table 1) of undoped $MnFe_2O_4$ compared to doped $MnFe_2O_4$ could be attributed to the formation of pores through the synthesis processes [2–5]. The variation in XRD and physical densities (d_{XRD} and d_p, Table 1) caused by small fluctuations in lattice constant can be explained by considering the changes in the cation distribution within the A and B sites [2–5,16,33,35]. The rapid densification during the annealing process leads to a slight decrease in porosity (P, Table 1) with the increase in annealing temperature [2–5]. The P value of doped $MnFe_2O_4$ is lower than that of undoped $MnFe_2O_4$. The decrease in P with the increase in d_p may be the consequence of the different grain sizes, while by annealing, the growth of the irregular shape grains occurs [2–5]. Concluding, D_{XRD}, DC, a, V, d_A, d_B and d_p increase, while d_{XRD} and P decrease with increasing annealing temperature.

3.3. Morphological Analysis

The sample morphology was investigated through the AFM microscopy, the nanoparticles disposal in the adsorbed thin film and their shape and size is better visible in the topographic images in Figure 3. The increase in the particle diameter with the annealing temperature evidences the evolution of the crystalline phase as a function of temperature. The diameter of pure, spherical $MnFe_2O_4$ nanoparticles (Figure 3a–c) is strongly influenced by the annealing temperature (15 nm at 500 °C, 18 nm at 800 °C and 70 nm at 1200 °C). These values are in good agreement with the crystallite size estimated by XRD data and demonstrate that the observed nanoparticles are mono–crystalline (each observed nanoparticle represents one crystallite). At low annealing temperatures, the nanoparticles are uniformly distributed onto the thin film, but at 800 °C the distribution is slightly changed due to the increase in the crystallite size domains and significantly changed at 1200 °C due to the growth of crystallite size. These results are in good agreement with previous data reported regarding $MnFe_2O_4$ nanoparticles [36].

At low annealing temperatures, Ag^+ doping has a low impact on the size and shape of the nanoparticles, i.e., the particle diameter is around 13 nm (Figure 3d) at 500 °C, while at 800 °C few big nanoparticles (35–40 nm) are surrounded by a uniform and compact film of 16 nm particles (Figure 3e). However, by annealing at 1200 °C, the diameter of spherical shape particles of 57 nm has a uniform distribution in the thin film (Figure 3f). Similarly, Na^+ doping (Figure 3g–i) does not significantly influence the particle diameter compared to that of undoped $MnFe_2O_4$. The diameter of particles at 500 °C (17 nm) and 800 °C (20 nm) and 1200 °C (45 nm) decreases compared to that of undoped $MnFe_2O_4$. The decrease in the ferrite particle size by doping with monovalent (Ag^+, Na^+) ions and annealing at 1200 °C is most probably due to the shrinkage of the crystal lattice and dependence of particle size on annealing temperature [37,38].

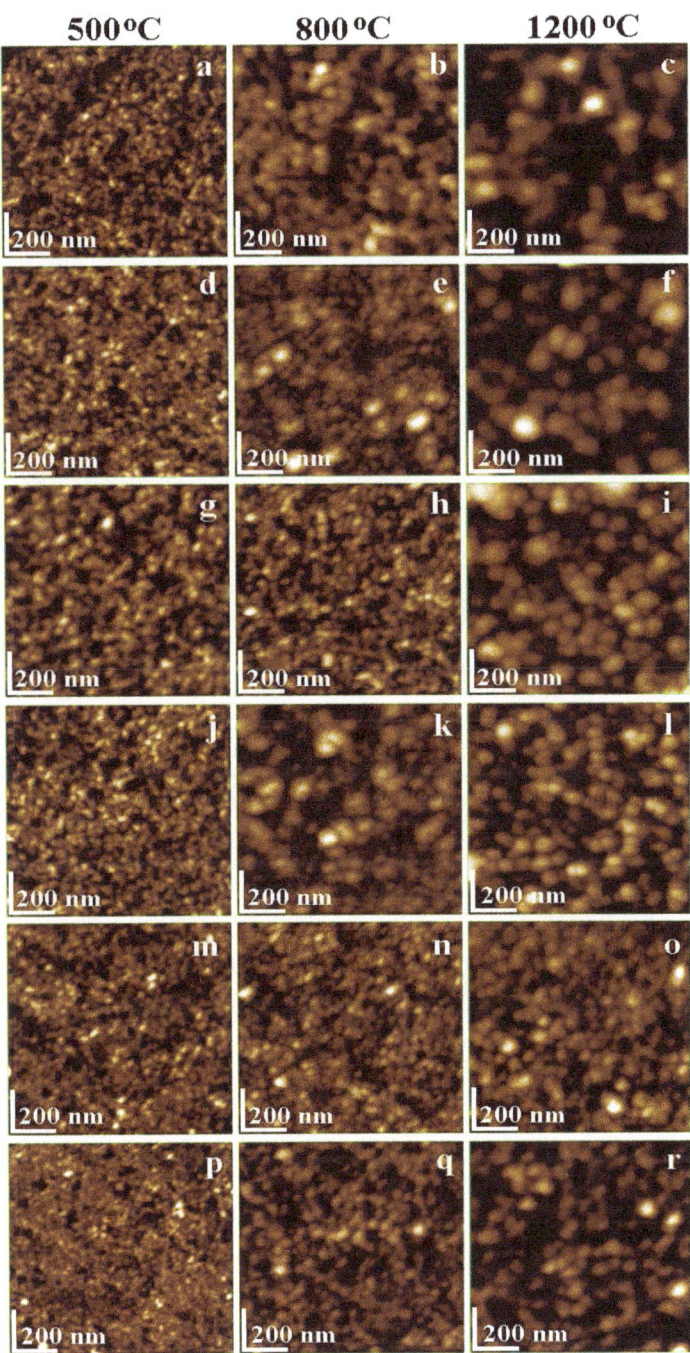

Figure 3. AFM topographic images of $MnFe_2O_4$ (**a–c**), $Ag_{0.1}Mn_{0.95}Fe_2O_4$ (**d–f**), $Na_{0.1}Mn_{0.95}Fe_2O_4$ (**g–i**), $Ca_{0.1}Mn_{0.90}Fe_2O_4$ (**j–l**), $Cd_{0.1}Mn_{0.90}Fe_2O_4$ (**m–o**), $La_{0.1}MnFe_{1.9}O_4$ (**p–r**), nanoparticles annealed at 500, 800 and 1200 °C.

Ca^{2+} doping leads to the formation of spherical nanoparticles with a diameter depending on the annealing temperature. Accordingly, low temperature annealing generates fine nanoparticles (12 nm), in good agreement with the crystallite sizes of about 10 nm (Figure 3j), while by increasing the annealing temperature to 800 °C (Figure 3k), the particle diameter also increases, resulting in nanoparticles of about 40–45 nm surrounded by smaller nanoparticles (18 nm). By annealing at 1200 °C, the particle diameter increases to 60 nm (Figure 3l), slightly lower than that of undoped $MnFe_2O_4$, most probably due to some nanostructure refinement induced by Ca^{2+} doping. At high annealing temperatures, Cd^{2+} doping influences the evolution of $MnFe_2O_4$ nanoparticles, but their shape is maintained spherical. This fact agrees with the XRD data indicating that Cd^{2+} doping leads to a decrease in crystallite size. A compact, thin film of well-individualized spherical nanoparticles with an average diameter of about 14 nm are observed at 500 °C (Figure 3m) and 8 nm at 800 °C (Figure 3n), while at 1200 °C the diameter of the particles is 58 nm (Figure 3o), due to better development of the crystalline phase within the ferrite grains.

La^{3+} doping has a moderate influence on the particle diameter compared to undoped $MnFe_2O_4$. Thermal treatment at 500 and 800 °C leads to forming small spherical nanoparticles of about 13 nm, respectively 19 nm (Figure 3p,q). However, higher temperatures (1200 °C) result in nanoparticles of about 55 nm (Figure 3r), the well-contoured and individualized nanoparticles proving a solid consolidation of crystalline phase, in good agreement with the XRD data.

Thin film topography is better evidenced by the tridimensional images in Figure 4, the revealed aspects being in close relation with roughness and the other surface parameters presented in Table 2. The thin film surface was measured on the AFM tridimensional profiles in Figure 4 and the values are shown in Table 2. The surface irregularities significantly increase at 1200 °C, as the surface of thin film is bigger for the gels annealed at this temperature.

All ferrite nanoparticles released into the aqueous solutions are well individualized and were adsorbed uniformly onto the glass slide forming well-structured compact thin films. Depending on the annealing temperature, different topographies were remarked. In the case of gels annealed at 500 °C, the nanoparticles form a very thin uniform film with heights of 8 to 12 nm (Figure 4a,d,g,j,m,p) and low (0.5–1.34 nm) surface roughness (Table 2). The roughest film was obtained by the undoped $MnFe_2O_4$ and the smoothest for La^{3+} doped $MnFe_2O_4$ nanoparticles. Thin films obtained by the adsorption of the Na^+, Cd^{2+}, and La^{3+}-doped $MnFe_2O_4$ nanoparticles annealed at 800 °C (Figure 4h,n,q) are compact and smooth, dominated by relatively small particles uniformly distributed on the film surface (Table 2). The other gels annealed at 800 °C present a few big particles mixed between the average-sized ones, generating a relatively irregular surface (Figure 4b,e,k) with increased roughness (1.6–3.8 nm, Table 2). Bigger nanoparticles resulted after annealing at 1200 °C form a uniform, compact and well-structured thin film with relatively irregular topography (Figure 4c,f,i,l,o,r) and relatively high surface roughness (2.28–4.78 nm, Table 2). Doped $MnFe_2O_4$ thin films with similar roughness were obtained by various methods, such as sputtering [39], spin coating [40], and spray pyrolysis [41].

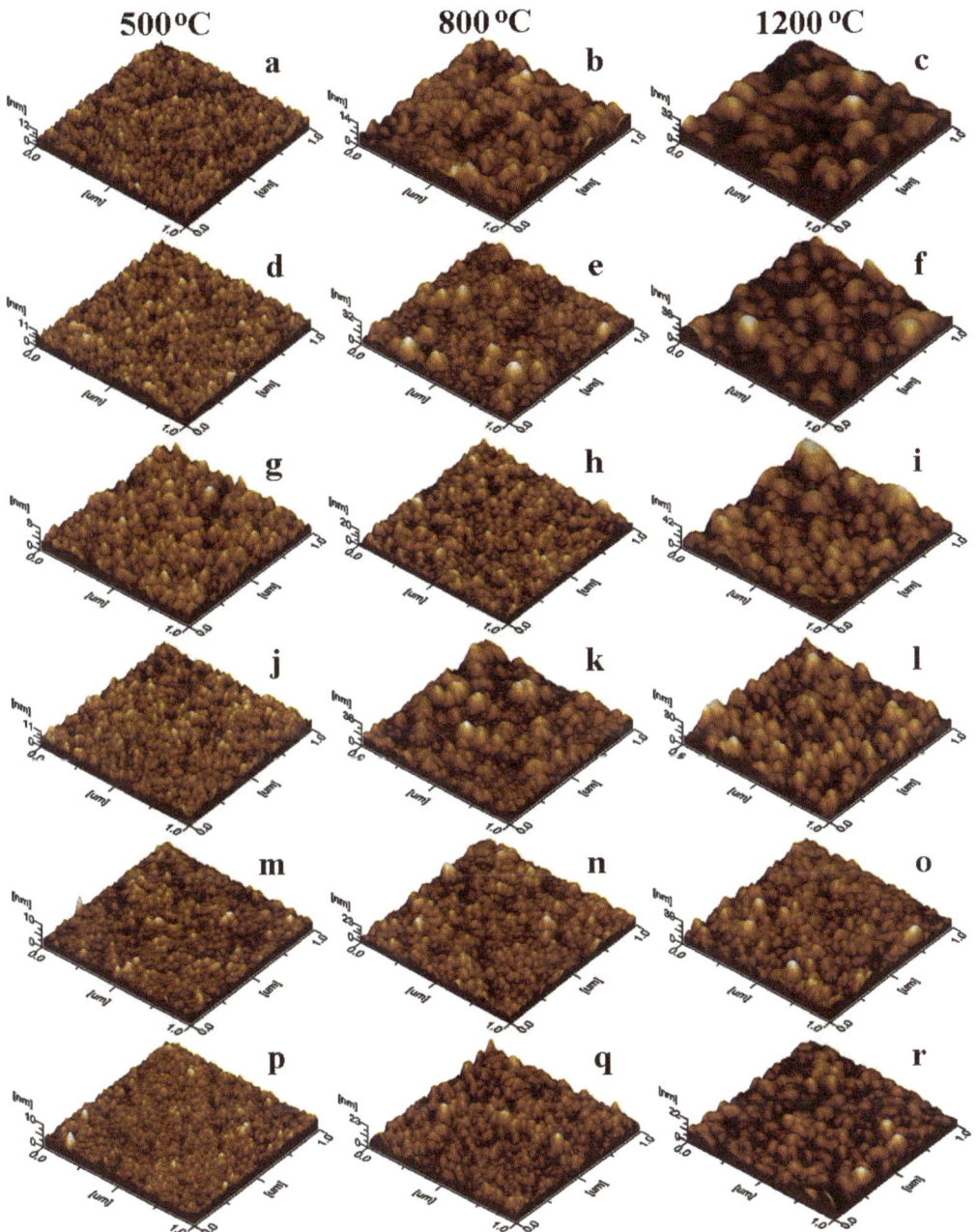

Figure 4. Tridimensional profiles of $MnFe_2O_4$ (**a–c**), $Ag_{0.1}Mn_{0.9}Fe_2O_4$ (**d–f**), $Na_{0.1}Mn_{0.95}Fe_2O_4$ (**g–i**), $Ca_{0.1}Mn_{0.95}Fe_2O_4$ (**j–l**), $Cd_{0.1}MnFe_{1.9}O_4$ (**m–o**), and $La_{0.1}Mn_{0.9}Fe_2O_4$ (**p–r**) nanoparticles annealed at 500, 800, and 1200 °C.

Table 2. Average nanoparticles diameter (D_{AFM}), thin film height (H), roughness (Rq) and surface of $MnFe_2O_4$, $Ag_{0.1}Mn_{0.95}Fe_2O_4$, $Na_{0.1}Mn_{0.95}Fe_2O_4$, $Ca_{0.1}Mn_{0.90}Fe_2O_4$, $Cd_{0.1}Mn_{0.9}Fe_2O_4$, and $La_{0.1}Mn_1Fe_{1.9}O_4$ gels annealed at 500, 800, and 1200 °C.

Gel	Temperature (°C)	H (nm)	Rq (nm)	D_{AFM} (nm)	Surface (nm²)
$MnFe_2O_4$	500	12	1.34	15	1012
	800	14	1.60	18	1017
	1200	32	3.73	70	1022
$Ag_{0.1}Mn_{0.95}Fe_2O_4$	500	11	1.28	13	1019
	800	32	3.08	16	1023
	1200	38	4.78	57	1035
$Na_{0.1}Mn_{0.95}Fe_2O_4$	500	8	0.92	17	1014
	800	20	2.29	20	1031
	1200	42	5.25	45	1044
$Ca_{0.1}Mn_{0.9}Fe_2O_4$	500	11	1.29	12	1017
	800	36	3.81	18	1027
	1200	30	4.17	60	1039
$Cd_{0.1}Mn_{0.9}Fe_2O_4$	500	10	0.50	14	1009
	800	23	2.41	18	1037
	1200	39	3.87	58	1054
$La_{0.1}MnFe_{1.9}O_4$	500	10	0.54	13	1014
	800	23	2.40	19	1030
	1200	22	2.28	55	1022
Error	-	± 1.0	± 0.20	± 5.0	±5.0

3.4. Magnetic Properties

For the ideal (containing no defects) $MnFe_2O_4$ ferrites, the magnetic properties are dictated by the antiferromagnetic interactions between the magnetic cations distributed between the A and B sites of the spinel structure. By doping with different ions, this distribution can be changed to manipulate the main magnetic parameters of the samples. The distribution of M^{2+} ions between the A and B sites in some ferrites can be modified by heat treatment, depending on whether the compounds are slowly cooling down from a high temperature or are quenched to a lower temperature [42]. The grain boundaries contain unreacted phases with isolate disordered magnetic moments for polycrystalline samples. In the case of nanoparticles, a large surface-to-volume ratio implies that many atoms are at the surface or near the surface with the associated spin distortion due to the surface effects [35,43,44]. This will make dominant the behavior of the surface atoms and from the close vicinity of the surface of the nanoparticle particle over those from the core as shown in Table 2, the annealing temperature strongly affecting the average nanoparticles diameter. The smaller the particles' size, the more different their magnetic behavior is compared to the bulk material behavior. Thus, magnetic properties are influenced by the crystalline structure, defects, and cationic distribution, which can be controlled by both ion doping and annealing temperature [42].

The magnetic hysteresis loops, $M(\mu_0 H)$ and the first derivatives $dM/d(\mu_0 H)$ of gels annealed at 800 °C (Figure 5) and 1200 °C (Figure 6) indicate a ferromagnetic behavior. The single maximum in the $dM/d(\mu_0 H)$ vs. $\mu_0 H$ curves close to the coercive field indicates the occurrence of a single magnetic phase, specific for crystalline samples [5]. The magnetic hysteresis loops indicate small coercive fields (H_C) ascribed to particle coalescence during annealing and to relatively low saturation magnetization (M_S) values [2–5,16,33,35]. For all doping ions, the particle sizes increase with increasing annealing temperature. For the gels annealed at 1200 °C, the magnetization saturates in low magnetic fields, while for those annealed at 800 °C, the saturation is reached in higher magnetic fields, suggesting a higher magnetic disorder. Smal particle size has larger magnetic disorder at the surface, containing isolated magnetic moments or canted spin, which require a higher magnetic field for saturation.

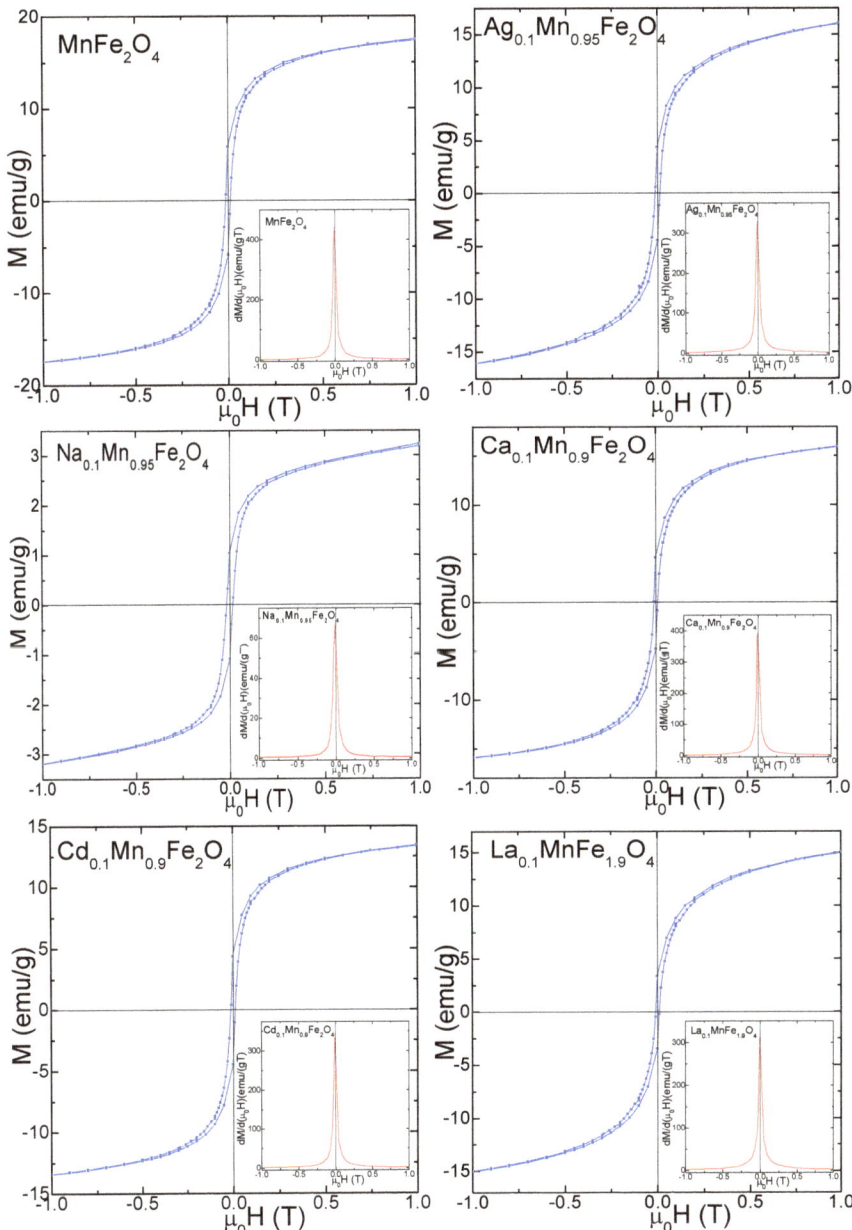

Figure 5. Magnetic hysteresis loops and the magnetization first derivatives of $MnFe_2O_4$, $Ag_{0.1}Mn_{0.95}Fe_2O_4$, $Na_{0.1}Mn_{0.95}Fe_2O_4$, $Ca_{0.1}Mn_{0.9}Fe_2O_4$, $Cd_{0.1}Mn_{0.9}Fe_2O_4$, and $La_{0.1}MnFe_{1.9}O_4$ gels annealed at 800 °C.

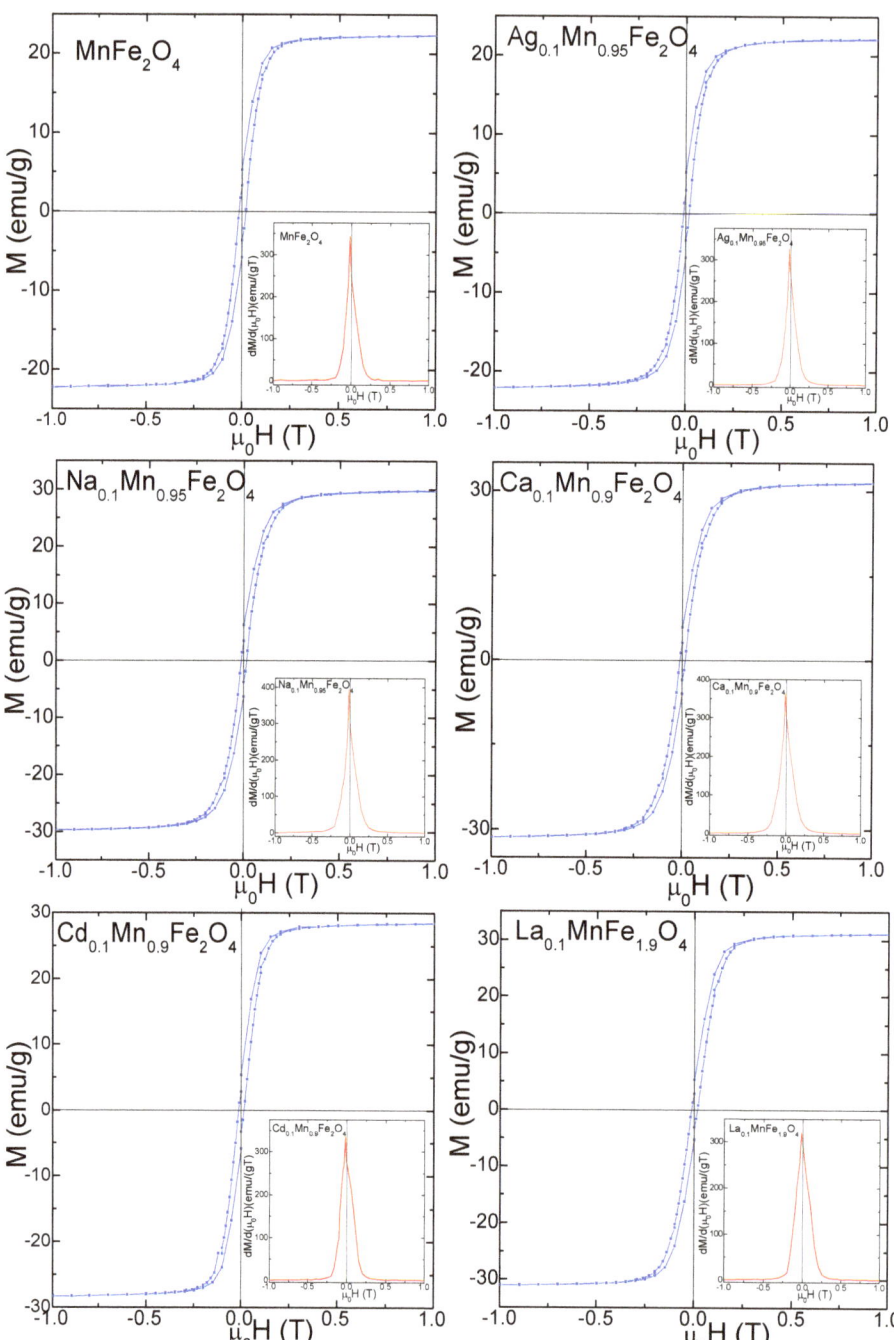

Figure 6. Magnetic hysteresis loops and the magnetization first derivatives $MnFe_2O_4$, $Ag_{0.1}Mn_{0.95}Fe_2O_4$, $Na_{0.1}Mn_{0.95}Fe_2O_4$, $Ca_{0.1}Mn_{0.9}Fe_2O_4$, $Cd_{0.1}Mn_{0.9}Fe_2O_4$, and $La_{0.1}MnFe_{1.9}O_4$, gels annealed at 1200 °C.

The M_S of gels annealed at 800 °C decreases from 21.5 to 6.4 emu/g for doped $MnFe_2O_4$ compared to the undoped $MnFe_2O_4$. The small-size particles have a large surface-to-volume ratio, indicating that the high surface defect density is responsible for the M_S 'depreciation [2–5]. Moreover, the magnetically dead layer at the particle surface containing broken chemical bonds, lattice defects, and isolated magnetic moments contributes to the low M_S value of the nanoparticles [2–5,16,33,35]. Increasing the fraction of this layer will make its behavior dominant over the behavior of the core.

For the gels annealed at 1200 °C, the M_S value increases with doping from 23.3 to 32.2 emu/g (Table 3) and with the annealing temperature owing to the increase in particle size. The small-size particles have a large surface-to-volume ratio, indicating that the high surface defect density is responsible for the M_S 'depreciation [2–5]. Moreover, the existence of magnetically dead layer at the particle surface containing broken chemical bonds, lattice defects and isolated magnetic moments contribute to the nanoparticle's low M_S [2–5,16,33,35]. Increasing the fraction of this layer will make its behavior overriding over the core. According to XRD, the annealing temperature dictates the gel's crystallinity, while the presence of secondary phases affects the magnetic properties of the nanoparticles. Moreover, the increase in particle sizes, improvement of crystallinity, increase in spin disorder, and change in the distribution of Fe^{3+} and Mn^{2+} ions between the A and B sites will result in a higher net magnetic moment [43,44].

Table 3. Magnetic parameters of $MnFe_2O_4$, $Ag_{0.1}Mn_{0.95}Fe_2O_4$, $Na_{0.1}Mn_{0.95}Fe_2O_4$, $Ca_{0.1}Mn_{0.9}Fe_2O_4$, $Cd_{0.1}Mn_{0.9}Fe_2O_4$, and $La_{0.1}MnFe_{1.9}O_4$ gels annealed at 800 and 1200 °C.

Parameter	Temp (°C)	$MnFe_2O_4$	$Ag_{0.1}Mn_{0.95}Fe_2O_4$	$Na_{0.1}Mn_{0.95}Fe_2O_4$	$Ca_{0.1}Mn_{0.9}Fe_2O_4$	$Cd_{0.1}Mn_{0.9}Fe_2O_4$	$La_{0.1}MnFe_{1.9}O_4$	Errors
M_s (emu/g)	800	21.5	21.0	6.4	19.5	17.5	20.1	±1.1
	1200	23.3	26.3	30.7	32.2	29.9	31.7	±2.8
M_R (emu/g)	800	5.9	4.4	1.0	4.6	4.4	3.4	±0.4
	1200	6.5	5.4	6.2	5.8	5.5	5.6	±0.6
H_c (Oe)	800	116	112	100	105	110	113	±10
	1200	160	145	120	125	140	156	±15
$K \cdot 10^3$ (erg/cm^3)	800	1.57	1.48	0.40	1.29	1.21	1.43	±0.10
	1200	2.34	2.39	3.68	2.53	2.63	3.11	±0.25

The remnant magnetization (M_R) decreases for the doped $MnFe_2O_4$ gels compared to undoped $MnFe_2O_4$, from 5.9 to 1.0 emu/g (at 800 °C) and from 6.5 to 5.4 emu/g (at 1200 °C). Generally, the behavior of M_S and M_R can be assessed using Neel's theory via cation distribution between the A and B sites. The Ag^+, Na^+, Ca^{2+}, Cd^{2+} and La^{3+} ions tend to occupy the A sites, Mn^{2+} ions occupy both A and B sites, while Fe^{3+} ions can be found mainly in B sites [2–5,16,33,35,44]. The net magnetization is given by the antiparallel coupling between the magnetic moments of the A and B sites in the presence of spin disorder at the surface and/or spin canting in the presence of secondary phases. The magnetic moments of A sites are antiparallel coupled with those from B sites and the net magnetization derives from the uncompensated magnetic moment between both sites [2–5,16,33,35,44].

Compared to undoped $MnFe_2O_4$ gel, the H_C of doped $MnFe_2O_4$ gels decreases from 116 to 100 Oe for the gels annealed at 800 °C and from 160 to 120 Oe for those annealed at 1200 °C, due to the decrease in crystallite size with a typical single magnetic domain behavior (i.e., H_C increases with increasing crystallite size). The lower H_C values also suggest a spin distortion due to the surface effects, which affect the magnetocrystalline anisotropy [35,44].

Similar to M_S, the magnetocrystalline anisotropy constant (K) for the doped $MnFe_2O_4$ gels annealed at 800 °C decreases (from 1.57×10^{-3} to 0.40×10^{-3} erg/cm^3), while for those annealed at 1200 °C increases (from 2.34×10^{-3} to 3.68×10^{-3} erg/cm^3) compared to undoped $MnFe_2O_4$. A possible explanation could be the decrease in particle size which results in increased surface spin disorder and related surface effects. K is affected by the shape and surface anisotropy of nanoparticles [45]. Since the AFM indicated that the particle shape and area do not change substantially by doping and annealing temperature, the surface anisotropy dictates the variation in K.

While the properties of the obtained doped $MnFe_2O_4$ nanoparticles can be further enhanced by adjusting the dopant ions content, annealing temperature, or the SiO_2 to ferrite ratio, this study introduces valuable information on the properties of doped $MnFe_2O_4/SiO_2$ nanocomposites. Metal ion doping and annealing temperature can have remarkable effects on the structure, morphology, and magnetic properties of the $MnFe_2O_4/SiO_2$ nanocomposites, which allow the control of the physical properties of these nanocomposites to make them potential candidates for various applications such as microwave and communication devices, information storage systems, ferrofluid technology, gas sensors, and medical applications for magnetic hyperthermia, magnetic resonance imaging, and photocatalytic activity.

4. Conclusions

The effect of doping with mono- (Ag^+, Na^+), di– (Ca^{2+}, Cd^{2+}), and trivalent (La^{3+}) ions and annealing temperature on the structural, morphological, and magnetic properties of $MnFe_2O_4$ were studied. Low-crystalline $MnFe_2O_4$ at low annealing temperatures and well-crystallized $MnFe_2O_4$ at high annealing temperatures were remarked. For the gels annealed at 1200 °C, in $MnFe_2O_4$ doped with divalent and trivalent metals, three crystalline phases associated with the SiO_2 matrix (cristobalite, quartz, and tridymite) were also observed. The crystallite size, degree of crystallinity, lattice constant, unit cell volume, hopping length, and density increased, while the porosity decreased with the annealing temperature. The crystallite size estimated by XRD was in good agreement with the particle size measured by AFM, suggesting that the observed nanoparticles contain a single ferrite crystallite. Doping slightly reduced the ferrite particle diameter depending on the doping ion radius. Uniformly self-assembling ferrite nanoparticles in thin films by adsorption from an aqueous solution may be a straightforward approach for doped ferrite nanocrystalline coatings. The saturation magnetization (M_S) of doped $MnFe_2O_4$ gels annealed at 800 °C decreased compared to the undoped $MnFe_2O_4$ (from 21.5 to 6.4 emu/g), while M_S of gels annealed at 1200 °C increased for the doped $MnFe_2O_4$ gels (from 23.3 to 32.2 emu/g). Similar behavior was found for the magnetocrystalline anisotropy constant (K). The coercive field (H_c) decreased by doping for the gels annealed at 800 °C (from 116 to 100 Oe) and 1200 °C (from 160 to 120 Oe). The magnetocrystalline anisotropy constant (K) of the doped $MnFe_2O_4$ gels was lower at 800 °C (from 1.57×10^{-3} to 0.40×10^{-3} erg/cm^3) and higher at 1200 °C (from 2.34×10^{-3} to 3.68×10^{-3} erg/cm^3) compared to the undoped $MnFe_2O_4$. The results obtained confirm that doping and annealing temperature play an important role in tailoring the structural, morphological, and magnetic properties of doped $MnFe_2O_4/SiO_2$ nanocomposites making them important for catalysts and biomedical applications, such as magnetic resonance imaging, biomolecule detection, and magnetic hyperthermia. Although their use in biomedical applications is still in the beginning stage, some challenges, such as tuning size, shape, and magnetic properties of nanoparticles, exploring additional dopants, and optimizing annealing conditions, require further study. Additionally, adjusting $MnFe_2O_4$' s surface as key performance for biomedical applications should be further explored as one of the most important challenges to obtaining ferrite nanoparticles.

Author Contributions: Conceptualization, T.D.; methodology, T.D.; investigation, E.A.L., I.P., I.G.D. and O.C.; formal analysis, T.D., E.A.L., I.G.D., I.P. and O.C.; resources, T.D., E.A.L., I.G.D., I.P. and O.C.; data curation, T.D.; writing—original draft preparation, T.D., O.C., I.P., I.G.D and E.A.L.; writing—review and editing, T.D., O.C. and E.A.L.; visualization, T.D.; supervision, T.D. All authors have read and agreed to the published version of the manuscript.

Funding: This research was funded by the Ministry of Research, Innovation and Digitization through Program 1—Development of the national research & development system, Subprogram 1.2—Institutional performance—Projects that finance the RDI excellence, Contract no. 18PFE/30.12.2021 (E.A.L. and O.C.).

Data Availability Statement: The data that support the findings of this study are available on request from the corresponding author.

Acknowledgments: The authors acknowledge the Research Centre in Physical Chemistry "CECHIF" for AFM assistance and Institute of Physics "Ioan Ursu" of the Faculty of Physics of Babes–Bolyai University for magnetic measurements.

Conflicts of Interest: The authors declare no conflict of interest.

References

1. Akhlaghi, N.; Najafpour-Darzi, G. Manganese ferrite ($MnFe_2O_4$) nanoparticles: From synthesis to application—A review. *J. Ind. Eng. Chem.* **2021**, *103*, 292–304. [CrossRef]
2. Dippong, T.; Deac, I.G.; Cadar, O.; Levei, E.A. Effect of silica embedding on the structure, morphology and magnetic behavior of $(Zn_{0.6}Mn_{0.4}Fe_2O_4)_\delta/(SiO_2)_{(100-\delta)}$ nanoparticles. *Nanomaterials* **2021**, *11*, 2232. [CrossRef]
3. Dippong, T.; Levei, E.A.; Deac, I.G.; Petean, I.; Cadar, O. Dependence of structural, morphological and magnetic properties of manganese ferrite on Ni-Mn substitution. *Int. J. Mol. Sci.* **2022**, *23*, 3097. [CrossRef] [PubMed]
4. Dippong, T.; Levei, E.A.; Cadar, O.; Deac, I.G.; Lazar, M.; Borodi, G.; Petean, I. Effect of amorphous SiO_2 matrix on structural and magnetic properties of $Cu_{0.6}Co_{0.4}Fe_2O_4/SiO_2$ nanocomposites. *J. Alloys Compd.* **2020**, *849*, 156695. [CrossRef]
5. Dippong, T.; Levei, E.A.; Cadar, O. Investigation of structural, morphological and magnetic properties of MFe_2O_4 (M = Co, Ni, Zn, Cu, Mn) obtained by thermal decomposition. *Int. J. Mol. Sci.* **2022**, *23*, 8483. [CrossRef]
6. Wang, Z.; Ma, H.; Zhang, C.; Feng, J.; Pu, S.; Ren, Y.; Wang, Y. Enhanced catalytic ozonation treatment of dibutyl phthalate enabled by porous magnetic Ag-doped ferrospinel $MnFe_2O_4$ materials: Performance and mechanism. *Chem. Eng. J.* **2018**, *354*, 42–52. [CrossRef]
7. Baig, M.M.; Zulfiqar, S.; Yousuf, M.A.; Touqeer, M.; Ullah, S.; Agboola, P.O.; Warsi, M.F.; Shakir, I. Structural and photocatalytic properties of new rare earth La^{3+} substituted $MnFe_2O_4$ ferrite nanoparticles. *Ceram. Int.* **2020**, *46*, 23208–23217. [CrossRef]
8. Kalaiselvan, C.R.; Laha, S.S.; Somvanshi, S.B.; Tabish, T.A.; Thorat, N.D.; Sahu, N.K. Manganese ferrite ($MnFe_2O_4$) nanostructures for cancer theranostics. *Coord. Chem. Rev.* **2022**, *473*, 214809. [CrossRef]
9. Sun, Y.; Zhou, J.; Liu, D.; Li, X.; Liang, H. Enhanced catalytic performance of Cu-doped $MnFe_2O_4$ magnetic ferrites: Tetracycline hydrochloride attacked by superoxide radicals efficiently in a strong alkaline environment. *Chemosphere* **2022**, *297*, 134154. [CrossRef]
10. Shayestefar, M.; Mashreghi, A.; Hasani, S.; Rezvan, M.T. Optimization of the structural and magnetic properties of $MnFe_2O_4$ doped by Zn and Dy using Taguchi method. *J. Magn. Magn. Mater.* **2022**, *541*, 168390. [CrossRef]
11. Angadi, J.V.; Shigihalli, N.B.; Batoo, K.M.; Hussain, S.; Sekhar, E.V.; Wang, S.S.; Kubrin, P. Synthesis and study of transition metal doped ferrites useful for permanent magnet and humidity sensor applications. *J. Magn. Magn. Mater.* **2022**, *564*, 170088. [CrossRef]
12. Debnath, S.; Das, R. Study of the optical properties of Zn doped Mn spinel ferrite nanocrystals shows multiple emission peaks in the visible range—A promising soft ferrite nanomaterial for deep blue LED. *J. Mol. Struct.* **2020**, *1199*, 127044. [CrossRef]
13. Mallesh, S.; Srinivas, V. A comprehensive study on thermal stability and magnetic properties of MnZn-ferrite nanoparticles. *J. Magn. Magn. Mater.* **2019**, *475*, 290–303. [CrossRef]
14. He, Q.; Liu, J.; Liang, J.; Huang, C.; Li, W. Synthesis and antibacterial activity of magnetic $MnFe_2O_4/Ag$ composite particles. *Nanosci. Nanotechnol. Lett.* **2014**, *6*, 385–391. [CrossRef]
15. Andrade, R.G.D.; Ferreira, D.; Veloso, S.R.S.; Santos-Pereira, C.; Castanheira, E.M.S.; Côrte-Real, M.; Rodrigues, L.R. Synthesis and cytotoxicity assessment of citrate-coated calcium and manganese ferrite nanoparticles for magnetic hyperthermia. *Pharmaceutics* **2022**, *14*, 2694. [CrossRef] [PubMed]
16. Dippong, T.; Levei, E.A.; Leostean, C.; Cadar, O. Impact of annealing temperature and ferrite content embedded in SiO_2 matrix on the structure, morphology and magnetic characteristics of $(Co_{0.4}Mn_{0.6}Fe_2O_4)_\delta(SiO_2)_{100-\delta}$ nanocomposites. *J. Alloys Compd.* **2021**, *868*, 159203. [CrossRef]
17. Kozerozhets, I.V.; Panasyuk, G.P.; Semenov, E.A.; Avdeeva, V.V.; Danchevskaya, M.N.; Simonenko, N.P.; Vasiliev, M.G.; Kozlova, L.O.; Ivakin, I.D. Recrystallization of nanosized boehmite in an aqueous medium. *Powder Technol.* **2023**, *416*, 118030. [CrossRef]
18. Kang, S.; Wang, C.; Chen, J.; Meng, T.; Jiaqiang, E. Progress on solvo/hydrothermal synthesis and optimization of the cathode materials of lithium-ion battery. *J. Energy Storage* **2023**, *67*, 107515. [CrossRef]
19. Ravindra, A.V.; Ju, S. Mesoporous $CoFe_2O_4$ nanocrystals: Rapid microwave-hydrothermal synthesis and effect of synthesis temperature on properties. *Mater. Chem. Phys.* **2023**, *303*, 127818. [CrossRef]
20. Baublytè, M.; Vailionis, A.; Sokol, D.; Skaudžius, R. Enhanced functionality of Scots pine sapwood by in situ hydrothermal synthesis of $GdPO_4 \cdot H_2O:Eu^{3+}$ Composites in woods matrix. *Ceram. Int.* **2023**, *in press*. [CrossRef]

21. Marques de Gois, M.; de Alencar Souza, L.W.; Nascimento Cordeiro, C.H.; Tavares da Silva, I.B.; Soares, J.M. Study of morphology and magnetism of $MnFe_2O_4$–SiO_2 composites. *Ceram. Int.* **2023**, *49*, 11552–11562. [CrossRef]
22. Dippong, T.; Levei, E.A.; Cadar, O. Formation, structure and magnetic properties of $MFe_2O_4@SiO_2$ (M = Co, Mn, Zn, Ni, Cu) nanocomposites. *Materials* **2021**, *14*, 1139. [CrossRef] [PubMed]
23. Yin, P.; Zhang, L.; Wang, J.; Feng, X.; Zhao, L.; Rao, H.; Wang, Y.; Dai, J. Preparation of SiO_2-$MnFe_2O_4$ composites via one-pot hydrothermal synthesis method and microwave absorption investigation in S-band. *Molecules* **2019**, *24*, 2605. [CrossRef] [PubMed]
24. Lu, J.; Ma, S.; Sun, J.; Xia, C.; Liu, C.; Wang, Z.; Zhao, X.; Gao, F.; Gong, Q.; Song, B.; et al. Manganese ferrite nanoparticle micellar nanocomposites as MRI contrast agent for liver imaging. *Biomaterials* **2009**, *30*, 2919–2928. [CrossRef] [PubMed]
25. Kurtan, U.; Admir, M.; Yildiz, A.; Baykal, A. Synthesis and magnetically recyclable MnFe2O4@SiO2@Ag nanocatalyst: Its high catalytic performances for azo dyes and nitro compunds reduction. *Appl. Surf. Sci.* **2016**, *376*, 16–25. [CrossRef]
26. Zhu, J.Q.; Zhang, X.J.; Wang, S.W.; Wang, G.S.; Yin, P.G. Enhanced microwave absorption material of ternary nanocomposites based on MnFe2O4@SiO2, polyaniline and polyvinylidene fluoride. *RSC Adv.* **2006**, *6*, 88104–88109. [CrossRef]
27. Shirzadi-Ahodashti, M.; Ebrahimzadeh, M.A.; Ghoreishi, S.M.; Naghizadeh, A.; Mortazavi-Derazkola, S. Facile and eco-benign synthesis of a novel $MnFe_2O_4@SiO_2@Au$ magnetic nanocomposite with antibacterial properties and enhanced photocatalytic activity under UV and visible-light irradiations. *Appl. Organomet. Chem.* **2020**, *34*, e5614.
28. Kavkhani, R.; Hajalilou, A.; Abouzari-Lotf, E.; Ferreira, L.P.; Cruz, M.M.; Yusefi, M.; Arvini, E.; Ogholbey, A.B.; Ismail, U.N. CTAB assisted synthesis of $MnFe_2O_4@$ SiO_2 nanoparticles for magnetic hyperthermia and MRI application. *Mater. Today* **2023**, *31*, 103412. [CrossRef]
29. Asghar, K.; Qasim, M.; Das, D. Preparation and characterization of mesoporous magnetic $MnFe_2O_4@mSiO_2$ nanocomposite for drug delivery application. *Mater. Today Proc.* **2020**, *26*, 87–93. [CrossRef]
30. Desai, H.B.; Hathiya, L.J.; Joshi, H.H.; Tanna, A.R. Synthesis and characterization of photocatalytic $MnFe_2O_4$ nanoparticles. *Mater. Today Proc.* **2020**, *21*, 1905–1910. [CrossRef]
31. Kozerozhets, V.; Semenov, E.A.; Avdeeva, V.V.; Ivakin, Y.D.; Kupreenko, S.Y.; Egorov, A.V.; Kholodkova, A.A.; Vasilev, M.G.; Kozlova, L.O.; Panasyuk, G.P. State and forms of water in dispersed aluminum oxides and hydroxides. *Ceram. Int.* **2023**, in press. [CrossRef]
32. Calvin, J.J.; Rosen, P.F.; Ross, N.L.; Navrotsky, A.; Woodfield, B.F. Review of surface water interactions with metal oxide nanoparticles. *J. Mater. Res.* **2019**, *34*, 416–427. [CrossRef]
33. Balarabe, B.Y.; Bowmik, S.; Ghosh, A.; Maity, P. Photocatalytic dye degradation by magnetic XFe_2O_3 (X: Co, Zn, Cr, Sr, Ni, Cu, Ba, Bi, and Mn) nanocomposites under visible light: A cost efficiency comparison. *J. Magn. Magn. Mater.* **2022**, *562*, 169823. [CrossRef]
34. Dippong, T.; Levei, E.A.; Toloman, D.; Barbu-Tudoran, L.; Cadar, O. Investigation on the formation, structural and photocatalytic properties of mixed Mn-Zn ferrites nanoparticles embedded in SiO_2 matrix. *J. Alloys Compd.* **2021**, *158*, 105281. [CrossRef]
35. Ajeesha, T.L.; Manikandan, A.; Anantharaman, A.; Jansi, S.; Durka, M.; Almessiere, M.A.; Slimani, Y.; Baykal, A.; Asiri, A.M.; Kasmery, H.A.; et al. Structural investigation of Cu doped calcium ferrite ($Ca_{1-x}Cu_xFe_2O_4$; x = 0, 0.2, 0.4, 0.6, 0.8, 1) nanomaterials prepared by co-precipitation method. *J. Mater. Res. Technol.* **2022**, *18*, 705–719. [CrossRef]
36. Liandi, A.R.; Cahyana, A.H.; Kusumah, A.J.F.; Lupitasari, A.; Alfariza, D.N.; Nuraini, R.; Sari, R.W.; Kusumasari, F.C. Recent trends of spinel ferrites (MFe_2O_4: Mn, Co, Ni, Cu, Zn) applications as an environmentally friendly catalyst in multicomponent reactions: A review. *Case Stud. Chem. Environ. Eng.* **2023**, *7*, 100303. [CrossRef]
37. Rout, J.; Choudhary, R.N.P.; Sharma, H.B.; Shannigrahi, S.R. Effect of co-substitutions (Ca–Mn) on structural, electrical and magnetic characteristics of bismuth ferrite. *Ceram. Int.* **2015**, *41*, 9078–9087. [CrossRef]
38. Torre, F.; Sanchez, T.A.; Doppiu, S.; Bengoechea, M.O.; Ergueta, P.L.A.; Palomo del Barrio, E. Effect of atomic substitution on the sodium manganese ferrite thermochemical cycle for hydrogen production. *Mater. Today Energy* **2022**, *29*, 101094. [CrossRef]
39. Yadav, B.S.; Vishwakarma, A.K.; Singh, A.K.; Kumar, N. Oxygen vacancies induced ferromagnetism in RF-sputtered and hydrothermally annealed zinc ferrite ($ZnFe_2O_4$) thin films. *Vacuum* **2023**, *207*, 111617. [CrossRef]
40. Sharma, A.D.; Sharma, H.B. Structural, optical, and dispersive parameters of (Gd, Mn) co-doped $BiFeO_3$ thin film. *Mater. Today Proc.* **2022**, *65*, 2837–2843. [CrossRef]
41. Sarıtaş, S.; Şakar, B.C.; Kundakci, M.; Gürbulak, B.; Yıldırım, M. Analysis of magnesium ferrite and nickel doped magnesium ferrite thin films grown by spray pyrolysis. *Mater. Today Proc.* **2021**, *46*, 6920–6923. [CrossRef]
42. Cullity, B.D.; Graham, C.D. *Introduction to Magnetic Materials*; Wiley: Hoboken, NJ, USA, 2011.
43. Goya, G.F.; Berquó, T.S.; Fonseca, F.C.; Morales, M.P. Static and dynamic magnetic properties of spherical magnetite nanoparticles. *J. Appl. Phys.* **2003**, *94*, 3520–3528. [CrossRef]
44. Chavarría-Rubio, J.A.; Cortés-Hernández, D.A.; Garay-Tapia, A.M.; Hurtado-López, G.F. The role of lanthanum in the structural, magnetic and electronic properties of nanosized mixed manganese ferrites. *J. Magn. Magn. Mater.* **2022**, *553*, 169253. [CrossRef]
45. Caruntu, D.; Caruntu, G.; O'Connor, C.J. Magnetic properties of variable-sized Fe_3O_4 nanoparticles synthesized from nonaqueous homogeneous solutions of polyols. *J. Phys. D Appl. Phys.* **2007**, *40*, 5801–5809. [CrossRef]

Disclaimer/Publisher's Note: The statements, opinions and data contained in all publications are solely those of the individual author(s) and contributor(s) and not of MDPI and/or the editor(s). MDPI and/or the editor(s) disclaim responsibility for any injury to people or property resulting from any ideas, methods, instructions or products referred to in the content.

Article

Effects of Lanthanum Substitution and Annealing on Structural, Morphologic, and Photocatalytic Properties of Nickel Ferrite

Thomas Dippong [1,*], Dana Toloman [2], Mihaela Diana Lazar [2] and Ioan Petean [3]

1 Faculty of Science, Technical University of Cluj-Napoca, 76 Victoriei Street, 430122 Baia Mare, Romania
2 National Institute for Research and Development of Isotopic and Molecular Technologies, 67-103 Donath Street, 400293 Cluj-Napoca, Romania; dana.toloman@itim-cj.ro (D.T.); diana.lazar@itim-cj.ro (M.D.L.)
3 Faculty of Chemistry and Chemical Engineering, Babes-Bolyai University, 11 Arany Janos Street, 400028 Cluj-Napoca, Romania; ioan.petean@ubbcluj.ro
* Correspondence: dippong.thomas@yahoo.ro

Citation: Dippong, T.; Toloman, D.; Lazar, M.D.; Petean, I. Effects of Lanthanum Substitution and Annealing on Structural, Morphologic, and Photocatalytic Properties of Nickel Ferrite. *Nanomaterials* **2023**, *13*, 3096. https://doi.org/10.3390/nano13243096

Academic Editor: Antonio Guerrero-Ruiz

Received: 13 November 2023
Revised: 29 November 2023
Accepted: 5 December 2023
Published: 7 December 2023

Copyright: © 2023 by the authors. Licensee MDPI, Basel, Switzerland. This article is an open access article distributed under the terms and conditions of the Creative Commons Attribution (CC BY) license (https://creativecommons.org/licenses/by/4.0/).

Abstract: Nanoparticles of $NiLa_xFe_{2-x}O_4$ ferrite spinel incorporated in a SiO_2 matrix were synthesized via a sol-gel method, followed by annealing at 200, 500, and 800 °C. The resulting materials were characterized via XRD, AFM, and BET techniques and evaluated for photocatalytic activity. The XRD diffractograms validate the formation of a single-phase cubic spinel structure at all temperatures, without any evidence of secondary peaks. The size of crystallites exhibited a decrease from 37 to 26 nm with the substitution of Fe^{3+} with La^{3+} ions. The lattice parameters and crystallite sizes were found to increase with the rise in La^{3+} content and annealing temperature. Isotherms were employed to calculate the rate constants for the decomposition of malonate precursors to ferrites and the activation energy for each ferrite. All nanocomposites have pores within the mesoporous range, with a narrow dispersion of pore sizes. The impact of La content on sonophotocatalytic activity was evaluated by studying Rhodamine B degradation under visible light irradiation. The results indicate that the introduction of La enhances nanocomposite performance. The prepared Ni-La ferrites may have potential application for water decontamination.

Keywords: nickel–lanthanum ferrite; crystalline phase; specific surface; sonophotocatalysis

1. Introduction

Nickel ferrite ($NiFe_2O_4$) stands out as one of the most prominent in the spinel ferrite class. In bulk, it exhibits a rhombohedrally distorted cubic structure, characterized by an inverse spinel structure; this arrangement involves an antiparallel spin alignment between Fe^{3+} and Ni^{2+} ions on octahedral sites, while equal Fe^{3+} ions are positioned at the tetrahedral sites [1–4]. The unit cell of $NiFe_2O_4$ is composed of 32 O^{2-}, 16 Fe^{3+}, and 8 Ni^{2+} ions. The oxygen ions form 32 octahedral sites (B-sites) and 64 tetrahedral ones (A-sites). These sites have the capacity to host a total of 24 cations. Within this inverse spinel structure, an exclusive occupation of A-sites by Fe^{3+} ions arises, whereas the B-site is shared by Ni^{2+} and Fe^{3+} ions, both demonstrating electron exchange at the octahedral site, highlighting their unique electrical and magnetic properties [5]. The superior characteristics of nickel ferrites, such as a low permittivity superparamagnetism and favorable optical band gap (Eg) values, make them suitable for high-frequency applications [2,4–6]. The incorporation of La^{3+} ions into spinel ferrites induces a strong spin–orbit coupling of their angular momentum, resulting in enhanced dielectric properties [7]. The presence of the rare earth ion in the spinel ferrite contributes to improved densification, electrical resistivity, and reduced eddy current losses [7]. Their unique properties make them suitable for a wide range of applications such as photoacoustic imaging, transformer cores, biosensors, high-density storage media, electron transport devices, hyperthermia, analog devices, imaging, biological field, radio frequency, microwave absorbing materials, water treatment, gas sensors, lithium-ion batteries, spin canting, surface anisotropy, and superparamagnetism [1–4,6].

The properties of Ni-La ferrite nanoparticles are intricately linked to the chosen synthesis method, which impacts both composition and microstructure [8]. Several synthesis methods are worth mentioning, including ball milling, sol–gel, solid-state reaction, spray pyrolysis, microemulsion, thermal pyrolysis, hydrothermal, solvothermal, citrate gel auto-combustion, self-propagating high-temperature synthesis, microwave, or chemical coprecipitation [1–9]. Sol–gel is arguably the most versatile method, as it requires less time and offers a reduced cost due to the low temperature requirement; it is highly reproducible, allowing for a good stoichiometry control that leads to a homogenous, single-phase final product under normal ambient conditions [1–6]. It is worth mentioning that the annealing process influences both phases and the increase in crystallite size [10]. The incorporation of $NiLa_xFe_{2-x}O_4$ into a mesoporous SiO_2 matrix plays a crucial role in enhancing water stability, improving biocompatibility, and mitigating the degradation of $NiLa_xFe_{2-x}O_4$ nanoparticles [11–14]. The SiO_2 coating not only prevents agglomeration by regulating dipolar attraction between nanoparticles but also facilitates the binding of biomolecules on the mesoporous SiO_2 surface, enabling targeted ligands and drug loading on the nanocarrier surface [11–14]. The synthesis of $NiLa_xFe_{2-x}O_4$ embedded in the SiO_2 matrix following a sol–gel method involves the mixing of reactants with tetraethyl orthosilicate (TEOS), forming strong networks with moderate reactivity that allow the incorporation of various inorganic and organic molecules [11–14]. Simple adjustments in synthesis conditions, such as pH, time, and annealing temperature, can provide more precise control over nucleation and particle growth [11–14].

Photocatalytic properties are highly dependent upon parameters such as surface area, particle size, and concentration of dopants [15]. The photochemical process occurs at the surface of metal oxides, involving two types of reactions, namely oxidation (resulting in positive holes) and reduction (producing negative electrons) [4]. By tuning the band gap energy (Eg) of ferrites below 3 eV, one can improve upon their photocatalytic properties [6]. In spite of its oxidation capacity, the use of a wide UV band gap, and noteworthy photocatalytic activity, to achieve better efficiency proves to be a challenge due to rapid recombination. Due to their use of visible light, ferrites with lower Eg values are suitable for applications like wastewater treatment in pollutant degradation [6]. According to Zhang et al. [16], conventional homogeneous photocatalysis is characterized by inherent drawbacks, including the easy recombination of photo-induced electron–hole (e^-/h^+) pairs and light absorption restricted to the ultraviolet region. This study proposes that the development of heterogeneous photocatalysis has proven to be an effective strategy for expanding the range of light absorption wavelengths and enhancing the separation of charge carriers [16]. Another study, by Shah et al. [17], highlights the inclusion of a new energy level in between the conduction and valence bands of TiO_2 and $NiFe_2O_4$, thus facilitating the separation of photoinduced electrons and holes. The investigation of Zhang et al. [16] emphasizes the stability of a $TiO_2/Ag/SnO_2$ photocatalyst following a methylene blue degradation over four cycles. Ghoneim [18] evaluates the potential of using $Cu_{0.3}Cd_{0.7}CrFeO_4$ nano-spinel for cost-effective wastewater treatment. Padmapriya et al. [19] note that the zinc–ferrite nanoparticle photocatalysis depends on surface area and particle size. Additionally, sonocatalysis, utilizing ultrasound for pollutant degradation, combines effectively with photocatalysis in the versatile sonophotocatalysis technique [20]. This study explores the synthesis, structural aspects (crystallite size and lattice parameter), surface characteristics (specific surface area and porosity), morphology (particle size, roughness, and height), and the sonophotocatalytic performance of Ni-La ferrites incorporated into SiO_2. These nanocomposites were prepared using a sol–gel method followed by thermal treatment. The crystalline phases, crystallite sizes, and lattice constants were examined via X-ray diffraction (XRD). The Ni-La-Fe ferrite composition was investigated via inductively coupled plasma optical emission spectrometry (ICP-OES). Specific surface area (SSA) and porosity were determined by analyzing N_2

adsorption–desorption isotherms. Particle attributes, including shape, size, size distribution, and agglomeration, were characterized using atomic force microscopy (AFM). The sonophotocatalytic degradation of the samples was evaluated under visible light exposure with concurrent sonication using a Rhodamine (RhB) solution.

2. Materials and Methods

$Ni(NO_3)_2 \cdot 6H_2O$, $La(NO_3)_2 \cdot 6H_2O$, $Fe(NO_3)_3 \cdot 9H_2O$, 1,3-propandiol (1,3PD), TEOS, and ethanol were used to synthesize Ni-La ferrites embedded in a SiO_2 matrix (50% wt. ferrite, 50% wt. SiO_2) using a sol–gel method. Ni:La:Fe molar ratios of 10:1:19 ($NiLa_{0.1}Fe_{1.9}O_4/SiO_2$), 10:3:17 ($NiLa_{0.3}Fe_{1.7}O_4/SiO_2$), 10:5:15 ($NiLa_{0.5}Fe_{1.5}O_4/SiO_2$), 10:7:13 ($NiLa_{0.7}Fe_{1.3}O_4/SiO_2$), 10:9:11 ($NiLa_{0.9}Fe_{1.1}O_4@SiO_2$), 10:11:9 ($NiLa_{1.1}Fe_{0.9}O_4@SiO_2$), and a nitrate:1.3PD:TEOS molar ratio of 1:1:1 were used. The as-produced sols were maintained at room temperature for gelation and the process was followed by grinding and drying at 200 °C (6 h), heating at 500 °C (6 h), and annealing at 800 °C (6 h).

The structural characterization was explored via X-ray diffraction using a Bruker D8 Advance diffractometer in normal temperature conditions (with CuKα radiation, λ = 1.5406 Å), running at 40 kV and 40 mA. The content of Ni, La, and Fe in the ferrites was confirmed through an inductively coupled plasma optical emission spectrometry (ICP-OES) using an Optima 5300 DV (Perkin Elmer, Waltham, MA, USA) spectrometer, after an aqua regia microwave digestion employing a Berghof Speedwave Xpert system. atomic force microscopy (AFM) was effectuated in AC mode with a JEOL Scanning Probe Microscope 4210 (Jeol Company, Akashima, Japan) using sharp probes (NSC 15 produced by Mikromasch Company, Watsonville, CA, USA) featuring a resonance frequency of 325 kHz and a spring constant of 40 N/m. The annealed powders were dispersed in deionized water to release nanoparticles which were adsorbed on the solid substrate (e.g., glass) as thin layers. Three different areas of 1 μm² were scanned for each specimen. The topographic characteristics were measured using JEOL WinSPM 2.0 processing software (Jeol Company, Akashima, Japan). The shape and clustering of the particles were examined by depositing dried sample suspensions onto a copper grid coated with a thin carbon film using a Hitachi HD-2700 transmission electron microscope (Hitachi, Tokyo, Japan). Specific surface area (SSA) and porosity parameters (mean pore size and pore size distribution) were calculated from N_2 adsorption–desorption isotherms using the BET method (for SSA) and the Dollimore–Heal model (for porosity). The isotherms were recorded on a Sorptomatic 1990 instrument (Thermo Fisher Scientific, Waltham, MA, USA). A V570 model UV-VIS-NIR spectrophotometer (JASCO, Oklahoma City, OK, USA) containing the absolute reflectivity accessory (ARN-475, JASCO) was used to register the UV–VIS absorption spectra. The optical band gap was determined from Tauc's relationship. The sonophotocatalytic efficacy was assessed using a Rhodamine (RhB) solution exposed to visible light within a Laboratory-Visible-Reactor system, using a 400 W halogen lamp (Osram, Munich, Germany) and an ultrasonic bath. In this experimental setup, 10 mg of catalyst was blended with a 20 mL solution of 1.0×10^{-5} mol/L RhB in water, and the resulting mixture was stirred in darkness until adsorption equilibrium was achieved on the catalyst surface. Each photodegradation test spanned 240 min, with 3.5 mL samples extracted every 60 min for subsequent analysis. Following catalyst removal, the absorbance of the RhB solution was measured at 554 nm. Sonophotocatalytic activity was determined based on the degradation rate. Prior to sonophotodegradation tests, the adsorption of RhB on the nanoparticle surface was assessed. This involved mixing the photocatalyst with the RhB solution in the dark for 60 min until adsorption–desorption equilibrium was reached. The photodegradation of RhB was modeled using the first-order kinetic model, relying on the absorbance data. To demonstrate the generation of reactive oxygen species (ROS) by the samples, we employed the EPR Bruker E-500 ELEXSYS X-band spectrometer (Bruker, Billerica, MA, USA) (9.52 GHz), coupled with the spin trapping technique. The spin trapping reagent used for this purpose was 5,5-Dimethyl-1-pyrroline N-oxide (DMPO).

3. Results and Discussion

The XRD patterns of NiLa$_x$Fe$_{2-x}$O$_4$/SiO$_2$ (x = 0.1–1.1) nanocomposites annealed at 200, 500, and 800 °C are presented in Figure 1. At 200 °C in case of low La content, the formation of the weakly crystalline Ni-La ferrite phase can be observed (NiFe$_2$O$_4$ (JCPDS card no 89-4927) and La$_{0.14}$Fe$_3$O$_4$ (JCPDS card no. 75-8137)). The NiFe$_2$O$_4$ (x = 0.1) crystalline phase originates from a combination of Ni's low oxidation capacity, low melting point, high electronegativity, high thermal expansion coefficient, and high specific heat capacity [1]. At 200 °C, the broad peak observed at 2θ =20–30° suggests a low level of crystallization in the nanocomposites. At 500 and 800 °C, the intensity of the diffraction peaks increases due to a better crystallization of ferrites. The existence of reflection planes such as (220), (311), (222), (400), (331), (422), (511), and (440) confirms the distinct phase of Ni-La ferrite, characterized by a face-centered cubic inverse spinel structure in the *Fd-3m* space group [1]. All the samples revealed a homogeneous phase spinel Ni-La ferrite without any impurities being registered in the XRD patterns. The peak intensity of the (311) peak decreased with the increase in La content and increased with the increase in thermal treatment temperature. The diffraction peaks sharpen, and their intensity increases with rising annealing temperatures, which are attributed to pronounced agglomeration without immediate recrystallization; this process leads to the formation of a single crystal rather than a polycrystal [21].

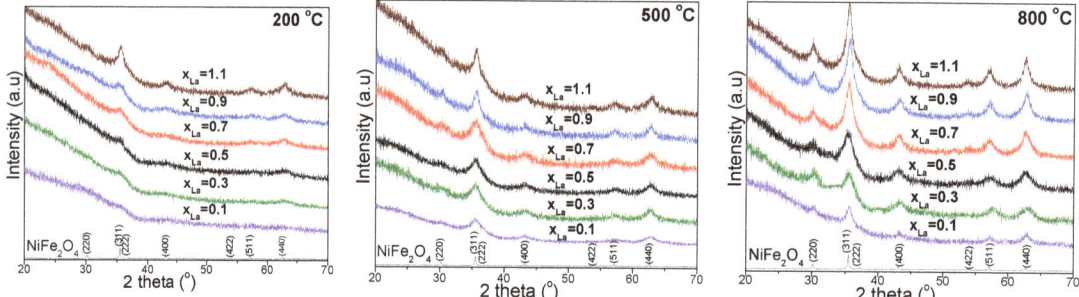

Figure 1. XRD patterns of NiLa$_x$Fe$_{2-x}$O$_4$/SiO$_2$ (x$_{La}$ = 0.1, 0.3, 0.5, 0.7, 0.9, 1.1) annealed at 200, 500, and 800 °C.

The average crystallite size (D$_C$) increases from 19.5 to 35.4 nm at 800 °C, with an increase in the Fe^{3+} substitution by La^{3+} ion (Table 1). The increase in D$_C$ with temperature is attributed to the agglomeration of crystallites without recrystallization, resulting in a transition from a polycrystalline structure to single crystals at elevated temperatures [11–14]. Another plausible explanation is the coalescence process, with small nanoparticles merging into larger ones at high annealing temperatures [1,11–14]. The substitution of La^{3+} for Fe^{3+} induces crystalline anisotropy following the substantial difference in their sizes, with the doping of La^{3+} ions acting as a kinetic barrier to further grain growth [7,22]. Increasing the temperature distinctly enhances the crystallinity of lanthanum–nickel ferrite [10]. The amorphous phase dominates at low annealing temperatures and undergoes partial transformation into various crystalline phases at higher annealing temperatures [1]. The increase in D$_C$ indicates a reduction in the densities of nucleation centers [1]. La–O bond energy is greater when compared with that of Fe–O; as such, the replacement of Fe^{3+} ions with La^{3+} ions at the octahedral site in NiFe$_2$O$_4$ causes La^{3+} ions to enter the interstitial location and hinders nickel ferrites from crystallizing [6]. The variation in D$_C$ may also be attributed to the peak broadening associated with lattice strains, the grain growth blocking effect induced by the SiO$_2$ matrix, as well as thermal and instrumental effects [1,11–14]. Following Vegard's law, the increase in lattice parameter (a) with the increase in lanthanum concentration could be explained on the basis of ionic radii of the La^{3+} ion substituted in the structure [1–3,9,22]. In nickel ferrite, Ni^{2+} ions predominantly occupy octahedral sites (B-site), whereas Fe^{3+} ions occupy both tetrahedral (A) and octahedral

(B) sites. The increase in the lattice constant (a) following the increase in La content (Table 1) can be attributed to the larger ionic radius of La^{3+} ions (oct: 1.06 Å) when compared with that of Fe^{3+} ions (oct: 0.67 Å); the La^{3+} ions of higher radii substitute Fe^{3+} ions of smaller radii at the octahedral sites [1–3,6,9,22]. The inverse spinel structures cause a partial migration of Ni^{2+} ions from A to B sites; the migration is consorted by an opposite relegation of the corresponding numerical values of Fe^{3+} ions from B to A sites in order to relax the strain [1–3,11–14].

Table 1. Lattice parameter (a), crystallite size (D_C), particle size from AFM (D_{AFM}), H (high), and Rq (roughness) of $NiLa_xFe_{2-x}O_4/SiO_2$ (x_{La} = 0.1, 0.3, 0.5, 0.7, 0.9, 1.1) annealed at 200, 500, and 800 °C.

$NiLa_xFe_{2-x}O_4/SiO_2$	Temp, °C	a, Å	D_C, nm	D_{TEM}, nm	D_{AFM}, nm	H, nm	Rq, nm	SSA m²/g
x_{La} = 0.1	200	-	-	-	10	12	1.23	283
	500	8.444	10.4	-	22	15	1.33	230
	800	8.458	19.5	26	33	11	1.23	3
x_{La} = 0.3	200	8.451	1.2	-	12	14	1.38	267
	500	8.462	12.9	-	25	12	1.17	216
	800	8.481	22.8	31	38	10	1.42	3
x_{La} = 0.5	200	8.469	3.4	-	15	15	1.43	257
	500	8.483	15.5	-	28	12	1.15	173
	800	8.497	25.3	35	42	15	2.41	2
x_{La} = 0.7	200	8.488	5.9	-	18	9	0.94	229
	500	8.501	17.8	-	31	13	1.23	208
	800	8.521	28.1	39	44	18	2.77	<0.5
x_{La} = 0.9	200	8.502	8.1	-	20	9	1.01	258
	500	8.517	20.1	-	35	11	1.11	198
	800	8.541	31.7	43	46	22	3.91	<0.5
x_{La} = 1.1	200	8.525	9.8	-	21	12	1.24	-
	500	8.538	22.6	-	37	11	1.14	241
	800	8.566	35.4	47	50	26	4.56	<0.5

Based on the content of Ni, La, and Fe within the samples, the Ni/La/Fe molar ratio was calculated and compared with the theoretic value for each sample (Table 2). A good agreement was observed between the theoretical and experimental molar ratios across all samples and calcination temperatures.

Table 2. Ni/La/Fe molar ratios of $NiLa_xFe_{2-x}O_4/SiO_2$ samples calcined at 200, 500, and 800 °C.

x_{La}	Ni/La/Fe Molar Ratio			
	Theoretical	200 °C	500 °C	800 °C
0.1	1.0/0.1/1.9	1.01/0.11/1.88	1.00/0.09/1.91	1.00/0.11/1.89
0.3	1.0/0.3/1.7	1.00/0.29/1.71	1.02/0.31/1.67	1.00/0.30/1.70
0.5	1.0/0.5/1.5	1.01/0.52/1.47	1.01/0.49/1.52	1.00/0.49/0.51
0.7	1.0/0.7/1.3	0.99/0.68/1.32	0.98/0.71/1.31	1.01/0.69/1.30
0.9	1.0/0.9/1.1	0.98/0.93/1.09	0.99/0.89/1.12	1.00/0.91/1.09
1.1	1.0/1.1/0.9	1.00/1.11/0.89	1.00/1.08/0.92	1.01/1.09/0.90

The nitrogen adsorption–desorption isotherms recorded at −196 °C are utilized to provide information about the porous structure and the surface area of the nanocomposites. The isotherm shapes observed in the composites annealed at 200 °C and 500 °C exhibit characteristics typical of mesoporous materials, all falling into the type IV category according to the IUPAC classification. Additionally, they display minimal hysteresis at high relative pressures [23]. For the materials thermally treated at higher temperatures (800 °C), the isotherms could only be recorded for the nanocomposite samples with lower lanthanum content: x_{La} between 0.1 and 0.5. For the composites with higher La concentration, the surface area is below the detection limit of the equipment (below 0.5 m^2/g). As can be observed in Figure 2, the shape of the isotherms is comparable for all nanocomposites annealed at 200 °C and 500 °C, suggesting a very similar porous structure of these materials. For the nanocomposite samples calcined at the same temperature, the SSA values vary only in moderate proportion to the lanthanum concentration. A direct, linear correlation between the lanthanum content and the specific surface area was not observed (Table 1). For the samples annealed at 200 °C, there is a general trend of a slow decrease in the SSA with the increase in La substitution within the ferrite lattice. However, no such trend can be observed for the samples calcined at 500 °C. In this case, after an initial decrease in SSA with the increase in La content, the trend reverses and the SSA started to increase again, with the sample with the lowest value for SSA being NiLa$_{0.5}$Fe$_{1.5}$O$_4$/SiO$_2$ (173 m^2/g compared with 230 m^2/g for NiLa$_{0.1}$Fe$_{1.9}$O$_4$/SiO$_2$ and 240 m^2/g for NiLa$_{1.1}$Fe$_{0.9}$O$_4$/SiO$_2$, respectively).

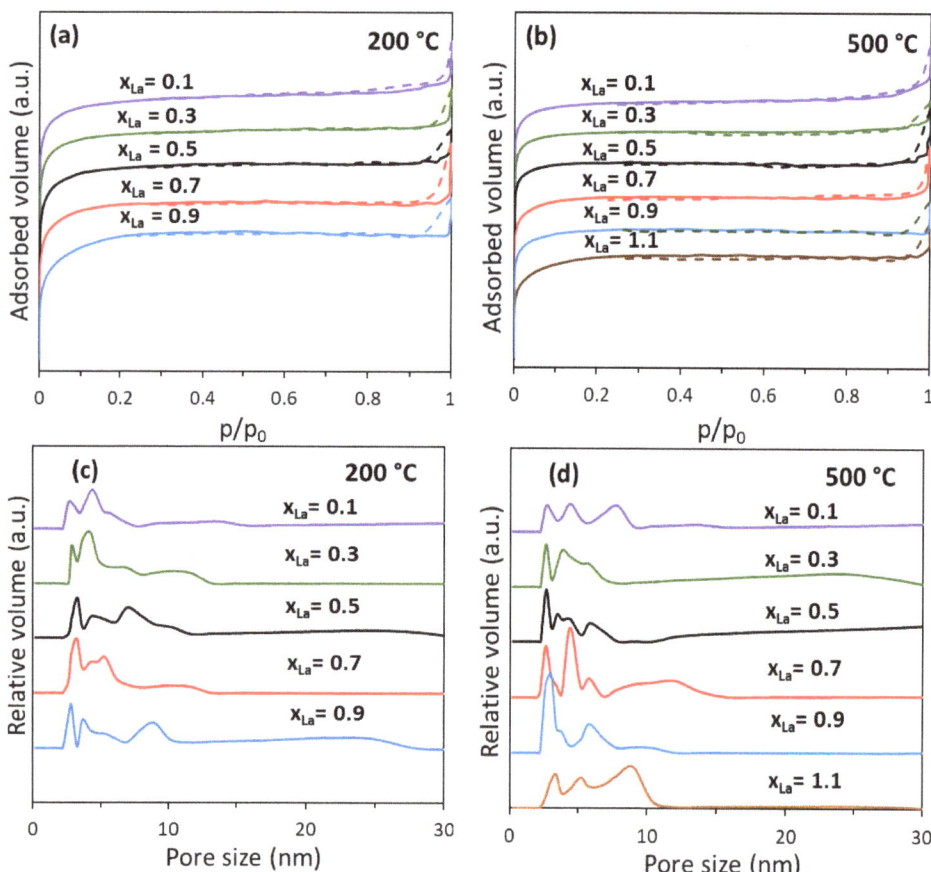

Figure 2. Nitrogen adsorption–desorption isotherms (**a**,**b**) and pore diameter distribution (**c**,**d**) for NiLa$_x$Fe$_{2-x}$O$_4$/SiO$_2$ (x_{La} = 0.1, 0.3, 0.5, 0.7, 0.9, 1.1) annealed at 200 °C and 500 °C. (full line—adsorption and dashed line—desorption branch of the isotherms).

This trend corresponds to the one observed for surface roughness (Rq) of the film prepared for AFM analysis. For the samples containing the same amount of La but thermally treated at different temperatures, the increase in temperature led to a decrease in SSA. This behavior was previously reported for nickel ferrite [24], ferrite-SiO$_2$ composite materials [25], as well as other oxides [26], being usually related to an increase in crystallinity due to crystallite growth (Table 1 and Figure 1) and/or to the prevalence of silica crystalline forms with lower surface areas. All the samples annealed at 800 °C present very low surface areas (less than 3 m^2/g) indicating that in the samples calcined at this temperature, the porous structure is no longer present. For the samples calcined at 200 °C and 500 °C, the distribution in pore sizes (Figure 2c,d) confirms the mesoporous structure of NiLa$_x$Fe$_{2-x}$O$_4$/SiO$_2$. The pore dimensions for all samples are low, with values less than 10 nm, and are thus situated in the lower region of the mesoporous domain. All the tested nanocomposites present a multimodal pore size distribution that is very similar across all samples. The size distribution is relatively narrow (3–10 nm), with the multiple pore dimensions inside this range being characteristic of composite materials in which the global porous structure is given by the combined effect of the porosity of each component

The ferrite powder obtained after annealing is slightly agglomerated. Therefore, each sample was dispersed into deionized water and transferred onto glass substrates via vertical adsorption. The intense Brownian movement within dispersed particles promotes their individualization, with these nanoparticles being attracted to the glass surface and subsequently adsorbing, forming a thin film [27,28]. The obtained ferrite thin films were investigated using AFM microscopy and the obtained topographic images are presented in Figure 3. A small substitution of Fe atoms with La (x_{La} = 0.1) within nickel ferrites has limited influence on particle size and shape and is mainly observed only at the crystalline lattice level as revealed by XRD patterns in Figure 1. Thus, the AFM image of the sample treated at 200 °C reveals very small rounded particles of about 10 nm in diameter, as shown in Figure 3a. As no crystallites were observed in the XRD analysis results, we assume that these nanoparticles are amorphous. Increasing the annealing temperature to 500 °C, a diameter enhancement to 22 nm was observed, as shown in Figure 3b. This observation is in good agreement with ferrite crystallite development. The ferrite core has a crystallite of 10.4 nm, which is covered with amorphous silica up to the observed diameter of nanoparticles.

The crystallization process is enhanced at 800 °C, developing a ferrite core of 19.5 nm that conducts to the formation of larger nanoparticles of 33 nm in diameter, as shown in Figure 3c. The crystalline core introduces certain sharp corners to the shape of the nanoparticles, but these corners are rounded by the amorphous silica glaze. This behavior aligns with AFM observations made on nickel ferrite [29,30]. Increasing the amount of La substitution with x_{La} = 0.3, less significant changes in the obtained nanoparticles after annealing at 200 °C and 500 °C were found. The exception here is a slight increase in their diameters, as shown in Figure 3d,e. The major change occurs after annealing at 800 °C, as shown in Figure 3f, where the nanoparticles have an increased size of about 38 nm with a ferrite crystalline core of 22.8 nm (as calculated from XRD patterns) and their shape becomes boulder-like with rounded corners due to the cubic FCC crystals' expansion. This tendency is progressively accentuated by increasing x_{La} from 0.5 to 0.9. Smaller rounded particles are observed after annealing at 200 °C, exhibiting a gradual increase in size from approximately 15 to 20 nm. This size evolution is attributed to the development of small ferrite crystallites and the presence of an amorphous silica coating, as shown in Figure 3g,j, and m. A similar enhancement from a diameter of about 28 nm to one of 35 nm is observed after annealing at 500 °C, with the development of a vigorous ferrite core crystallite (as determined from XRD patterns); however, it was not strong enough to alter the rounded shape of the nanoparticles, as shown in Figure 3h,k,n.

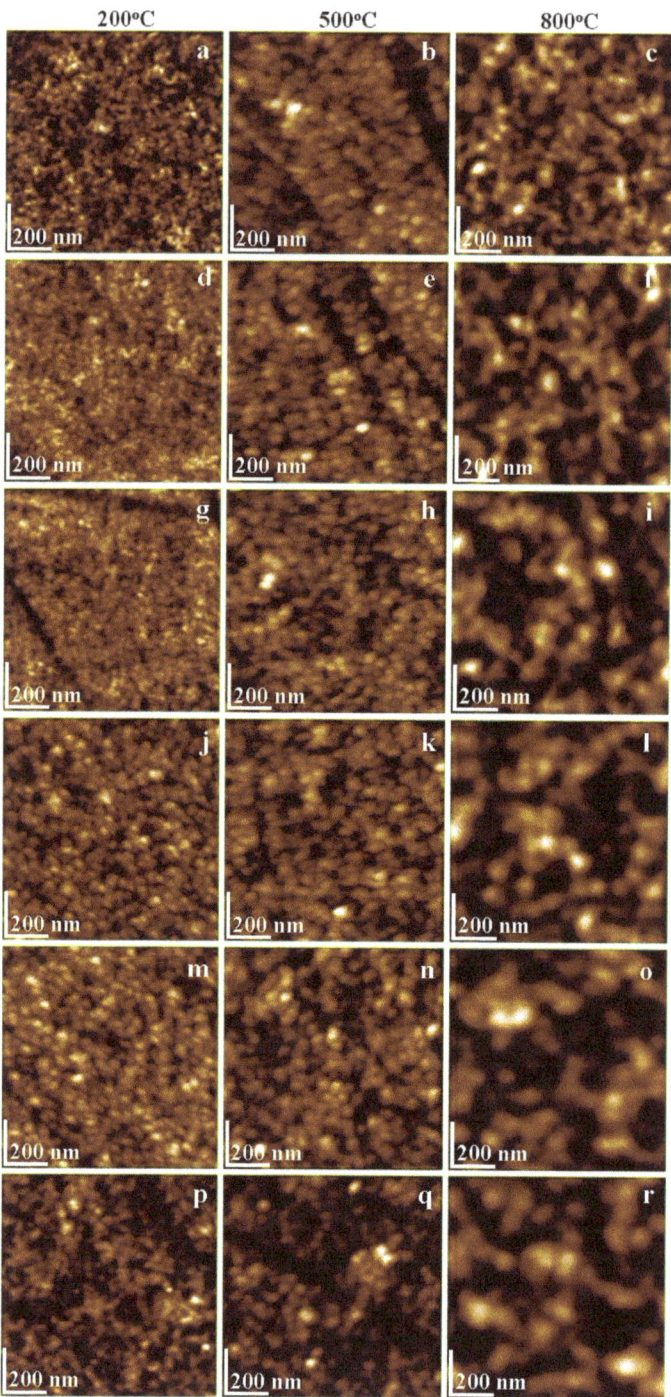

Figure 3. AFM topographical images of NiLa$_x$Fe$_{2-x}$O$_4$/SiO$_2$ annealed at 200, 500, and 800 °C. (**a**–**c**) x$_{La}$ = 0.1; (**d**–**f**) x$_{La}$ = 0.3; (**g**–**i**) x$_{La}$ = 0.5; (**j**–**l**) x$_{La}$ = 0.7; (**m**–**o**) x$_{La}$ = 0.9; (**p**–**r**) x$_{La}$ = 1.1.

The situation is more favorable after annealing at 800 °C because the nanoparticles are well developed, evidencing a small and constant increase in the diameter from 42 to 46 nm along with the accentuation of their boulder aspect as a consequence of cubic FCC single-phase development through the ferrite core, as shown in Figure 3i,l,o. The high amount of La substitution of Fe atoms within nickel ferrite (x_{La} = 1.1) has a major impact on the nanostructure of all particles. The shape remains rounded after the annealing at 200 °C. However, the diameter increases at 21 nm with a ferrite crystallite core of 9.8 nm, as shown in Figure 3p, and is comparable to x_{La} = 0.1 annealed at 500 °C, Figure 3b. The condition is further improved following annealing at 800 °C, resulting in nanoparticles with a diameter of approximately 37 nm.

These nanoparticles exhibit a crystalline core measuring 22.6 nm, and the initially rounded shape shows a slight alteration with the emergence of square corners, as depicted in Figure 3q. These features are not readily apparent due to the presence of the amorphous silica glaze, but they are clearly indicated by the XRD pattern. The nanoparticles exhibit significant elongation of the boulder shape, attributed to the strong development of the ferritic core following annealing at 800 °C, as illustrated in Figure 3r, resulting in a size of approximately 50 nm and a crystallite core measuring 35.4 nm. The substitution of Fe atoms with La appears to be facilitated by higher annealing temperatures, promoting the formation of topographically anisotropic nanoparticles. This distinctive assembly at the nanostructural level suggests the potential for special properties.

By examining the three-dimensional profiles of the resulting thin films in Figure 4, one can observe that nanoparticles annealed at 200 °C produce smooth, uniform, and compact layers characterized by low heights and minimal surface roughness, as indicated in Table 1. As the annealing temperature rises, the nanoparticle diameter expands, and the thin film becomes more agglomerated, resulting in localized unevenness that increases both height and surface roughness. Notably, nanoparticles obtained after annealing at 800 °C exhibit excellent individualization, and the adsorbed thin film is less compact. This phenomenon is linked to the augmented diameter, leading to a notable increase in surface roughness. These findings hold potential for the future application of customized surfaces through thin film deposition. Achieving desired properties may involve adjusting the nanoparticle range appropriately.

The morphological characteristics of $NiLa_xFe_{2-x}O_4/SiO_2$ samples annealed at 800 °C (x_{La} = 0.1, 0.3, 0.5, 0.7, 0.9, 1.1) were investigated using transmission electron microscopy (TEM), as illustrated in Figure 5. TEM images clearly reveal the spherical nature and uniform size distribution of particles. Also, the mean size of particles increases from 26 nm to 47 nm with rising La^{3+} content, which is likely influenced by higher surface tension in smaller nanoparticles, driving increased agglomeration [1–3,6,9]. Spherical particle shapes may be attributed to the synthesis method and surface properties, while agglomeration could result from interfacial surface tension phenomena [1–3,6,9]. Discrepancies between crystallite size (D_{XRD}), particle size from AFM data (D_{AFM}), and TEM-derived particle size (D_{TEM}) may be explained by interference from the amorphous SiO_2 matrix and large nanoparticles in diffraction patterns [11–14]. Agglomeration could be explained by the influence of thermal treatment temperature and potential surface defects [3].

The sonophotocatalytic performance of the samples was examined when exposed to visible irradiation using a RhB synthetic solution. Prior to the irradiation, the samples were subjected to 60 min of darkness to achieve the adsorption equilibrium. The results of this investigation are depicted in Figure 6.

For samples subjected to annealing at 500 °C, the adsorption capacity falls within the range of 12% to 22%, except for $NiLa_{0.5}Fe_{2.5}O_4/SiO_2$, which exhibits a notably higher adsorption capacity at around 45%. An interesting observation is the fact that this sample has the smallest specific surface area. The adsorption capacity increases for samples annealed at 800 °C, ranging from 22% to 45%, excluding $NiLa_{1.1}Fe_{0.9}O_4/SiO_2$ nanocomposites with an adsorption capacity of less than 5%. As expected, all samples annealed at 800 °C, with the exception of that with x_{La} = 1.1, have lower specific surface area and still have higher adsorption capacity. Based on this observation, it can be concluded that the adsorption capacity of these samples might be attributed to interactions between the surface functional groups of the prepared ferrites and the active functional groups of RhB [31]. Figure 6 illustrates the removal rates of the samples after 5 h of irradiation. Among samples annealed at 500 °C, NiLa$_{0.5}$Fe$_{1.5}$O$_4$/SiO$_2$ demonstrates the most effective removal rate at approximately 60%. However, as the annealing temperature rises to 800 °C, the removal rate of this sample decreases to 53%, which is attributed to its larger crystallite size dimension (D$_C$) and lower specific surface area (SSA). Notably, in the samples annealed at 800 °C, NiLa$_{0.5}$Fe$_{1.5}$O$_4$/SiO$_2$ is surpassed in removal rate by NiLa$_{0.3}$Fe$_{1.5}$O$_4$/SiO$_2$, reaching a maximum removal rate of 73%. This value is higher than the previously reported removal rate of Ni-ferrite of about 39% [32]. To assess the impact of ultrasound, this sample was kept in the dark for an additional 300 min, and the results are incorporated in Figure 6. It can be observed that ultrasound does not have a significant effect on the sample removal rate.

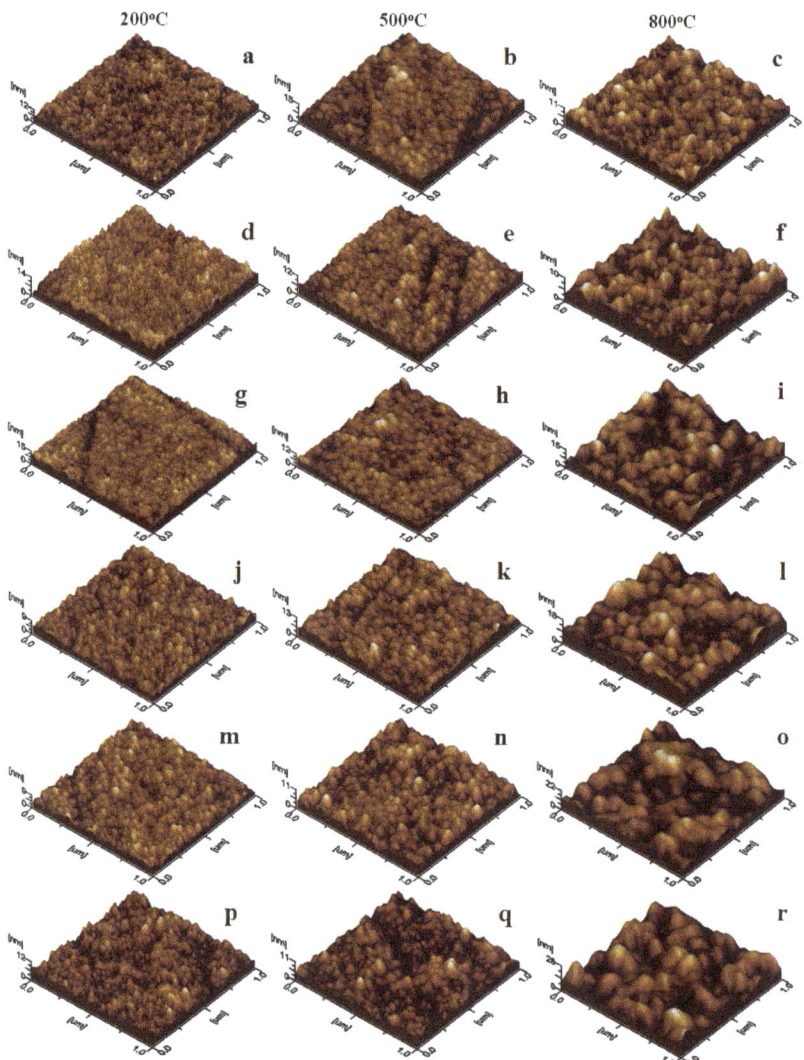

Figure 4. AFM tridimensional profiles of NiLa$_x$Fe$_{2-x}$O$_4$/SiO$_2$ annealed at 200, 500, and 800 °C. (**a**–**c**) x$_{La}$ = 0.1; (**d**–**f**) x$_{La}$ = 0.3; (**g**–**i**) x$_{La}$ = 0.5; (**j**–**l**) x$_{La}$ = 0.7; (**m**–**o**) x$_{La}$ = 0.9; (**p**–**r**) x$_{La}$ = 1.1.

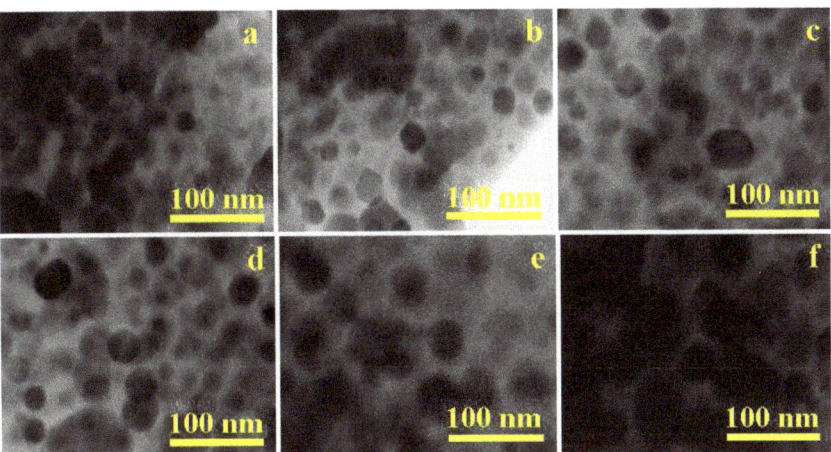

Figure 5. TEM images of NiLa$_x$Fe$_{2-x}$O$_4$/SiO$_2$ annealed at 800 °C. (a) x_{La} = 0.1; (b) x_{La} = 0.3; (c) x_{La} = 0.5; (d) x_{La} = 0.7; (e) x_{La} = 0.9; (f) x_{La} = 1.1.

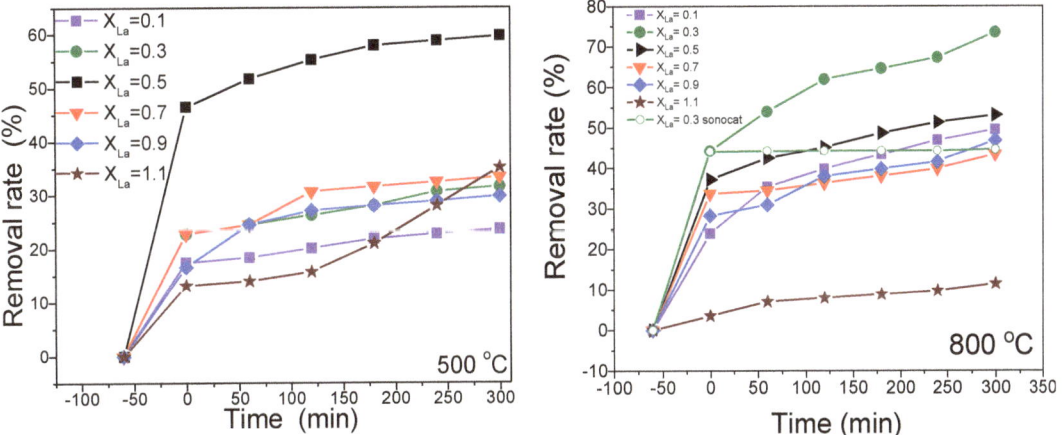

Figure 6. RhB solution degradation rate under visible irradiation in the presence of the 500 and 800 °C tested samples.

The first-order kinetic model was applied to describe the photocatalytic process (1):

$$-\ln\frac{A_t}{A_0^*} = k \times t, \tag{1}$$

where A_t represents RhB absorbance at t time; A_0^* is the absorbance of RhB after dark adsorption; t—irradiation time; and k—apparent kinetic constant.

The experimental data were fitted using the rate equation, and the resulting plots, demonstrating a linear correlation with irradiation time, are showcased in Figure 7.

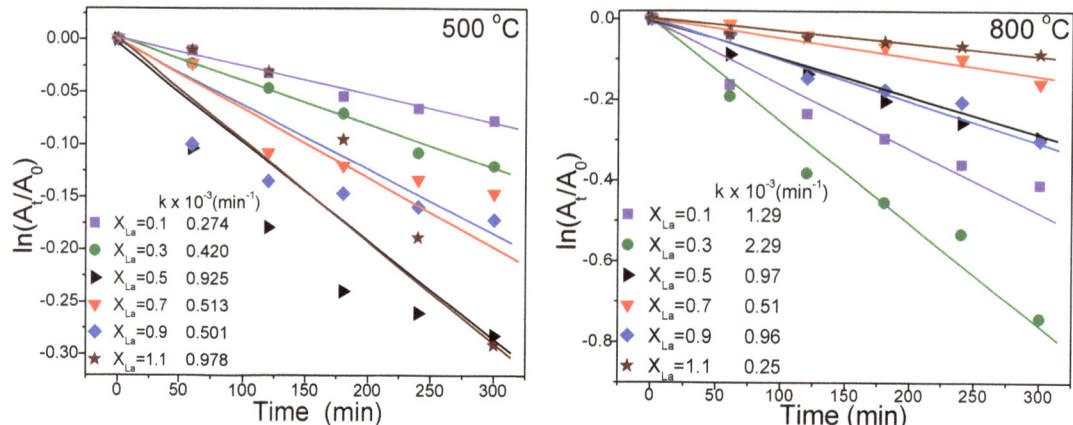

Figure 7. Photodegradation kinetics of RhB synthetic solution at 500 and 800 °C.

The inset of Figure 7 displays the rate constant values derived from the fitting process. Upon analyzing the obtained results, it is evident that the NiLa$_{0.3}$Fe$_{1.7}$O$_4$/SiO$_2$ sample, annealed at 800 °C, exhibited the most superior photocatalytic activity.

The photocatalytic activity is influenced by various factors, and one key factor is the band gap energy. To assess this, we determined the band gap energy using Tauc's equation based on the UV-Vis absorption spectra. In Figure 8a, the UV-Vis absorption profiles of all samples annealed at 800 °C are presented. The significant absorption is a result of electron excitation from the valence band to the conduction band. The variation in absorbance band edge is attributed to the presence of interface defects, point defects, and interactions involving photogenerated electrons [1]. The UV–visible spectrum results from electronic transitions, moving from a lower energy band to a higher energy band. In the case of nickel ferrite, this transition is attributed to electrons moving from the O 2p level to the Fe 3d level. This is explained by considering the O 2p orbital as the valence band and the Fe 3d orbital as the conduction band, as the band structure is primarily defined in this manner [33]. Through a substitution with La ions, the maximum absorption and the absorption edge vary; however, all samples exhibit a broad response across the entire visible range.

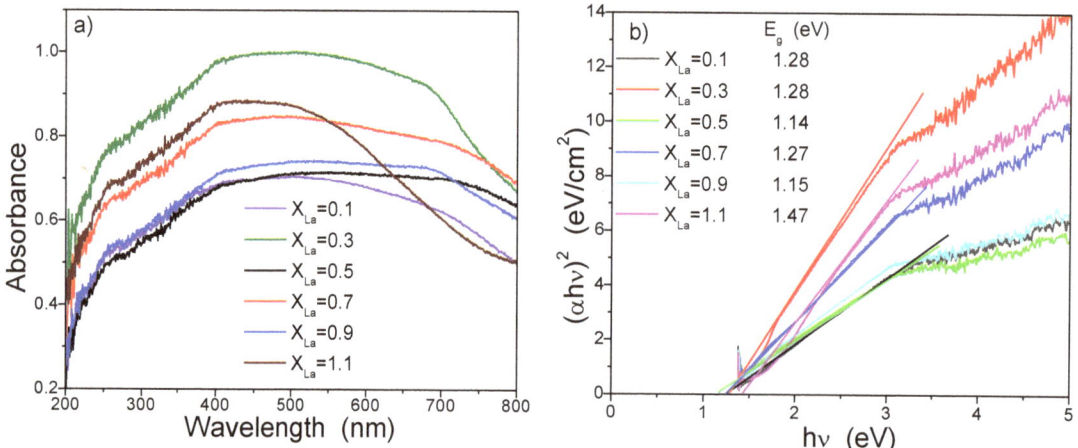

Figure 8. (a) UV-Vis absorption spectra of the samples. (b) Tauc's plot.

Figure 8b shows the Tauc plot of $(\alpha h \nu)^2$ versus $h\nu$ for the direct allowed transition of all samples. The extrapolation of this plot by linear region to the point = 0 gives the corresponding values of the direct band gap. The optical band gaps were calculated and are inserted in the inset of Figure 8b. The calculated values indicate a significant decrease in the band gap energies of $NiLa_{0.5}Fe_{1.5}O_4/SiO_2$ and $NiLa_{0.9}Fe_{1.1}O_4/SiO_2$ when compared with the 1.45 eV band gap of Ni-ferrite reported in a previous study [34]. This reduction is likely attributed to the introduction of additional dopant levels into the band gap of Ni ferrite. In accordance with Rajeshwari et al. [34], the band gap values for the prepared Lanthanum-doped manganese nanoferrite range from 1.89 to 2.35 eV, showing improvement compared with the 1.25–1.38 eV band gap values of Mn nanoferrite, owing to the influence of La^{3+} ions. The photocatalytic mechanism could be explained through the generation of electron–hole pairs when the ferrite surface is exposed to an energy equal to or greater than the band gap energy. Consequently, the photoexcited electron moves from the valence band to the conduction band, creating a hole in the valence band. Effective photocatalysis occurs when these generated pairs remain uncombined. In this scenario, the electrons engage with the adsorbed O_2 on the photocatalyst's surface, yielding superoxide radicals, while the holes interact with H_2O, forming hydroxyl radicals. Both types of radicals are classified as reactive oxygen species capable of breaking down organic pollutant molecules. A significant challenge in this process is the recombination of electron–hole pairs, which hampers the production of reactive oxygen species, thereby impeding the photodegradation process. In our study, the energy levels of dopants, derived from the Ni-ferrite band gap, effectively capture the generated electrons, thereby hindering the recombination of electron–hole pairs. More precisely, part of the photogenerated electrons undergo excitation and reach defect levels, while simultaneously, the photogenerated holes participate in photo-oxidation reactions. The enhanced photocatalytic activity observed in the $NiLa_{0.3}Fe_{1.7}O_4/SiO_2$ sample can be attributed to its increased adsorption capacity and the introduction of dopant energy levels into the band gap of Ni-ferrite through La substitution. These dopant energy levels serve as mediators for interfacial charge transfer [35], resulting in a high separation rate of photogenerated charge carriers. However, with a higher doping level, La ions become recombination centers, leading to the quenching of photocatalytic activity.

To confirm the generation of reactive oxygen species (ROS) by the $NiLa_{0.3}Fe_{1.7}O_4/SiO_2$ sample under visible irradiation, we utilized EPR spectroscopy coupled with the spin trapping technique. DMPO was employed as the spin trapping agent, and the resulting spectrum is depicted in Figure 9.

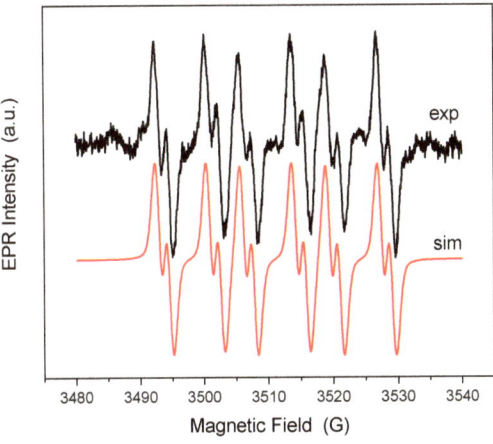

Figure 9. Experimental and simulated spectra of DMPO spin adducts generated by the $NiLa_{0.3}Fe_{1.5}O_4/SiO_2$ sample after 25 min of irradiation.

To discern the species accountable for this signal, a simulation was performed. The experimental spectrum fitted closely with the spectrum of the •DMPO-O_2^- spin adduct having the following spin Hamiltonian parameters g = 2.0098, ΔH = 1.38 G, a_N = 13.2474 G, a_H^β = 8.0109 G, and a_H^γ = 1.6051 G. Unexpectedly, the sample generates only O_2^-, meaning that the maximum valence band position has a lower potential than the oxidation one of the OH-/•OH and H_2O/•OH redox pair; consequently, these reactions cannot occur [36]. The photocatalytic activity of the sample is exclusively attributed to the generation of superoxide radicals when exposed to visible light.

The obtained photocatalytic performance results (removal rate and the first-order rate constant, k) for $NiLa_{0.3}Fe_{1.5}O_4/SiO_2$ annealed at 800 °C are in the same range with other Ni-ferrites previously reported in the literature. Table 3 provides a comparison of various Ni-ferrites, considering both reported work and the current study, with respect to the first-order rate constant.

Table 3. Comparison of various Ni-ferrite samples with respect to the reported first-order rate constant values.

Sample	Lights	Dyes	k × 10^{-3} (min^{-1})	Reference
$NiFe_2O_4$	Visible	Methylene blue	3.4	[4]
$NiFe_2O_4$	Visible	Methyl Orange	2.4	[37]
$NiFe_2O_4$	Visible	Methylene blue	2.3	[38]
$NiFe_2O_4$	Visible	Methylene blue	2.4	[39]
ZnO-$NiFe_2O_4$	Visible	RhB	2.5	[40]
$ZnO/NiFe_2O_4$	Visible	Methylene blue	1.7	[4]
$Ni_xZn_{1-x}Fe_2O_4$	Sun	Fluorescein	2.7	[41]
$Ni_xCu_{(1-x)}Fe_2O_4$	Visible	RhB	3.6	[42]
$Ni_{0.5}Zn_{0.5}Fe_2O_4$	Sun	Methylene blue	6.5	[43]
$TiO_{2-x}N_x/SiO_2/NiFe_2O_4$	Visible	Methyl Orange	4.7	[44]
$NiLa_{0.3}Fe_{1.5}O_4/SiO_2$	Visible	RhB	2.3	This work

The photostability of the $NiLa_{0.3}Fe_{1.7}O_4/SiO_2$ sample (annealed at 800 °C) was verified via reutilization tests in three consecutive trials. The results are depicted in Figure 10. The sample, extracted from the solution using a magnet, underwent a washing with water and ethyl alcohol before each run, followed by an overnight drying. As could be observed from the results, the removal rate shows minimal variation, signifying the robust stability of the photocatalyst.

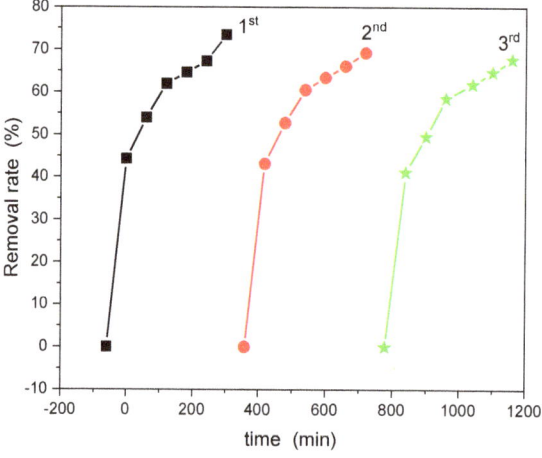

Figure 10. Photocatalyst stability test of $NiLa_{0.3}Fe_{1.7}O_4/SiO_2$ annealed at 800 °C for removal of RhB.

4. Conclusions

Nickel nanoferrite samples doped with La^{3+} ions, featuring various compositions, were synthesized using a sol–gel method. Single-phase nanostructures in the form of an inverse spinel were achieved for Ni-La ferrites across all concentrations at both 500 and 800 °C. The substitution of iron with lanthanum ions within the lattice revealed an expansion of the lattice parameter. This is attributed to the considerable difference in ionic radii between La^{3+} and Fe^{3+}, influencing both crystallite size and the fraction of A sites occupied by ferrite cations. Consequently, the degree of the inverse spinel structure experienced an increase. The crystallite size of the mixed Ni-La ferrites increases with the increase in La content and increased temperature, from 19.5 nm to 35.4 nm at 800 °C, from 10.4 nm to 22.6 nm at 500 °C, and from 1.2 nm to 9.8 nm at 200 °C. The particles have an asymmetric spherical shape. The results affirm that the preparation method effectively provided a straightforward means of achieving the desired morphology and microstructure for the ferrite nanocrystals. The specific surface area (SSA) values exhibit variation in accordance with the lanthanum content, showing a decrease as the heat treatment temperature increases. This decline is attributed to the augmentation of grain sizes and crystallinity during the heating process. All nanocomposites present pores in the mesoporous region, with narrow pore size dispersion. All samples show good optical response in the visible range. The best sonophotocatalytic performance was registered for $NiLa_{0.3}Fe_{1.7}O_4/SiO_2$; this result is most likely because of the La additional levels inserted in the band gap of Ni-ferrite and the equilibrium between La and Fe in Ni-La ferrite.

Author Contributions: Conceptualization, T.D.; methodology, T.D.; software, T.D., I.P., M.D.L., and D.T.; formal analysis, T.D., I.P., M.D.L., and D.T.; investigation, T.D., I.P., M.D.L., and D.T.; resources, T.D., I.P., M.D.L., and D.T.; data curation, T.D.; writing—original draft preparation, T.D., I.P., M.D.L., and D.T.; writing—review and editing, T.D., I.P., M.D.L., and D.T.; visualization, T.D.; supervision, T.D. All authors have read and agreed to the published version of the manuscript.

Funding: The APC was funded by the Technical University of Cluj-Napoca Grant Support CA106, 28 June 2022.

Data Availability Statement: Data are contained within the article.

Acknowledgments: M.D.L. and D.T. acknowledge support from the Romanian Ministry of Research, Innovation, and Digitalization, NUCLEU Program within the National Research Development and Innovation Plan 2022–2027, project numbers PN 23 24 01 01 and PN 23 24 01 03, and Program 1—Development of the national research and development system, Subprogram 1.2—Institutional performance—Projects that finance the RDI excellence, contract no. 37PFE/30.12.2021. The authors acknowledge the Research Centre in Physical Chemistry "CECHIF" for AFM assistance.

Conflicts of Interest: The authors declare no conflict of interest.

References

1. Deepapriya, S.; Vinosha, P.A.; Rodney, J.D.; Krishnan, S.; Jose, J.E.; Das, S.J. Effect of Lanthanum Substitution on Magnetic and Structural Properties of Nickel Ferrite. *J. Nanosci. Nantotechnol.* **2018**, *18*, 6987–6994. [CrossRef] [PubMed]
2. Deepapriya, S.; Vinosha, P.A.; Rodney, J.D.; Jose, M.; Krishnan, S.; Jose, J.E.; Das, S.J. Investigation on lanthanum substitution in magnetic and structural properties of group IV elements. *Vacuum* **2019**, *161*, 5–13. [CrossRef]
3. Al Angari, Y.M. Magnetic properties of La-substituted $NiFe_2O_4$ via egg-white precursor route. *J. Magn. Magn. Mater.* **2011**, *323*, 1835–1839. [CrossRef]
4. Munir, S.; Warsi, M.F.; Zulfiqar, S.; Ayman, I.; Haider, S.; Alsafari, I.A.; Agboola, P.O.; Shakir, I. Nickel ferrite/zinc oxide nanocomposite: Investigating the photocatalytic and antibacterial properties. *J. Saudi Chem. Soc.* **2021**, *25*, 101388. [CrossRef]
5. Priyadharshini, P.; Pushpanathan, K. Synthesis of Ce-doped $NiFe_2O_4$ nanoparticles and their structural, optical, and magnetic properties. *Chem. Phys. Impact.* **2023**, *6*, 100201. [CrossRef]
6. Shetty, P.B.; Maddani, K.I.; MahaLaxmi, K.S.; Lakshmi, C.S.; Sridhar, C.S.L.N. Studies on lanthanum-doped nickel ferrites for improved structural, magnetic and optical properties. *J. Mater. Sci. Mater. Electron.* **2023**, *34*, 1246. [CrossRef]
7. Lenin, N.; Rajesh Kanna, R.; Sakthipandi, K.; Senthil Kumar, A. Structural, electrical and magnetic properties of $NiLa_xFe_{2-x}O_4$ nanoferrites. *Mater. Chem. Phys.* **2018**, *212*, 385–393. [CrossRef]
8. Liu, X.M.; Gao, W.L. Preparation and Magnetic Properties of $NiFe_2O_4$ Nanoparticles by Modified Pechini Method. *Mater. Manuf. Process.* **2012**, *27*, 905–909. [CrossRef]

9. Gaba, S.; Rana, P.S.; Kumar, A.; Pant, R.P. Structural and paramagnetic resonance properties correlation in lanthanum ion doped nickel ferrite nanoparticles. *J. Magn. Magn. Mater.* **2020**, *508*, 166866. [CrossRef]
10. Ahmad, M.N.; Khan, H.; Islam, L.; Alnasir, M.H.; Ahmad, S.N.; Qureshi, M.T.; Khan, M.Y. Investigating Nickel Ferrite ($NiFe_2O_4$) Nanoparticles for Magnetic Hyperthermia Applications. *J. Mater. Phys. Sci.* **2023**, *4*, 32–45. [CrossRef]
11. Dippong, T.; Deac, I.G.; Cadar, O.; Levei, E.A. Effect of silica embedding on the structure, morphology and magnetic behavior of $(Zn_{0.6}Mn_{0.4}Fe_2O_4)_\delta/(SiO_2)_{(100-\delta)}$ nanoparticles. *Nanomaterials* **2021**, *11*, 2232. [CrossRef] [PubMed]
12. Dippong, T.; Levei, E.A.; Deac, I.G.; Petean, I.; Cadar, O. Dependence of structural, morphological and magnetic properties of manganese ferrite on Ni-Mn substitution. *Int. J. Mol. Sci.* **2022**, *23*, 3097. [CrossRef] [PubMed]
13. Dippong, T.; Levei, E.A.; Cadar, O.; Deac, I.G.; Lazar, M.; Borodi, G. Effect of amorphous SiO_2 matrix on structural and magnetic properties of $Cu_{0.6}Co_{0.4}Fe_2O_4/SiO_2$ nanocomposites. *J. Alloys Comp.* **2020**, *849*, 156695. [CrossRef]
14. Dippong, T.; Levei, E.A.; Cadar, O. Investigation of structural, morphological and magnetic properties of MFe_2O_4 (M = Co, Ni, Zn, Cu, Mn) obtained by thermal decomposition. *Int. J. Mol. Sci.* **2022**, *23*, 8483. [CrossRef] [PubMed]
15. Jadhav, S.A.; Somvanshi, S.B.; Khedkar, M.V.; Patade, S.R.; Jadhav, K.M. Magneto-structural and photocatalytic behavior of mixed Ni–Zn nano-spinel ferrites: Visible light-enabled active photodegradation of rhodamine B. *J. Mater. Sci. Mater. Electron.* **2020**, *31*, 11352–11365. [CrossRef]
16. Zhang, Z.; Ma, Y.; Bu, X.; Wu, Q.; Hang, Z.; Dong, Z.; Wu, X. Facile one-step synthesis of $TiO_2/Ag/SnO_2$ ternary heterostructures with enhanced visible light photocatalytic activity. *Sci. Rep.* **2018**, *8*, 10532. [CrossRef]
17. Shah, P.; Joshi, K.; Shah, M.; Unnarkat, A.; Patel, F.J. Photocatalytic dye degradation using nickel ferrite spinel and its nanocomposite. *Environ. Sci. Poll. Res.* **2022**, *29*, 78255–78264. [CrossRef]
18. Ghoneim, A.I. High Surface Area and Photo-catalysis of $Cu_{0.3}Cd_{0.7}CrFeO_4$ Nanocrystals in Degradation of Methylene Blue (MB). *Egypt. J. Solids* **2021**, *43*, 211–226. [CrossRef]
19. Padmapriya, G.; Manikandan, A.; Krishnasamy, V.; Jaganathan, S.K.; Antony, S.A. Enhanced catalytic activity and magnetic properties of spinel $Mn_xZn_{1-x}Fe_2O_4$ (0.0≤ x≤ 1.0) nano-photocatalysts by microwave irradiation route. *J. Supercond. Nov. Magn.* **2016**, *29*, 2141–2149. [CrossRef]
20. Mapukata, S.; Ntsendwana, B.; Mokhena, T.; Sikhwivhilu, L. Advances on sonophotocatalysis as a water and wastewater treatment technique: Efficiency, challenges and process optimisation. *Front. Chem.* **2023**, *23*, 1252191. [CrossRef]
21. Munoz, M.; De Pedro, Z.M.; Casas, J.A.; Rodriguez, J.J. Preparation of magnetite-based catalysts and their application in heterogeneous Fenton oxidation—A review. *Appl. Catal. B Environ.* **2015**, *176–177*, 249–265. [CrossRef]
22. Hossen, M.M.; Hossen, M.B. Structural, electrical and magnetic properties of $Ni_{0.5}Cu_{0.2}Cd_{0.3}La_xFe_{2-x}O_4$ nano-ferrites due to lanthanum doping in the place of trivalent iron. *Phys. B* **2020**, *585*, 412116. [CrossRef]
23. Thommes, M.; Kaneko, K.; Neimark, A.V.; Olivier, J.P.; Rodriguez-Reinoso, F.; Rouquerol, J.; Sing, K.S.W. Physisorption of gases, with special reference to the evaluation of surface area and pore size distribution (IUPAC Technical Report). *Pure Appl. Chem.* **2015**, *87*, 1051–1069. [CrossRef]
24. Simon, C.; Zakaria, M.B.; Kurz, H.; Tetzlaff, D.; Blösser, A.; Weiss, M.; Timm, J.; Weber, B.; Apfel, U.P.; Marschall, R. Magnetic $NiFe_2O_4$ Nanoparticles Prepared via Non-Aqueous Microwave-Assisted Synthesis for Application in Electrocatalytic Water Oxidation. *Chem. Eur. J.* **2021**, *27*, 16990–17001. [CrossRef] [PubMed]
25. Dippong, T.; Lazar, M.D.; Deac, I.G.; Palade, P.; Petean, I.; Borodi, G.; Cadar, O. The effect of cation distribution and heat treatment temperature on the structural, surface, morphological and magnetic properties of $Mn_xCo_{1-x}Fe_2O_4@SiO_2$ nanocomposites. *J. Alloys Compd.* **2022**, *895*, 162715. [CrossRef]
26. Sivaraj, C.; Contescu, C.; Schwarz, J.A. Effect of calcination temperature of alumina on the adsorption/impregnation of Pd(II) compounds. *J. Catal.* **1991**, *132*, 422–431. [CrossRef]
27. Samuel, Z.; Ojemaye, M.O.; Okoh, O.O.; Okoh, A.I. Adsorption of simazine herbicide from aqueous solution by novel pyrene functionalized zinc oxide nanoparticles: Kinetics and isotherm studies. *Mater. Today Commun.* **2023**, *34*, 105435. [CrossRef]
28. Ahmad, R.; Ejaz, M.O. Synthesis of new alginate-silver nanoparticles/mica (Alg-AgNPs/MC) bionanocomposite for enhanced adsorption of dyes from aqueous solution. *Chem. Eng. Res. Des.* **2023**, *197*, 355–371. [CrossRef]
29. Abdelghani, G.M.; Al-Zubaidi, A.B.; Ahmed, A.B. Synthesis, characterization, and study of the influence of energy of irradiation on physical properties and biologic activity of nickel ferrite nanostructures. *J. Saudi Chem. Soc.* **2023**, *27*, 101623. [CrossRef]
30. Sapkota, B.; Martin, A.; Lu, H.; Mahbub, R.; Ahmadi, Z.; Azadehranjbar, S.; Mishra, E.; Shield, J.E.; Jeelani, S.; Rangari, V. Changing the polarization and mechanical response of flexible PVDF-nickel ferrite films with nickel ferrite additives. *Mater. Sci. Eng. B* **2022**, *283*, 115815. [CrossRef]
31. Pogacean, F.; Ştefan, M.; Toloman, D.; Popa, A.; Leostean, C.; Turza, A.; Coros, M.; Pana, O.; Pruneanu, S. Photocatalytic and Electrocatalytic Properties of NGr-ZnO Hybrid Materials. *Nanomaterials* **2020**, *10*, 1473. [CrossRef] [PubMed]
32. Dippong, T.; Levei, E.A.; Cadar, O.; Goga, F.; Toloman, D.; Borodi, G. Thermal behavior of Ni, Co and Fe succinates embedded in silica matrix. *J. Therm. Anal. Calorim.* **2019**, *136*, 1587–1596. [CrossRef]
33. Xu, S.; Feng, D.; Shangguan, W. Preparations and Photocatalytic Properties of Visible-Light-Active Zinc Ferrite-Doped TiO_2 Photocatalyst. *J. Phys. Chem. C* **2019**, *113*, 2463–2467. [CrossRef]
34. Rajeshwari, A.; Punithavthy, I.K.; Jeyakumar, S.J.; Lenin, N.; Vigneshwaran, B. Dependance of lanthanum ions on structural, magnetic and electrical of manganese based spinel nanoferrites. *Ceram. Int.* **2020**, *46*, 6860–6870. [CrossRef]

35. Yin, Q.; Qiao, R.; Li, Z.; Zhang, X.L.; Zhu, L.J. Hierarchical nanostructures of nickel-doped zinc oxide: Morphology controlled synthesis and enhanced visible-light photocatalytic activity. *J. Alloys Compd.* **2015**, *618*, 318–325. [CrossRef]
36. Zhang, R.; Zhao, C.; Yu, J.; Chen, Z.; Jiang, J.; Zeng, K.; Cai, L.; Yang, Z. Synthesis of dual Z-scheme photocatalyst $ZnFe_2O_4$/PANI/Ag_2CO_3 with enhanced visible light photocatalytic activity and degradation of pollutants. *Adv. Powder Technol.* **2022**, *33*, 103348. [CrossRef]
37. Gebreslassie, G.; Bharali, P.; Chandra, U.; Sergawie, A. Hydrothermal synthesis of g-C_3N_4/$NiFe_2O_4$ nanocomposite and its enhanced photocatalytic activity. *Appl. Organomet. Chem.* **2019**, *33*, e5002. [CrossRef]
38. Jadhav, S.A.; Khedkar, M.V.; Somvanshi, S.B.; Jadhav, K.M. Magnetically retrievable nanoscale nickel ferrites: An active photocatalyst for toxic dye removal applications. *Ceram. Int.* **2021**, *47*, 28623–28633. [CrossRef]
39. Tripta, R.; Suman, P.S. Tuning the morphological, optical, electrical, and structural properties of $NiFe_2O_4$@CdO nanocomposites and their photocatalytic application. *Ceram. Int.* **2023**, *49*, 18735–18744. [CrossRef]
40. Fujishima, A.; Honda, K. Electrochemical photolysis of water at a semiconductor electrode. *Nature* **1972**, *238*, 37–38. [CrossRef]
41. Nimisha, O.K.; Akshay, M.; Mannya, S.; Reena Mary, A.P. Synthesis and photocatalytic activity of nickel doped zinc ferrite. *Mater. Today Proc.* **2022**, *66*, 2370–2373. [CrossRef]
42. Azevedoa, I.G.D.D.; Rodrigues, M.V.; Gomes, Y.R.; de Araújo, C.P.B.; de Souza, C.P.; Moriyama, A.L.L. Photocatalytic Degradation of the Rhodamine B Dye Under Visible Light Using $Ni_xCu_{(1-x)}Fe_2O_4$ Synthesized by EDTA-Citrate Complexation Method. *Mater. Res.* **2023**, *26*, e20230061. [CrossRef]
43. Dhiman, P.; Rana, G.; Dawi, E.A.; Kumar, A.; Sharma, G.; Kumar, A.; Sharma, J. Tuning the Photocatalytic Performance of Ni-Zn Ferrite Catalyst Using Nd Doping for Solar Light-Driven Catalytic Degradation of Methylene Blue. *Water* **2023**, *15*, 187. [CrossRef]
44. Rauf, A.; Ma, M.; Kim, S.; Shah, M.S.A.S.; Chung, C.-H.; Park, J.H.; Yoo, P.J. Mediator-and co-catalyst-free direct Z-scheme composites of Bi_2WO_6–Cu_3P for solar-water splitting. *Nanoscale* **2018**, *10*, 3026–3036. [CrossRef]

Disclaimer/Publisher's Note: The statements, opinions and data contained in all publications are solely those of the individual author(s) and contributor(s) and not of MDPI and/or the editor(s). MDPI and/or the editor(s) disclaim responsibility for any injury to people or property resulting from any ideas, methods, instructions or products referred to in the content.

Article

Magnetic Properties and Magnetocaloric Effect of Polycrystalline and Nano-Manganites $Pr_{0.65}Sr_{(0.35-x)}Ca_xMnO_3$ ($x \leq 0.3$)

Roman Atanasov [1], Dorin Ailenei [1], Rares Bortnic [1], Razvan Hirian [1], Gabriela Souca [1], Adam Szatmari [1], Lucian Barbu-Tudoran [2] and Iosif Grigore Deac [1,*]

[1] Faculty of Physics, Babes-Bolyai University, Str. Kogalniceanu 1, 400084 Cluj-Napoca, Romania; atanasov.roman@ubbcluj.ro (R.A.); dorin.ailenei@stud.ubbcluj.ro (D.A.); rares.bortnic@ubbcluj.ro (R.B.); razvan.hirian@ubbcluj.ro (R.H.); gabriela.souca@ubbcluj.ro (G.S.); adam.szatmari@ubbcluj.ro (A.S.)
[2] National Institute for Research and Development of Isotopic and Molecular Technologies, Str. Donath 67-103, 400293 Cluj-Napoca, Romania; lucian.barbu@itim-cj.ro
* Correspondence: iosif.deac@phys.ubbcluj.ro

Citation: Atanasov, R.; Ailenei, D.; Bortnic, R.; Hirian, R.; Souca, G.; Szatmari, A.; Barbu-Tudoran, L.; Deac, I.G. Magnetic Properties and Magnetocaloric Effect of Polycrystalline and Nano-Manganites $Pr_{0.65}Sr_{(0.35-x)}Ca_xMnO_3$ ($x \leq 0.3$). Nanomaterials 2023, 13, 1373. https://doi.org/10.3390/nano13081373

Academic Editor: Julian Maria Gonzalez Estevez

Received: 19 March 2023
Revised: 10 April 2023
Accepted: 11 April 2023
Published: 14 April 2023

Copyright: © 2023 by the authors. Licensee MDPI, Basel, Switzerland. This article is an open access article distributed under the terms and conditions of the Creative Commons Attribution (CC BY) license (https://creativecommons.org/licenses/by/4.0/).

Abstract: Here we report investigations of bulk and nano-sized $Pr_{0.65}Sr_{(0.35-x)}Ca_xMnO_3$ compounds ($x \leq 0.3$). Solid-state reaction was implemented for polycrystalline compounds and a modified sol–gel method was used for nanocrystalline compounds. X-ray diffraction disclosed diminishing cell volume with increasing Ca substitution in Pbnm space group for all samples. Optical microscopy was used for bulk surface morphology and transmission electron microscopy was utilized for nano-sized samples. Iodometric titration showed oxygen deficiency for bulk compounds and oxygen excess for nano-sized particles. Measurements of resistivity of bulk samples revealed features at temperatures associated with grain boundary condition and with ferromagnetic (FM)/paramagnetic (PM) transition. All samples exhibited negative magnetoresistivity. Magnetic critical behavior analysis suggested the polycrystalline samples are governed by a tricritical mean field model while nanocrystalline samples are governed by a mean field model. Curie temperatures values lower with increasing Ca substitution from 295 K for the parent compound to 201 K for x = 0.2. Bulk compounds exhibit high entropy change, with the highest value of 9.21 J/kgK for x = 0.2. Magnetocaloric effect and the possibility of tuning the Curie temperature by Ca substitution of Sr make the investigated bulk polycrystalline compounds promising for application in magnetic refrigeration. Nano-sized samples possess wider effective entropy change temperature (ΔT_{fwhm}) and lower entropy changes of around 4 J/kgK which, however, puts in doubt their straightforward potential for applications as magnetocaloric materials.

Keywords: manganites; nanoparticle perovskites; crystallography; magnetic behavior; phase transition; critical behavior; magnetocaloric effect

1. Introduction

Organic life is possible in a certain range of temperature because chemical exchange is destroyed above and below that range [1]. Throughout history, while artists were driven by the need to stay "cool", the engineering part of the human brain kept searching for more practical solutions [2]. It is obvious that the best method for cooling our environment is the one that does not harm it. As such, magnetocaloric materials present a viable alternative to harmful gasses [3,4].

Magnetocaloric effect was reported in 1881 [5], but only recently has its potential for everyday use been discussed. Gadolinium Gd has been a forerunner in excellent magnetocaloric effect for a while, until its alloys were found to have higher values of entropy change [3]. They are expensive and require fields of over 5 T for operational use [4], and so the search for even more effective compounds has begun.

Compounds of the type $A_{1-x}B_xMnO_3$ (where A is a trivalent rare earth cation and B is a divalent alkaline earth cation [6] (pp. 1–153) have a perovskite structure. They allow for

Mn^{3+}-O-Mn^{4+} interaction where an electron from Mn^{3+} can hop to Mn^{4+}, thus aligning the magnetic moments. Such interaction is named "double exchange" and is the main reason for their electric and magnetic properties [7] (pp. 18–21), [8] (pp. 167–293).

Recent work has shown that compounds such as $La_{1-x}Sr_xMnO_3$ and $Pr_{1-x}Ba_xMnO_3$ [9,10] exhibit large magnetic entropy change. Structurally, a divalent Sr^{2+} and Ba^{2+} are doped in place of trivalent La^{3+} and Pr^{3+}. The strongest "double exchange" interaction results are achieved at the doping level of about x = 0.3 [8]. Further introduction of divalent elements leads to antiferromagnetic (AFM) arrangement and can cause localization of charge—the so-called charge ordered (CO) state [11]. Because of this, further substitution of trivalent or divalent ions with different size ions can change the structure, volume and the magnetic properties.

In this work, we have prepared polycrystalline and nanocrystalline samples of $Pr_{0.65}Sr_{0.35-x}Ca_xMnO_3$ (x = 0.02, 0.05, 0.1, 0.2, 0.3). Magnetic properties of the parent bulk compound $Pr_{0.65}Sr_{0.35}MnO_3$ have been reported previously in the literature [12]. Ca^{2+} ions act as the substitute for divalent Sr^{2+} ions in order to bring T_C down from 295 K of the parent compound $Pr_{0.65}Sr_{0.35}MnO_3$ [12] in bulk and T_C = 257 K of nano-sized compounds. Systems were crystallographically and morphologically investigated by X-ray diffraction (XRD), optical microscopy and transmission electron microscopy (TEM), and Rietlveld refinement analysis of XRD. Stoichiometry [13] and structure [8,9] of the samples greatly affects properties. Preparation method can affect stoichiometry of the compounds, sometimes causing accidental vacancies in Pr^{3+} ions. Some changes in magnetic and electrical behavior, therefore, could be observed, but are unquantifiable by this experiment. Additionally, oxygenation plays an important role in final measurements. Oxygen deficiency and excess change stoichiometry of the compounds by changing Mn^{3+}/Mn^{4+} ratio electrical conductivity and overall magnetization. In our work, oxygen content was investigated by chemical analysis of Iodometry. Critical behavior was analyzed by construction of modified Arrott plots (MAP), which was also confirmed by the Kouvel–Fisher (KF) method. Nanocrystalline compounds have lower maximum entropy change values but much wider effective temperature range. Electrical measurements, in the bulk samples, revealed colossal magnetoresistance behavior [11,14,15].

This paper is organized as follows: in Section 2 we describe the preparation methods, as well as all the characterization methods we used in structural, morphological, oxygen stoichiometric, electrical, and magnetic investigation. In Section 3 we present the results of our investigation, analysis of data, and we discuss the magnetic critical behavior, electrical and magnetic properties of our samples. Finally, we summarize our results in Section 4.

2. Materials and Methods

Polycrystalline samples were prepared by conventional solid-state reaction. High purity oxides of principal elements Pr_6O_{11}, SrO, MnO_2, and carbonate $CaCO_3$ were purchased from Alfa Aesar (Heysham, UK). After being mixed by hand in a mortar for 3 h, the powders were calcinated at 1100 °C for 24 h in air. Later, the powder was pressed into a pellet at 3 tons and sintered in air at 1350 °C for 30 h.

Nanocrystalline samples were prepared by sucrose sol-gel method. Nitrates $Pr(NO_3)_3 \cdot 6H_2O$, $Sr(NO_3)_2$, $Ca(NO_3)_2 \cdot 4H_2O$, and $Mn(NO_3)_2 \cdot 6H_2O$ were dissolved in pure water (18.2 MΩ × cm at 25 °C) for up to 1 h at 60 °C after which, 10 g of sucrose was added. This enables positive ions to attach themselves to OH hubs of the sucrose chain. After stirring for 45–60 min temperature was turn down to room temperature. Two grams of pectin was added to expand the xero-gel and mixed for further 20 min. Solutions were dried at 200 °C for 24 h or until all water evaporated. Finally, gels were burned at 1000 °C for 2 h to obtain the nano-sized particles. If the reaction does not happen, as was the case in this experiment, some small amount of Acetic acid can be added to the mix at the last stage before drying in order to bring the chains closer together. However, such action usually results in somewhat bigger particles.

X-ray diffraction (XRD) was implemented for structural categorization of all samples. Optical microscopy was used on bulk samples for grain size determination and surface defects. Transmission electron microscopy (TEM) allowed for nanocrystalline particle size determination. Analysis of XRD data was performed by Rietveld refinement method, as well as Williamson–Hall (W-H) method.

Iodometric analytical titration was applied in order to detect deficiency or excess of oxygen in the samples. A small amount of a sample was placed in a closed vessel containing hydrochloric acid (HCl) where positive Mn^+ ions react with Cl^- to produce Cl_2. An inert gas pushes Cl_2 into another vessel containing potassium iodine (KI) where it reacts to produce I_2. This mixture was titrated with sodium thiosulfate and the ratio of Mn^{3+} to Mn^{4+} was calculated by the stoichiometry of balanced chemical equations [16].

Electrical properties were found using the four-point technique in a cryogen-free superconducting setup. Four-point chips, measuring voltage and current separately were placed in applied magnetic fields of up to 5 T with a varied temperature between 10 K and 300 K. Resistance of the samples was recorded, resistivity was calculated using sample dimensions.

Magnetic measurements were made using a Vibrating Sample Magnetometer (VSM) in the temperature range of 4–300 K in external magnetic fields of up to 4 T.

3. Results

3.1. Structural Analysis

Visual inspection of stack XRD patterns, as shown in Figure 1, suggests possible decrease in cell dimensions due to the shift to the right of patterns with increasing Ca content. All samples are single phase with less than 5% levels of impurities. Wider peaks for nanocrystalline samples are indicative of smaller crystallites size in the compounds [17] which is further confirmed by Rietveld refinement analysis and Williamson–Hall method in Tables 1 and 2.

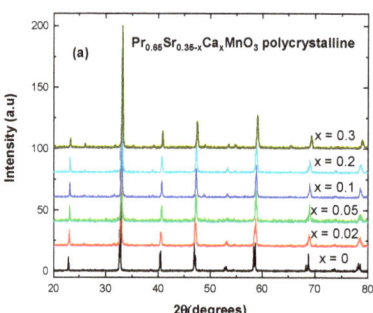
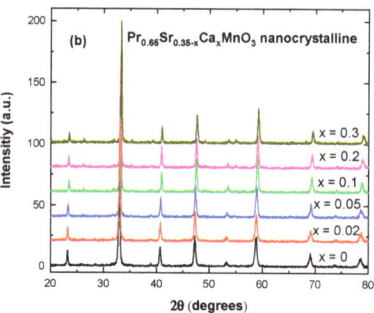

Figure 1. X-ray diffraction patterns for (**a**) $Pr_{0.65}Sr_{0.35-x}Ca_xMnO_3$ polycrystalline bulk samples and (**b**) $Pr_{0.65}Sr_{0.35-x}Ca_xMnO_3$ nano-sized samples.

Table 1. Calculated tolerance factors, Mn-O lengths, and crystallite sizes for polycrystalline samples using the Williamson–Hall and Rietveld methods and including particle diameters by optical microscopy.

Ca Content (Bulk)	t (Tolerance Factor)	Mn-O (Å)	Mn-O-Mn (°)	Average Particle Diameter (μm)	Williamson–Hall Size (nm)	Average Rietveld Size (nm)	Strain
x = 0.02	0.925	1.964	157.73	15	111.33	96.67	0.0022
x = 0.05	0.924	1.962	157.74	11	156.63	113.68	0.0019
x = 0.1	0.921	1.961	157.73	13	132.88	192.59	0.0024
x = 0.2	0.917	1.956	157.69	10	144.85	125.64	0.0023
x = 0.3	0.912	1.954	157.72	11	123.56	86.79	0.0021

Table 2. Calculated Mn-O lengths and crystallite sizes for nanocrystalline samples using the Williamson–Hall and Rietveld methods and including grain diameters from TEM.

Ca Content (Nano)	Mn-O (Å)	Mn-O-Mn (°)	Average Particle Diameter (nm)	Williamson–Hall Size (nm)	Average Rietveld Size (nm)	Strain
x = 0	1.972	157.74	68.1	72.45	54.33	0.0018
x = 0.02	1.964	157.75	64.8	71.15	45.64	0.0019
x = 0.05	1.958	157.69	78.2	84.59	57.21	0.0016
x = 0.1	1.957	157.69	87.5	82.95	39.75	0.0019
x = 0.2	1.955	157.7	71.5	66.73	45.62	0.0017
x = 0.3	1.951	157.71	65.7	69.69	51.11	0.002

Investigation of XRD data by Rietveld refinement analysis established orthorhombic (Pbnm) space group no. 62 for all samples, with decreasing cell dimensions and volume for each subsequent substitution of Sr. The results are illustrated in Figure 2. Stability of the structure in orthorhombic perovskite materials is assessed by the Goldschmidt tolerance factor. It was calculated using following relation [18,19]:

$$t = \frac{R_A + R_0}{\sqrt{2}(R_B + R_0)}, \quad (1)$$

where R_A is the radius of A cation, R_B is the radius of B cation, and R_0 is the radius of the anion. All samples fall within orthorhombic/rhombohedral tolerance range of 0.7–1 [20] (pp. 707–714).

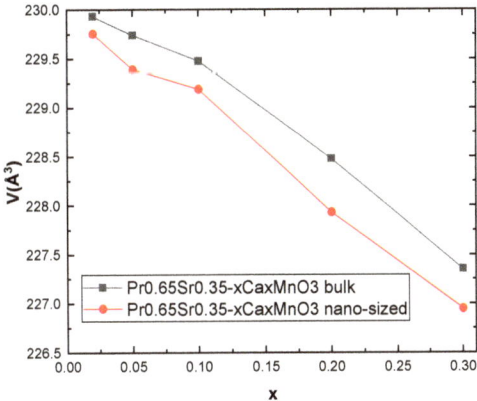

Figure 2. Plot of cell volume change for polycrystalline and nanocrystalline samples.

An addition of smaller crystal radius Ca^{2+} (1.32 Å) atoms in place of Sr^{2+} (1.45 Å) into the structure increases disorder [8,19] and decreases average Mn–O bond length, as seen in Tables 1 and 2, which plays a crucial role in "double exchange" interaction [8]. The average angle between Mn–O–Mn stays relatively the same throughout the range of substitutions at 157.72(3)°.

The Williamson–Hall (W-H) method is advantageous in that it takes into account strain between crystallites, as opposed to the Scherrer method, which does not [21]. Thus, results form W-H analysis are expected to be closer to the real values. A comparison between Rietveld refinement analysis and Williamson–Hall method for approximation of crystallite size can be performed by contrasting them with TEM and optical microscopy results. As can be observed in Table 1 for bulk samples, W-H and Rietveld results are similar, but are two orders of magnitude different from real values. This can be attributed to each grain containing many crystallites. In Table 2, for nano-sized samples, it can be seen that values from W-H calculations are on average larger than those performed by Rietveld analysis and closer to values from TEM. Real sizes for nanocrystalline particles are in the upper limit of nano dimensions at approximately 70–80 nm. Selected samples of optical microscope pictures are presented in Figure 3 and selected TEM pictures in Figure 4.

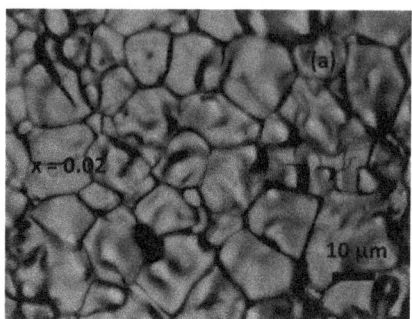 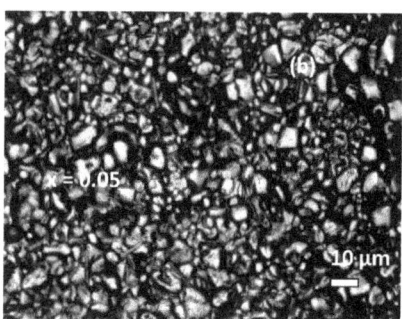

Figure 3. Selected optical microscope pictures for $Pr_{0.7}Sr_{0.3-x}Ca_xMnO_3$ bulk samples (a) for x = 0.02, (b) x = 0.05.

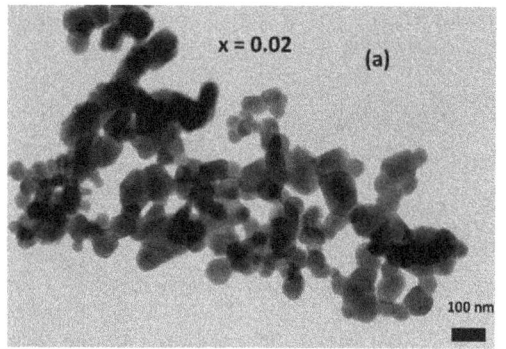 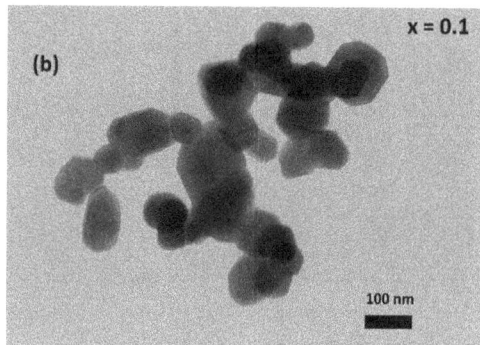

Figure 4. Selected TEM pictures for $Pr_{0.7}Sr_{0.3-x}Ca_xMnO_3$ nano-sized samples for x = 0.02 (a), x = 0.01 (b).

3.2. Oxygen Content

Implementation of iodometric titration analysis for samples is a reliable method for determining deficiency or excess of oxygen in manganites [13,16]. All results are presented in Table 3.

Titration of bulk compounds revealed oxygen deficiency for all samples. An average of $O_{2.93\pm0.02}$ for x = 0.02, similar to other samples, shows larger than expected deficiency of approximately $O_{2.98}$ [22]. This is attributed to preparation methods and difficulty in oxygenating the sample during calcination and sintering. Lower levels of oxygen would affect magnetic and electrical properties as it affects the stoichiometry and changes Mn^{3+}/Mn^{4+} ratio which would lower the amount of "double exchange" interaction [13].

All nanocrystalline compounds exhibited small excess of oxygen stoichiometry. The level of excess did not exceed $O_{3.02\pm0.01}$ as is with x = 0.02 sample. We suggest that the culprit for such result should be found in the high surface to volume ratio of nano-sized particle. Broken bonds on the surface will increase the ratio of Mn^{4+} to Mn^{3+} by attracting oxygen in the air. This, in turn, can create vacancies and affect its magnetic and electrical properties.

Table 3. Average oxygen content calculated using iodometry for bulk and nanocrystalline samples.

Ca Content	Average Mn^{3+}/Mn^{4+} Ratio	Standard Deviation	Relative Standard Deviation (%)	Average Oxygen Content
x = 0.02 bulk	0.788	0.0167	2.12	$O_{2.93\pm0.02}$
x = 0.05 bulk	0.783	0.0121	1.55	$O_{2.93\pm0.01}$
x = 0.1 bulk	0.778	0.0139	1.79	$O_{2.94\pm0.02}$
x = 0.2 bulk	0.781	0.0182	2.33	$O_{2.94\pm0.02}$
x = 0.3 bulk	0.765	0.0226	2.95	$O_{2.94\pm0.02}$
x = 0 nano	0.624	0.008	1.28	$O_{3.01\pm0.01}$
x = 0.02 nano	0.612	0.009	1.47	$O_{3.02\pm0.01}$
x = 0.05 nano	0.622	0.011	1.33	$O_{3.01\pm0.01}$
x = 0.1 nano	0.615	0.009	1.77	$O_{3.02\pm0.01}$
x = 0.2 nano	0.633	0.008	1.26	$O_{3.01\pm0.01}$
x = 0.3 nano	0.638	0.012	1.88	$O_{3.01\pm0.01}$

Finally, it can be observed that standard deviation, or deviation from the "mean", is larger for bulk samples than the nano-sized samples. This is explained by the variation of oxygenation within the bulk. Nevertheless, all relative standard deviations do not exceed 3% and can be considered reliable.

3.3. Electrical Measurements

An investigation of electrical behavior for polycrystalline samples x = 0.02; 0.05; 0.1 and 0.2 was performed using the four-point probe method in external fields of $\mu_0 H = 0$ T; 1 T; 2 T in temperature range of 10–290 K. For the sample with x = 0.3 fields of up to 5 T were used. Data were used to estimate metal–insulator transition temperature T_p and magnetoresistivity MR according to the following equation [23]:

$$MR\% = [(\varrho(H) - \varrho(0))/\varrho(0)] \times 100, \quad (2)$$

The first observation to be made is that all samples, except x = 0.3, exhibit a transition temperature T_{p1} typically associated with grain boundary conditions [23–25]. Visual inspection shows that it is a wide and smooth transition, different from the usual sharp peaks associated with ferromagnetic–paramagnetic transition (and consequently metal-insulator) at T_{p2} (inside of the grains). The addition of Ca ions increases disorder, which tends to spread itself toward the edges of the grains in order to conserve energy [26,27]. Boundaries act as insulators or semi-conductors [27,28]. As can be seen in Figure 5, the application of external field shifts T_p to the right. This is caused by lowering spin fluctuations and delocalization of charge carriers [19].

A separation between T_{p1} and T_{p2} (which is associated with T_C) is reported when grain boundary conditions are strong enough due to mismatch in ionic radius at the A-site [26]. Samples x = 0.02 and 0.05 do not show a peak T_{p2} because it is beyond the temperature range. Starting with x = 0.1 a sharp peak appears which lies higher in temperature than T_C value. With application of a field, T_{p2} tends to smooth out, as can be seen with the sample x = 0.2 in Figure 5b.

Special attention should be paid to the sample with x = 0.3. Its graph is presented in Figure 5c. Magnetic fields of up to 5 T were applied. For $\mu_0 H = 0$ T; 1 T the sample conductivity is below the detection level for temperatures below 50 K. With the application of 2 T resistivity drops significantly. Its curve can be seen in the main portion of the figure. Curves for fields of 3–5 T can be seen in the inset and show further drop in resistivity and a shift to higher temperatures for the peak. This suggests that higher field overcomes intergrain resistance and further increase in the field induces a parallel alignment of manganese ions. In higher applied magnetic fields, the magnetic moments of the neighboring grains tend to be parallel, enhancing the tunneling of the conduction electrons similarly with a giant magnetoresistance effect (GMR) [26].

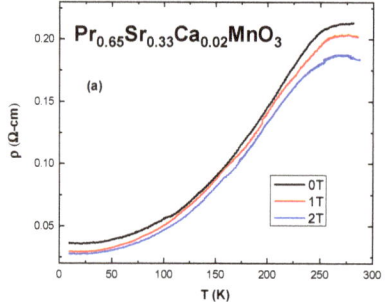
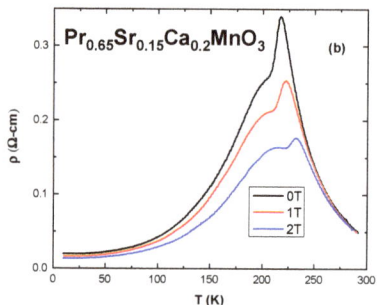
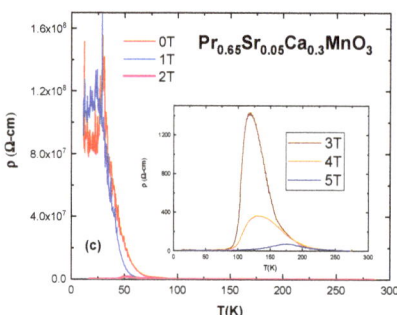

Figure 5. Graphs of resistivity vs. temperature for bulk samples: (**a**) x = 0.02, (**b**) x = 0.2, (**c**) x = 0.3; The inset shows plot of resistivity at higher fields which are not visible in the main plot.

Magnetoresistivity is negative in all samples. Table 4 shows an increasing value of MR with increasing level of Ca substitution. The highest MR_{Max} (2 T) is 51.96% and 29.08% for 1 T for x = 0.2 at 217 K. Sample with x = 0.02 exhibits the lowest MR_{Max} of 11.9% and 4.45% for 2 T and 1 T at 289 K, respectively. A sample with x = 0.3 has MR of 77.62% between 3 T and 4 T at 118 K and 99.99% for 2–3 T at 130 K (shown in brackets in Table 4). In order to calculate MR, both the peak value of resistivity from the lower field data and an isothermal value from the higher field data are taken and then inserted into Equation (2). The highest peak resistivity ϱ_{peak} of about 3 Ωcm at 275 K is observed in sample x = 0.05. The lowest ϱ_{peak} for x = 0.3 at 5 T is 73 Ωcm (in brackets) at 174 K. All relevant data, including T_C and T_p, are presented in Table 4.

Table 4. Experimental values for $Pr_{0.65}Sr_{0.35-x}Ca_xMnO_3$ bulk materials: electrical properties.

Compound (Bulk)	T_C (K)	T_{p1} (K) (T_{p2} (K))	ϱ_{peak} (Ωcm) in 0 T	MR_{Max} (%) (1 T)	MR_{Max} (%) (2 T)
$Pr_{0.65}Sr_{0.33}Ca_{0.02}MnO_3$	273	274	0.213	4.45	11.99
$Pr_{0.65}Sr_{0.3}Ca_{0.05}MnO_3$	261	273	3.094	12.28	22.75
$Pr_{0.65}Sr_{0.25}Ca_{0.1}MnO_3$	244	256 (258)	1.602	23.28	33.27
$Pr_{0.65}Sr_{0.15}Ca_{0.2}MnO_3$	201	210 (217)	0.341	29.08	51.96
$Pr_{0.65}Sr_{0.05}Ca_{0.3}MnO_3$		-	>100 × 10^8 (72.985 in 5 T)	77.62 (between 3 and 4 T)	99.99 (between 2 and 3 T)

3.4. Magnetic Properties

Investigation of magnetization vs. temperature (M vs. T), with the samples cooled in zero and in an applied magnetic field of $\mu_0 H = 0.05$ T (ZFC-FC) was carried out in a vibrating sample magnetometer (VSM). All samples except x = 0.3 exhibit strong ferromagnetic behavior. Samples with x = 0.3 for both systems exhibit antiferromagnetic charge-exchange-type (CE-AFM) and charged ordered (CO) state. In CE-AFM structure, Mn^{3+} and Mn^{4+} are arranged like a checkerboard in the ab-plane ($Mn-O_2$); exchange interaction causes Mn^{3+} e_g electron to occupy either $d_{3x^2-r^2}$ or $d_{3y^2-r^2}$ orbital. As a result, there are FM zig-zag chain arrangements which are AFM to each other and also stack antiferromagnetically in the c-plane. Figure 6 presents selected graphs of M vs. T with insets representing a derivative dM/dT of the plots for which the minimum is associated with T_C. The addition of Ca^{2+} ions causes smaller cell dimensions; increased disorder and it lowers the Curie temperature [17]. The increase in Ca substitution lowers T_C further. Nano-sized particles show lower values of T_C than their bulk counterpart due to their size and surface effect, including broken bonds, canted spins, and reduced magnetization [29,30]. In addition, the curves for nano-sized samples are "smoother", covering a wider temperature range in their ferromagnetic–paramagnetic change but have lower maximum value of magnetization M(T).

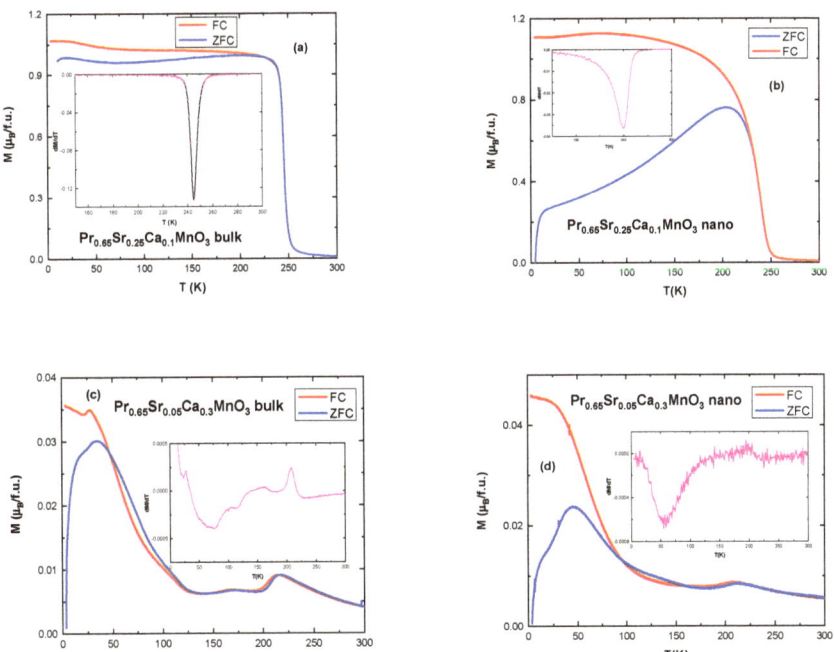

Figure 6. ZFC-FC plots in external field of 0.05 T (**a**) ZFC-FC curves and derivative of magnetization (inset) for the bulk sample $Pr_{0.65}Sr_{0.25}Ca_{0.1}MnO_3$; (**b**) ZFC-FC curves and derivative (inset) for nanocrystalline sample $Pr_{0.65}Sr_{0.25}Ca_{0.1}MnO_3$. (**c**) ZFC-FC curves and derivative shown as inset, for polycrystalline sample $Pr_{0.65}Sr_{0.05}Ca_{0.3}MnO_3$; (**d**) ZFC-FC curves and derivative shown in the inset for nanocrystalline sample $Pr_{0.65}Sr_{0.05}Ca_{0.3}MnO_3$.

An observation of ZFC-FC curves, especially for bulk samples, reveals an upturn in magnetization at lower temperatures at around 70–100 K, an example of which is in Figure 6a. This can be attributed to the magnetization of praseodymium Pr ions [10]. Samples with x = 0.3 also exhibit increased magnetization at low temperatures starting at around 100 K, seen in Figure 6c,d. Maximum value is not as high as for the rest of the samples, approximately 0.04 μ_B/f.u. for x = 0.3 vs. 1 μ_B/f.u. for x = 0.1. Other CO compounds, such as $La_{0.4}Ca_{0.6}Mn_{0.9}Ga_{0.1}O_3$ [11] and $Pr_{0.75}Na_{0.25}MnO_3$ [14] have been reported with similar increase in magnetization due to suppression of CO state. The difference between La^{3+} and Pr^{3+} is their crystal ionic size (1.356 Å for La^{3+} and 1.319 Å for Pr^{3+} [20]) and their electron configuration: La^{3+} has no 4f electrons while Pr^{3+} has the configuration = [Xe]4f^2. The size difference affects the bandwidth, with Pr-based manganites being referred to as narrow bandwidth manganites. Appearance of the charged ordered state is influenced by Mn^{3+}/Mn^{4+} ratio as well as the bandwidth [31], therefore Pr based manganites enter CO state more easily. The low temperature increase in magnetization is associated with phase separation (PS), i.e., existence of FM clusters, forming among AFM matrix [11]. The graphs for x = 0.3 at higher temperatures show behavior typical of charged ordered state. A small feature at around 170 K represents Néel temperature, T_N (most visible in the bulk sample), and the peak at around 210 K representing the onset of the charge ordering phase, T_{CO}. Such a behavior was revealed in several manganites systems such as $La_{0.250}Pr_{0.375}Ca_{0.375}MnO_3$ [32], $Pr_{0.7}Ca_{0.3}MnO_3$ [33], $La_{0.3}Ca_{0.7}Mn_{0.8}Cr_{0.2}O_3$ [34], $Pr_{0.57}Ca_{0.41}Ba_{0.02}MnO_3$ [35], $Nd_{0.5}Ca_{0.5}MnO_3$ [36], and $La_{0.4}Ca_{0.6}MnO_3$ [37], and it arises as a result of doping and/or due to the reducing of the sizes of the particles to the range of nanometers.

The feature from 45 K, in bulk sample, could be the signature of the blocking of isolated spins between FM clusters as seen in spin glass materials [38,39].

According to Landau's mean field theory [40], Gibbs free energy of the system around a critical point can be expanded in a Taylor series as:

$$G(T,M) = G\_0 + MH + aM^2 + bM^4 + \ldots, \quad (3)$$

where coefficients a and b depend on temperature. A derivative of the energy with respect to magnetization, to find the minima, results in an expression:

$$H/M = 2a + 4bM^2, \quad (4)$$

An easy way to confirm the correctness of the mean field theory approach for given samples is the Arrott plot, i.e., M^2 vs. H/M [41]. If isotherms are straight and parallel to each other around T_C then the assumption is correct. Observation of Arrott plots for our samples reveals curved, non-parallel lines for bulk compounds (Figure 7a,c) and almost straight lines for nano-sized compounds (Figure 7b,d). Furthermore, Banerjee criterion is a useful tool for determining the order of the phase transition [42]. According to the criterion, positive slope represents second order phase transition and negative slope corresponds to first order transition. Samples with x = 0, 0.02, 0.05, 0.1 show only positive slope. As can be seen in Figure 7c,d, samples with x = 0.2 show negative slope in the first stage of the magnetization at temperatures above T_C suggesting first order transition. Samples with x = 0.3, not pictured, show mostly negative slope plots.

Inexactness of the exponents in the Equation (4) can be solved by constructing an Arrott–Noakes plot, i.e., $M^{1/\beta}$ vs. $\mu_0 H/M^{1/\gamma}$ [43] with proper exponents β and γ. The exponent β relates to the spontaneous magnetization below T_C and γ relates to the inverse susceptibility above T_C [44] (pp. 7–10). An additional exponent δ relates magnetization and external field at T_C. In mean field model: β = 0.5 and γ = 1, but there exist more models with different exponent values. The value of the exponents relates to the system dimensionality, spin, and range of the interaction $J(r)$. In renormalization group theory [45], exchange interaction is defined as $J(r) = 1/r^{d+\sigma}$ where d—dimensionality of the system and σ—range of interaction. For σ greater than 2: β = 0.355, γ = 1.366 and δ = 4.8—is the case of the 3D Heisenberg model. If $\sigma < 3/2$, long-range interactions occur, according to mean field theory. For the tricritical point, the critical exponents are: β = 0.25, γ = 1, and δ = 5 [46]. The tricritical point sets a boundary between two different ranges of order phase transitions (first order and second order). These exponents can be generalized into following equations [43]:

$$M_S(T) = M_0 (-\varepsilon)^\beta, T < T_C, \quad (5)$$

$$\chi^{-1}(T) = \left(\frac{h_0}{M_0}\right)\varepsilon^\gamma, T > T_C, \quad (6)$$

$$M = D (\mu_0 H^{1/\delta}), T = T_C, \quad (7)$$

where ε is the reduced temperature $(T - T_C)/T_C$ and M_0, h_0/M_0, and D are critical amplitudes.

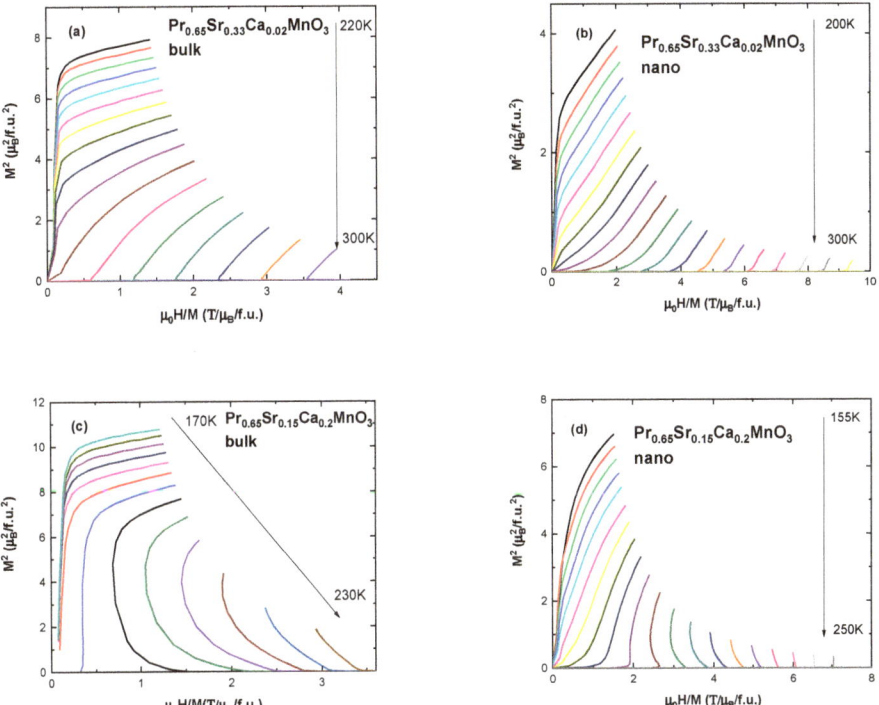

Figure 7. Arrott plot (M^2 vs. H/M) for (**a**) the bulk sample with x = 0.02; (**b**) the nanocrystalline sample with x = 0.02; (**c**) bulk sample with x = 0.2; (**d**) nanocrystalline sample x = 0.2.

To find the proper exponents, we constructed Arrott–Noakes plots by implementing the Modified Arrott plot (MAP) method [43]. It is an iterative method. It includes, at first, construction of Arrott–Noakes plots by choosing exponents which produce straight parallel lines, with the line at T_C crossing at the origin. Secondly, we find intercepts of lines around T_C; on the abscissa, above Curie temperature, to obtain the values of χ_0^{-1} and on the ordinate, below T_C, to find the values of M_s. Finally, these values are plugged into the Equations (5) and (6) to obtain new values of β and γ. This process is repeated until results stabilize. The exponent δ is found using the Widom relation: $\beta + \gamma = \beta \delta$ [10]. Selected MAP graphs are shown in Figure 8. Full results for MAP are presented in Table 5.

Remarkably, polycrystalline samples are more closely governed by tricritical mean field model (β = 0.212, γ = 1.057, δ = 5.986 for x = 0.02), rather than 3D Heisenberg model as reported for other manganites in the literature [10,47]. Sample x = 0.3 does not exhibit ferromagnetic behavior in the high temperature range, only showing small magnetization at low temperatures due to FM clusters and Praseodymium ions, therefore, no critical values are presented in this work. Alternatively, all critical exponents for nano-sized particles fall within the mean field model values ($\beta = 0.541$, $\gamma = 1.01$, $\delta = 2.867$ for x = 0.02) which is comparable to reported values in $La_{0.7}Ba_{0.3-x}Ca_xMnO_3$ compounds [22]. Similarly, to the bulk, critical values for nano-sized sample x = 0.3 are not presented.

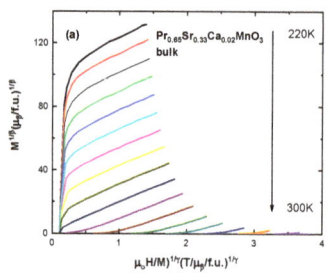 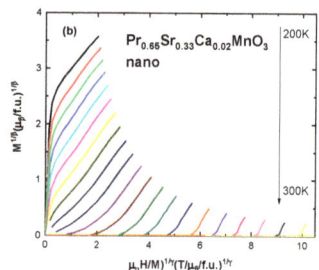

Figure 8. Modified Arrott plots for (**a**) the bulk sample x = 0.02 with $\beta = 0.212$, $\gamma = 1.057$, $\delta = 5.986$ and for (**b**) the nanocrystalline sample x = 0.02 with $\beta = 0.541$, $\gamma = 1.01$, $\delta = 2.867$.

Table 5. Critical exponent values for all samples from modified Arott plot method.

Compound		γ	β	δ	T_C (K)
x = 0.02	bulk	1.057	0.212	5.986	273
x = 0.05	bulk	0.981	0.232	5.228	261
x = 0.1	bulk	0.985	0.25	4.94	244
x = 0.2	bulk	1.036	0.217	5.774	201
x = 0.3	bulk	-	-	-	-
x = 0	nano	1.023	0.552	2.853	257
x = 0.02	nano	1.01	0.541	2.867	252
x = 0.05	nano	0.986	0.508	2.941	249
x = 0.1	nano	0.977	0.512	2.908	239
x = 0.2	nano	0.967	0.531	2.821	191
x = 0.3	nano	-	-	-	-
Mean field model		1	0.5	3	
3D Heisenberg model		1.366	0.355	4.8	
Ising model		1.24	0.325	4.82	
Tricritical mean field model		1	0.25	5	

The Kouvel–Fisher (KF) method for determining critical exponents is a widely implemented tool for ferromagnetic materials [47–49]. We used KF to confirm the results from MAP as the two methods are regarded to be very accurate and reliable [47]. Akin to MAP, it is also an iterative method. It requires the construction of Arrott–Noakes plot, finding the intercepts on ordinate and abscissa and fitting these values in the equations [47,49]:

$$M_s \{dM_s/dT\}^{-1} = (T - T_C)/\beta \tag{8}$$

$$\chi_0^{-1} \{d\chi_0^{-1}/dT\}^{-1} = (T - T_C)/\gamma \tag{9}$$

Ideally, plot of $M_s \{dM_s/dT\}^{-1}$ vs. T is a straight line with slope = $1/\beta$ and the intercept giving T_C/β. Same logic applies to the plot of $\chi_0^{-1} \{d\chi_0^{-1}/dT\}^{-1}$ vs. T which gives slope equal to $1/\gamma$. We present, in Figure 9, selected examples of results from KF calculations compared to results from MAP for the same compounds. It is evident, that the lines in KF plots are close to being straight and their slope values result in β and γ being close to results from MAP.

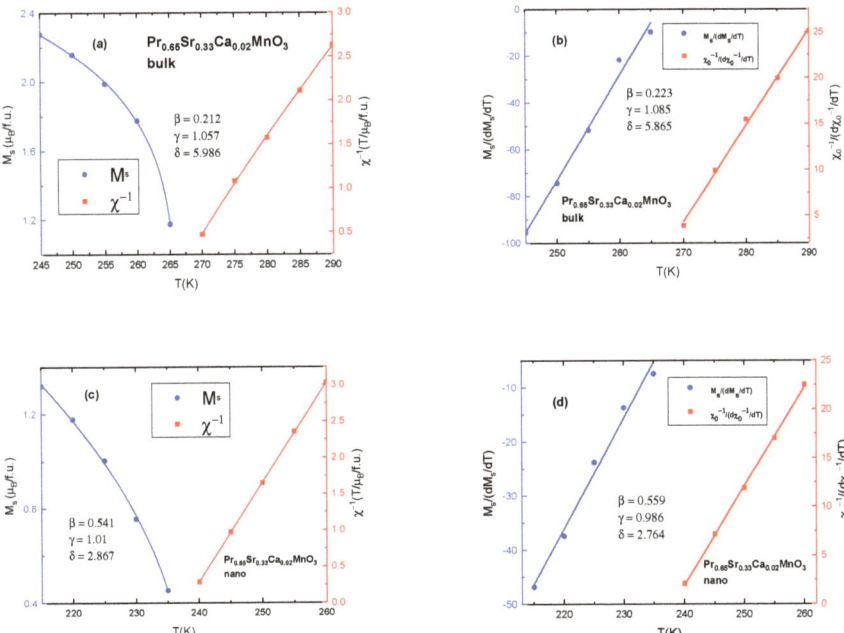

Figure 9. Calculated values for critical exponents for (**a**) MAP analysis for the bulk sample with x = 0.02 (**b**) KF analysis for the bulk sample with x = 0.02. (**c**) MAP analysis for the nanocrystalline sample x = 0.02 (**d**) KF analysis for nanocrystalline sample x = 0.02.

Magnetic entropy change was calculated using the following formula [50,51]:

$$\Delta S_m(T, H_0) = S_m(T, H_0) - S_m(T, 0) = \frac{1}{\Delta T} \int_0^{H_0} [M(T + \Delta T, H) - M(T, H)] dH, \quad (10)$$

where magnetization $M(\mu_0 H)$ is taken from isothermal data at fields of 1–4 T.

Plots of $-\Delta S_M$ vs. T (temperature) were constructed in order to better estimate magnetocaloric effect of the compounds, including calculations of relative cooling power RCP [51,52]:

$$RCP(S) = -\Delta S_m(T, H) \times \delta T_{FWHM} \quad (11)$$

where δT_{FWHM} is the range of temperature at full width half maximum.

Selected graphs of samples' entropy change are presented in Figure 10. It is noteworthy that, generally, polycrystalline samples exhibit higher maximum entropy change $-\Delta S_M$ compared to their nano-sized counterparts, as seen in Figure 10a vs. 10b. Bulk compound with x = 0.05 shows $-\Delta S_M$ (max) = 5.56 J/kgK at 4 T and nano-sized sample shows $-\Delta S_M$ (max) = 3.25 J/kgK at 4 T. Bulk samples with x = 0.1 and x = 0.2 exhibit high maximum entropy change at 6.9 J/kgK and 9.2 J/kgK respectively. For all samples, maximum entropy change occurs at around their Curie temperature T_C. It is also important to note that nanocrystalline compounds exhibit wider range of effective temperature δT_{FWHM} at around 40–70 K while bulk samples show range of 20–30 K.

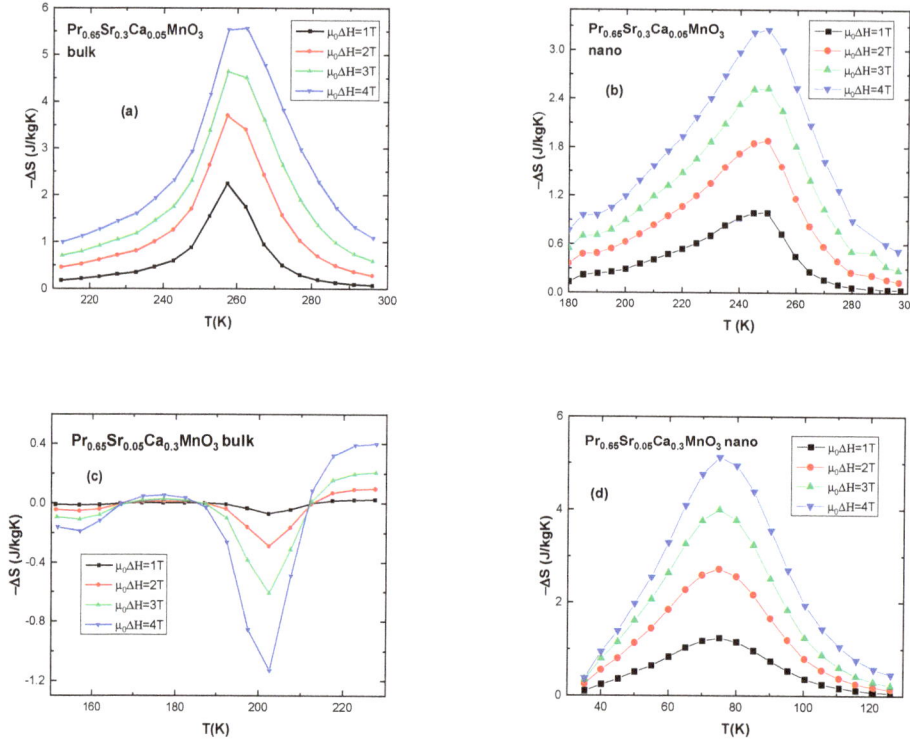

Figure 10. Magnetic entropy change vs. temperature for selected samples: (**a**) x = 0.05 bulk; (**b**) x = 0.05 nanocrystalline sample; (**c**) x = 0.3 bulk at AFM-FM; (**d**) x = 0.3 nanocrystalline at low temperature.

Construction of cooling equipment should not be "blindly" reliant on reported values of RCP [53]; nevertheless, for a long time it has been a useful tool in estimating the applicability of the magnetic material. Although it is the width of the temperature change in nano-sized compounds that is criticized as unreliable, in this work, nanocrystalline compounds deserve attention for their satisfactory level of entropy change, greater or close to 4 J/(kgK) for $\Delta\mu_0 H = 4$ T. Both nano-sized and bulk systems exhibit values of entropy change and RCP comparable with other manganites reported in the literature [10,52,54,55].

Delving deeper, unfortunately, RCP tends to overestimate the merits of the materials which have a large temperature range of magnetocaloric effect (δT_{FWHM}) but small entropy changes [55,56] in no small part because materials with same relative cooling power can behave differently in a magnetic cooling simulation [57]. Moreover, one of the last printed books about magnetic cooling ignores this figure of merit [58].

In order to be of use in magnetic refrigeration applications the chosen materials have to show a large magnetocaloric effect—a large magnetic entropy change. Besides this, there is a list of requirements related with heat transfer, heat capacity, heat conductivity, chemical and mechanical stability, hysteresis losses, eddy currents losses, etc. Therefore, magnetic measurements can be a trusted guide in deciding if a material deserves the effort of complete characterization. Some compounds of the type $A_{1-x}B_xMnO_3$ can be considered to be of interest for magnetic refrigeration because some of these compounds show $-\Delta S_m$ values ($-\Delta S_m \approx$ 4–6 J·kg^{-1}·K^{-1} for $\Delta H = 5$ T) comparable to intermetallic alloys [56]. This is the case of our investigated compounds.

The best way to decide the practicality of a material in magnetic cooling applications is by testing directly in a magnetic refrigerator, working on the principle of the AMRR (Active Magnetic Regenerative Refrigeration) cycle [12]. Recent investigations confirmed the magnetocaloric performance and the potential in magnetic refrigeration of the perovskite oxide $Pr_{0.65}Sr_{0.35}MnO_3$ [12,59], which has the stoichiometry of our parent compound. When Ca substitutes for Sr in this compound, the magnetic parameters and magnetocaloric effect can be tuned for various temperature applications without significant changes in the magnitude of magnetic entropy change. Our investigations support the bulk polycrystalline $Pr_{0.65}Sr_{(0.35-x)}Ca_xMnO_3$ compounds to be promising for magnetic cooling technology, while lower magnitude of magnetocaloric effect in nanocrystalline samples, unfortunately, seems to somewhat hinder their direct practical applications [53,56,57].

Special mention should be made of some feature for samples with x = 0.3. Both bulk (Figure 10c) and nano-sized (not pictured) compounds exhibit positive entropy change at temperatures associated with antiferromagnetic (CO)/ferromagnetic/paramagnetic phase transition. This phase transition can be observed in the ZFC-FC plots in Figure 6 at around 200–210 K for both samples. As can be observed in Figure 10c, bulk sample exhibits inverse (materials cool down when a magnetic field is adiabatically applied) to normal MCE change suggesting AFM/FM transition, but it seems the changes in the lattice parameters can also modify the magnetic exchange interaction to give rise to a such behavior [60,61]. Additionally, Figure 10d shows a negative entropy change of 5.1 J/kgK at 4 T for nano-sized samples at 75 K. Similar behavior and even higher entropy change is exhibited by the bulk sample at around 40 K (not pictured) at the suppression of the AFM state. These materials can be used to produce both cooling and heating when they are adiabatically demagnetized. Unfortunately, the large values of the magnetic entropy changes can be reach only at low temperatures, far from the range of domestic refrigeration, in the case of these samples.

Low coercivity is of utmost importance in application of cooling materials [51]. All samples exhibit low coercive fields as measured in a hysteresis loop at external fields of up to 4 T at 4 K. Largest coercive field exhibited by a bulk sample is of 190 Oe for x = 0.2; with the smallest being 130 Oe for x = 0.05. Nanocrystalline samples carry larger coercive field compared to polycrystalline samples. It increases with decreasing size of the particles until single domain size. Particles become superparamagnetic with further decrease in size [62]. Largest coercive field is produced by the parent nano-sized with x = 0 at 810 Oe. All values for coercive field and magnetic saturation M_s are presented in Tables 6 and 7. Nanocrystalline sample x = 0.3 does not reach saturation at 4 T.

Table 6. Experimental values for $Pr_{0.65}Sr_{0.35-x}Ca_xMnO_3$ bulk materials: magnetic measurements.

Compound (Bulk)	T_C (K)	M_s (μ_B/f.u.)	H_{ci} (Oe)	$|\Delta S_M|$ (J/kgK) $\mu_0\Delta H = 1\,T$	$|\Delta S_M|$ (J/kgK) $\mu_0\Delta H = 4\,T$	RCP (S) (J/kg) $\mu_0\Delta H = 1\,T$	RCP (S) (J/kg) $\mu_0\Delta H = 4\,T$	Refs.
$Pr_{0.65}Sr_{0.35}MnO_3$	295			2.3				[12]
$Pr_{0.65}Sr_{0.33}Ca_{0.02}MnO_3$	273	3.71	180	2.04	5.53	34	166	This work
$Pr_{0.65}Sr_{0.3}Ca_{0.05}MnO_3$	261	3.78	170	2.24	5.56	44	167	This work
$Pr_{0.65}Sr_{0.25}Ca_{0.1}MnO_3$	244	3.94	190	3.03	6.91	45	186	This work
$Pr_{0.65}Sr_{0.15}Ca_{0.2}MnO_3$	201	3.97	160	4.48	9.21	60	270	This work
$Pr_{0.65}Sr_{0.05}Ca_{0.3}MnO_3$		3.81	170	4.6 (40 K)	15.3 (40 K)	92 (40 K)	380 (40 K)	This work
$La_{0.7}Ca_{0.3}MnO_3$	256			1.38		41		[10]
$La_{0.7}Sr_{0.3}MnO_3$	365			-	4.44 (5 T)		128 (5 T)	[10]
$La_{0.6}Nd_{0.1}Ca_{0.3}MnO_3$	233			1.95		37		[10]
$Gd_5Si_2Ge_2$	276			-	18 (5 T)	-	535 (5 T)	[10]
Gd	293			2.8		35		[10]

Table 7. Experimental values for $Pr_{0.65}Sr_{0.35-x}Ca_xMnO_3$ nano materials.

Compound (Nano)	T_C (K)	M_s (μ_B/f.u.)	H_{ci} (Oe)	$|\Delta S_M|$ (J/kgK) $\mu_0\Delta H = 1\,T$	$|\Delta S_M|$ (J/kgK) $\mu_0\Delta H = 4\,T$	RCP(S) (J/kg) $\mu_0\Delta H = 1\,T$	RCP(S) (J/kg) $\mu_0\Delta H = 4\,T$	Refs.
$Pr_{0.65}Sr_{0.35}MnO_3$	257	3.6	810	1.62	4.55	54	263	This work
$Pr_{0.65}Sr_{0.33}Ca_{0.02}MnO_3$	252	3.08	720	0.69	2.5	38	175	This work
$Pr_{0.65}Sr_{0.3}Ca_{0.05}MnO_3$	249	3.16	540	0.99	3.25	48	178	This work
$Pr_{0.65}Sr_{0.25}Ca_{0.1}MnO_3$	239	3.49	510	1.41	4.37	39	215	This work
$Pr_{0.65}Sr_{0.15}Ca_{0.2}MnO_3$	191	3.39	620	1.08	3.9	41	185	This work
$Pr_{0.65}Sr_{0.05}Ca_{0.3}MnO_3$		-	600	1.23(75 K)	5.12(75 K)	48(75 K)	204(75 K)	This work
$La_{0.67}Ca_{0.33}MnO_3$	260				0.97 (5 T)		27 (5 T)	[63]
$Pr_{0.65}(Ca_{0.6}Sr_{0.4})_{0.35}MnO_3$	220			0.75		21.8		[64]
$La_{0.6}Sr_{0.4}MnO_3$	365			1.5		66		[17]

4. Conclusions

Two sample systems of $Pr_{0.65}Sr_{0.35-x}Ca_xMnO_3$ (x = 0.02, 0.05, 0.1, 0.2, 0.3) were prepared. Solid state reaction as a method of preparation of polycrystalline compounds resulted in samples with an average 12,000 nm grain size. Sucrose based sol–gel method for production of nanocrystalline samples resulted in particles of an average size of approximately 70 nm. X-ray diffraction measurements revealed a single crystallographic phase for all samples. Rietveld refinement analysis confirmed single phase, orthorhombic (Pbnm) symmetry, diminishing cell volume and Mn–O bond length with increasing Ca substitution. Iodometric titration was implemented on both systems to determine their oxygen content. All bulk samples show oxygen deficiency close to $O_{2.93\pm0.02}$ attributed to the preparation method. All nano-sized samples exhibit small oxygen excess of $O_{3.02\pm0.01}$ or less, credited to the effects of the surface/volume ratio of the particles. Bulk compounds were measured for their electrical properties and revealed a feature at a temperature T_{p1} associated with high disorder at grain boundaries due to Ca substitution. Additionally, samples with Ca levels x = 0.1, 0.2 exhibit a peak at a temperature T_{p2} associated with the ferromagnetic-paramagnetic transition. With the increase in applied magnetic field, T_{p2} tends to become higher and smooth out. The sample with x = 0.3 possesses "infinite" resistance for $\mu_0H = 0\,T$ and 1 T at around 50 K. Further increase in applied field up to 5 T results in overcoming intergrain resistance and shift in the resistance peak. Negative magnetoresistivity was observed for all bulk samples. The smallest value of MR was 11.99% for x = 0.02 at 2 T at T_{p1}, while the largest value of MR for x = 0.2 of 51.96% for T_{p2}. Zero-field cooled–field cooled graphs reveal ferromagnetic behavior for x = 0.02, 0.05, 0.01, 0.2 samples. Both x = 0.3 show ferromagnetic-like behavior at low temperatures and antiferromagnetic behavior at higher temperatures than 80 K, with transition to ferromagnetic to paramagnetic behavior at 210 K. Curie temperature T_C lowers with each Ca substitution: 273 K for bulk x = 0.02 compared to 295 K for parent compound; 252 K for nano-sized x = 0.02 vs. 257 K for parent compound. Low coercivity was found for all samples. Nanocrystalline samples exhibiting larger coercivity compared to their bulk counterparts: 720 Oe vs. 180 Oe for x = 0.02. Arrott plots confirm second order phase transition for all samples except for x = 0.3. Modified Arrott plot (MAP) analysis disclosed critical behavior. Critical exponents for polycrystalline samples belong to tricritical mean field model while exponents for nano-sized particles belong to mean field model. The Kouvel–Fisher (KF) method for analyzing critical exponents confirmed MAP results. Bulk compounds reveal higher magnetic saturation M_s and maximum entropy change ΔS_M than nano-sized compounds. Bulk sample with x = 0.2 exhibits highest entropy change $\Delta S_M = 9.21$ J/kgK (4 T) at 204 K. Relative cooling powers (RCP) for equivalent bulk and nano-sized compounds are comparable in value: 166 J/kg vs. 175 J/kg (4 T) for bulk and nano-sized with x = 0.02 This is due to the wide effective temperature range of entropy change δT_{FWHM} in nanocrystalline samples. All values of RCP are high and comparable with relevant compounds reported in the literature. Bulk samples with x = 0.3 exhibits positive entropy change at antiferromagnetic-ferromagnetic transition temperature Magnetic entropy changes and magnetic parameters of the bulk polycrystalline samples indicate them as potential candidates for magnetocaloric materials. Possibly, they can be combined in construction of multistep refrigeration processes to increase their temperature range and effectiveness. The nanocrystalline samples, which show comparatively lower maximum entropy change are moved to the immediate background of the picture of usable magnetocaloric compounds but could, possibly, be still useful in future developments.

Author Contributions: R.A., conceptualization, investigation, methodology, writing—original draft, writing—review and editing, visualization, supervision; D.A., R.H., R.B., G.S., A.S. and L.B.-T., methodology, investigation, review and editing; I.G.D., conceptualization, investigation, methodology, visualization, supervision, review and editing. All authors have read and agreed to the published version of the manuscript.

Funding: This research received no external funding.

Data Availability Statement: Data presented in this study are available in this article.

Acknowledgments: The authors acknowledge the Institute of Physics "Ioan Ursu" of Faculty of Physics from Babes Bolyai University for assistance.

Conflicts of Interest: The authors declare no conflict of interest.

References

1. Mckay, C.P. Requirements and limits for life in the context of exoplanets. *Proc. Natl. Acad. Sci. USA* **2014**, *111*, 12628–12633. [CrossRef] [PubMed]
2. Briley, G.C. A History of Refrigeration. *Ashrae J.* **2004**, *46*, S31–S34.
3. Gschneidner, K.A.; Pecharsky, V.K. Thirty Years of near Room Temperature Magnetic Cooling: Where We Are Today and Future Prospects. *Int. J. Refrig.* **2008**, *31*, 945–961. [CrossRef]
4. Salazar Munoz, V.E.; Lobo Guerrero, A.; Palomares-Sanchez, S.A. Review of magnetocaloric properties in lanthanum manganites. *J. Magn. Magn. Mater.* **2022**, *562*, 169787. [CrossRef]
5. Smith, A. Who discovered the magnetocaloric effect? *EPJH* **2013**, *38*, 507–517. [CrossRef]
6. Dagotto, E.; Hotta, T.; Moreo, A. Colossal Magnetoresistant Materials: The Key Role of Phase Separation. *Phys. Rep.* **2001**, *344*, 1–153. [CrossRef]
7. Pavarini, E.; Koch, E.; Anders, F.; Jarrell, M. *Correlated Electrons: From Models to Materials Modeling and Simulation*; Forschungszentrum Julich: Jülich, Germany, 2012; Chapter 7, Volume 2, pp. 18–21, ISBN 978-3-89336-796-2.
8. Coey, J.M.D.; Viret, M.; Von Molnár, S. Mixed-Valence Manganites. *Adv. Phys.* **1999**, *48*, 167–293. [CrossRef]
9. Rostamnejadi, A.; Venkatesan, M.; Alaria, J.; Boese, M.; Kameli, P.; Salamati, H.; Coey, J.M.D. Conventional and Inverse Magnetocaloric Effects in $La_{0.45}Sr_{0.55}MnO_3$ Nanoparticles. *J. Appl. Phys.* **2011**, *110*, 043905. [CrossRef]
10. Varvescu, A.; Deac, I.G. Critical Magnetic Behavior and Large Magnetocaloric Effect in $Pr_{0.67}Ba_{0.33}MnO_3$ Perovskite Manganite. *Phys. B Condens. Matter* **2015**, *470–471*, 96–101. [CrossRef]
11. Badea, C.; Tetean, R.; Deac, I.G. Suppression of Charge and Antiferromagnetic Ordering in Ga-doped $La_{0.4}Ca_{0.6}MnO_3$. *Rom. J. Phys.* **2018**, *63*, 604.
12. Guillou, F.; Legait, U.; Kedous-Lebouc, A.; Hardy, V. Development of a new magnetocaloric material used in a magnetic refrigeration device. *EPJ Web Conf.* **2012**, *29*, 21. [CrossRef]
13. Licci, F.; Turilli, G.; Ferro, P. Determination of Manganese Valence in Complex La-Mn Perovskites. *J. Magn. Magn. Mater.* **1996**, *164*, L268–L272. [CrossRef]
14. Zhang, X.H.; Li, Z.Q.; Song, W.; Du, X.W.; Wu, P.; Bai, H.L.; Jiang, H.Y. Magnetic properties and charge ordering in $Pr_{0.75}Na_{0.25}MnO_3$ manganite. *Solid State Commun.* **2005**, *135*, 356. [CrossRef]
15. Rao, C.N.R. Charge, Spin, and Orbital Ordering in the Perovskite Manganates, $Ln_{1-x}A_xMnO_3$ (Ln = Rare Earth, A = Ca or Sr). *J. Phys. Chem. B* **2000**, *104*, 5877–5889. [CrossRef]
16. Tali, R. Determination of Average Oxidation State of Mn in $ScMnO_3$ and $CaMnO_3$ by Using Iodometric Titration. *Damascus Univ. J. Basic Sci.* **2007**, *23*, 9–19.
17. Ehsani, M.H.; Kameli, P.; Ghazi, M.E.; Razavi, F.S.; Taheri, M. Tunable magnetic and magnetocaloric properties of $La_{0.6}Sr_{0.4}MnO_3$ nanoparticles. *J. Appl. Phys.* **2013**, *114*, 223907. [CrossRef]
18. Dagotto, E. *Nanoscale Phase Separation and Colossal Magnetoresistance*, 1st ed.; Springer Science & Business Media: New York, NY, USA, 2002; pp. 271–284.

19. Raju, K.; Manjunathrao, S.; Venugopal Reddy, P. Correlation between Charge, Spin and Lattice in La-Eu-Sr Manganites. *J. Low Temp. Phys.* **2012**, *168*, 334–349. [CrossRef]
20. Rao, C.N.R. Perovskites. In *Encyclopedia of Physical Science and Technology*; Elsevier: Amsterdam, The Netherlands, 2003; pp. 707–714.
21. Nath, D.; Singh, F.; Das, R. X-ray Diffraction Analysis by Williamson-Hall, Halder-Wagner and Size-Strain Plot Methods of CdSe Nanoparticles—A Comparative Study. *Mater. Chem. Phys.* **2020**, *239*, 2764–2772. [CrossRef]
22. Atanasov, R.; Bortnic, R.; Hirian, R.; Covaci, E.; Frentiu, T.; Popa, F.; Iosif Grigore Deac, I.G. Magnetic and Magnetocaloric Properties of Nano- and Polycrystalline Manganites La$_{(0.7-x)}$Eu$_x$Ba$_{0.3}$MnO$_3$. *Materials* **2022**, *15*, 7645. [CrossRef]
23. Saw, A.K.; Channagoudra, G.; Hunagund, S.; Hadimani, R.L.; Dayal, V. Study of transport, magnetic and magnetocaloric properties in Sr^{2+} substituted praseodymium manganite. *Mater. Res. Express* **2020**, *7*, 016105. [CrossRef]
24. Deac, I.G.; Tetean, R.; Burzo, E. Phase Separation, Transport and Magnetic Properties of La$_{2/3}$A$_{1/3}$Mn$_{1-x}$Co$_x$O$_3$, A = Ca, Sr ($0.5 \leq x \leq 1$). *Phys. B Condens. Matter* **2008**, *403*, 1622–1624. [CrossRef]
25. Panwar, N.; Pandya, D.K.; Agarwal, S.K. Magneto-Transport and Magnetization Studies of Pr$_{2/3}$Ba$_{1/3}$MnO$_3$:Ag$_2$O Composite Manganites. *J. Phys. Condens. Matter* **2007**, *19*, 456224. [CrossRef]
26. Panwar, N.; Pandya, D.K.; Rao, A.; Wu, K.K.; Kaurav, N.; Kuo, Y.K.; Agarwal, S.K. Electrical and Thermal Properties of Pr$_{2/3}$(Ba$_{1-x}$Cs$_x$)$_{1/3}$MnO$_3$ Manganites. *Eur. Phys. J. B* **2008**, *65*, 179–186. [CrossRef]
27. Ibrahim, N.; Rusop, N.A.M.; Rozilah, R.; Asmira, N.; Yahya, A.K. Effect of grain modification on electrical transport properties and electroresistance behavior of Sm$_{0.55}$Sr$_{0.45}$MnO$_3$. *Int. J. Eng. Technol.* **2018**, *7*, 113–117.
28. Kubo, K.; Ohata, N. A Quantum Theory of Double Exchange. *J. Phys. Soc. Jpn.* **1972**, *33*, 21–32. [CrossRef]
29. Arun, B.; Suneesh, M.V.; Vasundhara, M. Comparative Study of Magnetic Ordering and Electrical Transport in Bulk and Nano-Grained Nd$_{0.67}$Sr$_{0.33}$MnO$_3$ Manganites. *J. Magn. Magn. Mater.* **2016**, *418*, 265–272. [CrossRef]
30. Peters, J.A. Relaxivity of Manganese Ferrite Nanoparticles. *Prog. Nucl. Magn. Reson. Spectrosc.* **2020**, *120*, 72–94. [CrossRef] [PubMed]
31. Maheswar Repaka, D.V. Magnetic Control of Spin Entropy, Thermoelectricity and Electrical Resistivity in Selected Manganites. Ph.D. Thesis, National University of Singapore, Singapore, 2014.
32. Deac, I.G.; Diaz, S.V.; Kim, B.G.; Cheong, S.-W.; Schiffer, P. Magnetic relaxation in La$_{0.250}$Pr$_{0.375}$Ca$_{0.375}$MnO$_3$ with varying phase separation. *Phys. Rev. B* **2002**, *64*, 174426. [CrossRef]
33. Deac, I.G.; Mitchell, J.; Schiffer, P. Phase Separation and the Low-Field Bulk Magnetic Properties of Pr$_{0.7}$Ca$_{0.3}$MnO$_3$. *Phys. Rev. B* **2001**, *63*, 172408. [CrossRef]
34. Sudyoadsuka, T.; Suryanarayananb, R.; Winotaia, P.; Wengerc, L.E. Suppression of charge-ordering and appearance of magnetoresistance in a spin-cluster glass manganite La$_{0.3}$Ca$_{0.7}$Mn$_{0.8}$Cr$_{0.2}$O$_3$. *J. Magn. Magn. Mater.* **2004**, *278*, 96–106. [CrossRef]
35. Anuradha, K.N.; Rao, S.S.; Bhat, S.V. Complete 'melting' of charge order in hydrothermally grown Pr$_{0.57}$Ca$_{0.41}$Ba$_{0.02}$MnO$_3$ nanowires. *J. Nanosci. Nanotechnol.* **2007**, *7*, 1775–1778. [CrossRef]
36. Rao, S.S.; Tripathi, S.; Pandey, D.; Bhat, S.V. Suppression of charge order, disappearance of antiferromagnetism, and emergence of ferromagnetism in Nd$_{0.5}$Ca$_{0.5}$MnO$_3$ nanoparticles. *Phys. Rev. B* **2006**, *74*, 144416. [CrossRef]
37. Lu, C.L.; Dong, S.; Wang, K.F.; Gao, F.; Li, P.L.; Lv, L.Y.; Liu, J.M. Charge-order breaking and ferromagnetism in La$_{0.4}$Ca$_{0.6}$MnO$_3$ nanoparticles. *Appl. Phys. Lett.* **2007**, *91*, 032502. [CrossRef]
38. Hüser, D.; Wenger, L.E.; van Duynevelt, A.J.; Mydosh, J.A. Dynamical behavior of the susceptibility around the freezing temperature in (Eu,Sr)S. *Phys. Rev. B* **1983**, *27*, 3100. [CrossRef]
39. Cao, G.; Zhang, J.; Wang, S.; Yu, J.; Jing, C.; Cao, S.; Shen, X. Reentrant spin glass behavior in CE-type AFM Pr$_{0.5}$Ca$_{0.5}$MnO$_3$ manganite. *J. Magn. Magn. Mater.* **2007**, *310*, 169. [CrossRef]
40. Jeddi, M.; Gharsallah, H.; Bejar, M.; Bekri, M.; Dhahri, E.; Hlil, E.K. Magnetocaloric Study, Critical Behavior and Spontaneous Magnetization Estimation in La$_{0.6}$Ca$_{0.3}$Sr$_{0.1}$MnO$_3$ Perovskite. *RSC Adv.* **2018**, *8*, 9430–9439. [CrossRef]
41. Arrott, A. Criterion for Ferromagnetism from Observations of Magnetic Isotherms. *Phys. Rev.* **1957**, *108*, 1394–1396. [CrossRef]
42. Banerjee, B.K. On a Generalised Approach to First and Second Order Magnetic Transitions. *Phys. Lett.* **1964**, *12*, 16–17. [CrossRef]
43. Arrott, A.; Noakes, J.E. Approximate Equation of State for Nickel Near Its Critical Temperature. *Phys. Rev. Lett.* **1967**, *19*, 786–789. [CrossRef]
44. Stanley, H.E. *Introduction to Phase Transitions and Critical Phenomena*; Oxford University Press: Oxford, UK, 1987; pp. 7–10.
45. Fisher, M.E.; Ma, S.K.; Nickel, B.G. Critical Exponents for Long-Range Interactions. *Phys. Rev. Lett.* **1972**, *29*, 917. [CrossRef]
46. Pathria, R.K.; Beale, P.D. Phase Transitions: Criticality, Universality, and Scaling. In *Statistical Mechanics*; Elsevier: Amsterdam, The Netherlands, 2022; pp. 417–486.
47. Vadnala, S.; Asthana, S. Magnetocaloric effect and critical field analysis in Eu substituted La$_{0.7-x}$Eu$_x$Sr$_{0.3}$MnO$_3$ ($x = 0.0, 0.1, 0.2, 0.3$) manganites. *J. Magn. Magn. Mater.* **2018**, *446*, 68–79. [CrossRef]
48. Kim, D.; Revaz, B.; Zink, B.L.; Hellman, F.; Rhyne, J.J.; Mitchell, J.F. Tri-critical Point and the Doping Dependence of the Order of the Ferromagnetic Phase Transition of La$_{1-x}$Ca$_x$MnO$_3$. *Phys. Rev. Lett.* **2002**, *89*, 227202. [CrossRef] [PubMed]
49. Pelka, R.; Konieczny, P.; Fitta, M.; Czapla, M.; Zielinski, P.M.; Balanda, M.; Wasiutynski, T.; Miyazaki, Y.; Inaba, A.; Pinkowicz, D.; et al. Magnetic systems at criticality: Different signatures of scaling. *Acta Phys. Pol.* **2013**, *124*, 977. [CrossRef]
50. Souca, G.; Iamandi, S.; Mazilu, C.; Dudric, R.; Tetean, R. Magnetocaloric Effect and Magnetic Properties of Pr$_{1-x}$Ce$_x$Co$_3$ Compounds. *Stud. Univ. Babeș-Bolyai Phys.* **2018**, *63*, 9–18. [CrossRef]

51. Zverev, V.; Tishin, A.M. Magnetocaloric Effect: From Theory to Practice. In *Reference Module in Materials Science and Material Engineering*; Elsevier: Amsterdam, The Netherlands, 2016; pp. 5035–5041. [CrossRef]
52. Deac, I.G.; Vladescu, A. Magnetic and magnetocaloric properties of $Pr_{1-x}Sr_xCoO_3$ cobaltites. *J. Magn. Magn. Mater.* **2014**, *365*, 1–7. [CrossRef]
53. Griffith, L.D.; Mudryk, Y.; Slaughter, J.; Pecharsky, V.K. Material-based figure of merit for caloric materials. *J. Appl. Phys.* **2018**, *123*, 034902. [CrossRef]
54. Naik, V.B.; Barik, S.K.; Mahendiran, R.; Raveau, B. Magnetic and Calorimetric Investigations of Inverse Magnetocaloric Effect in $Pr_{0.46}Sr_{0.54}MnO_3$. *Appl. Phys. Lett.* **2011**, *98*, 112506. [CrossRef]
55. Gong, Z.; Xu, W.; Liedienov, N.A.; Butenko, D.S.; Zatovsky, I.V.; Gural'skiy, I.A.; Wei, Z.; Li, Q.; Liu, B.; Batman, Y.A.; et al. Expansion of the multifunctionality in off-stoichiometric manganites using post-annealing and high pressure: Physical and electrochemical studies. *Phys. Chem. Chem. Phys.* **2002**, *24*, 21872–21885. [CrossRef]
56. Sandeman, K.G. Magnetocaloric materials: The search for new systems. *Scr. Mater.* **2012**, *67*, 566–571. [CrossRef]
57. Smith, A.; Bahl, A.C.R.H.; Bjørk, R.; Engelbrecht, K.; Kaspar, K.; Nielsen, K.K.; Pryds, N. Materials challenges for high performance magnetocaloric refrigeration devices. *Adv. Energy Mater.* **2012**, *2*, 1288–1318. [CrossRef]
58. Kitanovski, A.; Tusek, J.; Tomc, U.; Plaznik, U.; Ozbolt, M.; Poredos, A. *Magentocaloric Energy Conversion: From Theory to Applications*, 1st ed.; Springer: Cham, Switzerland, 2015. [CrossRef]
59. Legait, U.; Guillou, F.; Kedous-Lebouc, A.; Hardy, A.V.; Almanza, M. An experimental comparison of four magnetocaloric regenerators using three different materials. *Int. J. Refrig.* **2014**, *37*, 147–155. [CrossRef]
60. Krishnamoorthi, C.; Barik, S.K.; Siu, Z.; Mahendiran, R. Normal and inverse magnetocaloric effects in $La_{0.5}Ca_{0.5}Mn_{1-x}Ni_xO_3$. *Solid State Commun.* **2010**, *150*, 1670–1673. [CrossRef]
61. von Ranke, P.J.; de Oliveira, N.A.; Alho, B.P.; Plaza, E.J.R.; de Sousa, V.S.R.; Caron, L.; Reis, M.S. Understanding the inverse magnetocaloriceffect in antiferro- and ferrimagnetic arrangements. *J. Phys. Condens. Matter* **2009**, *21*, 056004. [CrossRef] [PubMed]
62. Majumder, D.D.; Majumder, D.D.; Karan, S. Magnetic Properties of Ceramic Nanocomposites. In *Ceramic Nanocomposites*; Banerjee, R., Manna, I., Eds.; Woodhead Publishing Series in Composites Science and Engineering; Elsevier B.V.: Amsterdam, The Netherlands, 2013; pp. 51–91. [CrossRef]
63. Hueso, L.E.; Sande, P.; Miguéns, D.R.; Rivas, J.; Rivadulla, F.; López-Quintela, M.A. Tuning of the magnetocaloric effect in nanoparticles synthesized by sol–gel techniques. *J. Appl. Phys.* **2002**, *91*, 9943–9947. [CrossRef]
64. Anis, B.; Tapas, S.; Banerjee, S.; Das, I. Magnetocaloric properties of nanocrystalline $Pr_{0.65}(Ca_{0.6}Sr_{0.4})_{0.35}MnO_3$. *J. Appl. Phys.* **2008**, *103*, 013912. [CrossRef]

Disclaimer/Publisher's Note: The statements, opinions and data contained in all publications are solely those of the individual author(s) and contributor(s) and not of MDPI and/or the editor(s). MDPI and/or the editor(s) disclaim responsibility for any injury to people or property resulting from any ideas, methods, instructions or products referred to in the content.

Article

Facile Fabrication of Highly Active CeO$_2$@ZnO Nanoheterojunction Photocatalysts

Xiaoqian Ai [1,†], Shun Yan [2,†], Chao Lin [2,†], Kehong Lu [1], Yujie Chen [1] and Ligang Ma [2,*]

[1] School of Physics and Information Engineering, Jiangsu Province Engineering Research Center of Basic Education Big Data Application, Jiangsu Second Normal University, Nanjing 210013, China; xqai186@jssnu.edu.cn (X.A.)
[2] School of Electronic Engineering, Nanjing Xiaozhuang University, Nanjing 211171, China
* Correspondence: lgma@njxzc.edu.cn
† These authors contributed equally to this work.

Abstract: Photocatalyst performance is often limited by the poor separation and rapid recombination of photoinduced charge carriers. A nanoheterojunction structure can facilitate the separation of charge carrier, increase their lifetime, and induce photocatalytic activity. In this study, CeO$_2$@ZnO nanocomposites were produced by pyrolyzing Ce@Zn metal–organic frameworks prepared from cerium and zinc nitrate precursors. The effects of the Zn:Ce ratio on the microstructure, morphology, and optical properties of the nanocomposites were studied. In addition, the photocatalytic activity of the nanocomposites under light irradiation was assessed using rhodamine B as a model pollutant, and a mechanism for photodegradation was proposed. With the increase in the Zn:Ce ratio, the particle size decreased, and surface area increased. Furthermore, transmission electron microscopy and X-ray photoelectron spectroscopy analyses revealed the formation of a heterojunction interface, which enhanced photocarrier separation. The prepared photocatalysts show a higher photocatalytic activity than CeO$_2$@ZnO nanocomposites previously reported in the literature. The proposed synthetic method is simple and may produce highly active photocatalysts for environmental remediation.

Keywords: zinc oxide; cerium oxide; nanocomposites; photocatalysis; heterojunction

Citation: Ai, X.; Yan, S.; Lin, C.; Lu, K.; Chen, Y.; Ma, L. Facile Fabrication of Highly Active CeO$_2$@ZnO Nanoheterojunction Photocatalysts. Nanomaterials 2023, 13, 1371. https://doi.org/10.3390/nano13081371

Academic Editors: Thomas Dippong and Marco Stoller

Received: 11 March 2023
Revised: 7 April 2023
Accepted: 13 April 2023
Published: 14 April 2023

Copyright: © 2023 by the authors. Licensee MDPI, Basel, Switzerland. This article is an open access article distributed under the terms and conditions of the Creative Commons Attribution (CC BY) license (https:// creativecommons.org/licenses/by/ 4.0/).

1. Introduction

The continued use of fossil fuels has resulted in a global energy crisis, environmental pollution, and climate change [1,2], thus, more sustainable energy resources are essential to combat these issues [3]. Photocatalytic technologies [4,5] for exploiting light energy have drawn particular attention [6–8]. Such catalysts can be used to produce hydrogen from water for energy storage and convert the excess carbon dioxide in the atmosphere to valuable chemical feedstocks and fuels, such as methane and methanol [9–14]. Furthermore, photocatalytic materials can be used to degrade organic pollutants in contaminated water [15–17]. In particular, compared with traditional methods for water treatment, photocatalysis requires less energy and can achieve a complete degradation of pollutants. Therefore, photocatalytic systems are highly promising for clean energy production and environmental remediation. To achieve high photocatalytic activity, photoinduced charge carriers (i.e., electrons and holes) must be effectively generated and separated in the photocatalysts.

Several photocatalytic mechanisms for the degradation of organic pollutants such as dyes have been reported. Briefly, upon irradiation with ultraviolet light, electrons in the valence band (VB) are excited to the conduction band (CB); thus, holes are created in the VB. The electrons in the CB react with adsorbed oxygen to form superoxide radicals (•O$_2^-$), whereas the holes in the VB react with water to form hydroxyl radicals (•OH), and these two radicals react with organic pollutants and degrade them. To date, many photocatalysts have been reported, including CdS [18,19], ZnO [20], CeO$_2$ [21], TiO$_2$ [22], WO$_3$ [23], and graphitic carbon nitride (g-C$_3$N$_4$) [24]. However, the charge carriers generated by these single-phase catalysts can easily recombine, resulting in short carrier lifetimes and low catalytic efficiencies. To address this problem, multiphase catalysts, such as nanocomposites, i.e., Fe_2O_3/Cu_2O [25], ZnO/TiO_2 [26], GQDs/NiSe-NiO [27], $g-C_3N_4$/Ni-ZnO [28], and MoS_2/TiO_2 nanocomposites [29], have been prepared. These nanocomposites can increase the lifetimes of the charge carriers by restricting the generated electrons and holes in different phases and reducing their recombination.

Considering the matched band gaps of CeO_2 and ZnO, we have reported on a CeO_2@ZnO nanocomposite as an efficient photocatalyst [30]. In our previous study, a Ce@Zn-bimetallic metal–organic framework (Ce@Zn-MOF) precursor was prepared with a Zn:Ce atomic ratio of 1; subsequently, the Ce@Zn-MOF precursor was subjected to thermal decomposition to obtain photocatalytic CeO_2@ZnO nanocomposites. The optimal pyrolysis temperature was identified as 450 °C based on the structure, morphology, and photocatalytic degradation performance of the nanocomposites. However, the effects of the Zn:Ce ratio have not been studied.

Therefore, in this study, we fabricated Ce/Zn-MOF precursors with various Zn:Ce atomic ratios (0:10, 2:8, 4.5:5.5, 6.7:3.3, 8:2, and 10:0). CeO_2@ZnO nanocomposites were then obtained via thermal decomposition at the previously identified optimal temperature (450 °C). The structure, morphology, and optical properties of CeO_2, ZnO, and the CeO_2@ZnO nanocomposites were investigated via X-ray diffractometry (XRD), scanning electron microscopy (SEM), transmission electron microscopy (TEM), and UV-vis absorption spectroscopy. In addition, the prepared nanocomposites were employed for photocatalytic water remediation using rhodamine B (RhB) as a model organic pollutant. Finally, the photocatalytic degradation mechanism was determined.

2. Experimental Method
2.1. Precursor and Photocatalyst Synthesis

Ce-MOF, Zn-MOF, and Ce/Zn-MOF were prepared according to our previously reported method. Briefly, $Ce(NO_3)_2·6H_2O$ (30 mmol) and 2-methylimidazole (63 mmol) were dissolved in methanol (500 mL), and the mixture was stirred, precipitated, and centrifuged to obtain Ce-MOF. To prepare the bimetallic Ce/Zn-MOF precursors $Zn(NO_3)_2·6H_2O$ and $Ce(NO_3)_2·6H_2O$ in atomic ratios of 0:10, 2:8, 4.5:5.5, 6.7:3.3, 8:2, and 10:0 were added to methanol. The samples with Zn:Ce ratios of 0:10 and 10:0 yielded Ce-MOF and Zn-MOF, respectively. Ce-MOF, Zn-MOF, and Ce/Zn-MOF were obtained via sequential precipitation, washing, centrifugation, and drying. Finally, the Ce-MOF, Zn-MOF, and Ce/Zn-MOF precursors were pyrolyzed at 450 °C in a tubular sintering furnace for 3 h to produce the CeO_2, ZnO, and the CeO_2@ZnO nanocomposites [30]. The annealed samples are denoted as CeO_2@ZnO-x, where x is the ratio of Zn to Ce; for example, the sample with a ratio of 2:8 is denoted as CeO_2@ZnO-0.2.

2.2. Characterization

The effect of different ratios of ZnO and CeO_2 on the lattice structure of the CeO_2@ZnO nanocomposites was determined using XRD (Dmax-rB, Rigaku; Tokyo, Japan, Cu-K_α λ = 1.5418 Å) with a tube voltage and current of 40 kV and 80 mA, respectively. The changes in the morphology, microstructure, and elemental distribution of the CeO_2@ZnO-x nanocomposites were observed using field emission SEM (ZEISS Gemini 500) and TEM (FEITecnai G2 F30). For TEM analysis, the CeO_2@ZnO-x nanocomposites were ultrasonically dispersed in ethanol for 10 min and then dropped onto a Cu grid, and TEM observation was carried out at an acceleration voltage of 200 kV. The electronic structures and valence states of the elements were characterized using X-ray photoelectron spectroscopy (XPS). The optical band gaps of the nanocomposites were determined using UV-vis spectroscopy (UH4150, Hitachi). The spectrometer was equipped with an integrating sphere.

The photocatalytic activity of the prepared CeO_2@ZnO nanocomposites was evaluated by measuring the degradation of RhB as a model organic pollutant using a multi-channel photochemical reaction system (PCX-50C). The light source was ultraviolet light at 365 nm with a real power density of 320 mWcm^{-2}. For the degradation tests, the nanocomposites (50 mg) were ultrasonically dispersed in an aqueous RhB solution (50 mL, 10 mg/L) for 10 min. The suspension was then placed in the dark for 60 min with continuous magnetic agitation until it reached dynamic adsorption–desorption equilibrium. Before irradiation, an aliquot (3 mL) of the degradation solution was extracted and centrifuged to determine the degree of degradation. Subsequently, during light irradiation, aliquots (3 mL) were collected every 5 min and analyzed. Note that no precious metal catalyst was added during the degradation process. In addition, electron paramagnetic resonance (EPR, Bruker EMX-plus) spectroscopy was used to identify the free radicals produced upon light irradiation to investigate the degradation mechanism. For these measurements, the nanomaterials were added to a 5,5-dimethyl-1-pyridine-N-oxide (DMPO) solution and mixed with deionized water or CH_3OH to detect the concentrations of hydroxyl ($\bullet OH$) and superoxide ($\bullet O_2{}^-$) radicals, respectively. The transient photocurrent and electrochemical impedance spectroscopy (EIS) measurements were conducted on an electrochemical workstation (CHI660E) with three electrodes. A Pt wire and Ag/AgCl were used as the counter and reference electrodes, respectively. For the photocurrent measurement, indium tin oxide glass coated with the photocatalyst was used as the working electrode, whereas a glass-carbon electrode coated with the photocatalyst was the working electrode for the EIS measurement. A Na_2SO_4 solution (0.5 M) was used as the electrolyte. Photoluminescence (PL) spectra were recorded on a spectrofluorometer (Hitachi F7000) equipped with a 250 nm excitation source.

3. Results and Discussion

Figure 1a shows the XRD patterns of the nanomaterials formed by the pyrolysis of the MOF precursors at 450 °C. The diffraction peaks of the nanomaterial formed by pyrolyzing Ce-MOF are indexed to cubic CeO_2 (JCPDS Card No. 81-0792) [31]. When the Zn:Ce ratio increases to 0.2, the intensity of the CeO_2 diffraction peaks decreases slightly (Figure 1a, $CeO_2@ZnO$-0.2). With a further increase in the Zn:Ce ratio, these CeO_2 peaks decrease in intensity, and new diffraction peaks corresponding to hexagonal wurtzite ZnO (JCPDS Card No. 36-1451) appear (Figure 1a, $CeO_2@ZnO$-0.45) [32], suggesting the formation of a $CeO_2@ZnO$ nanocomposite. When the Zn:Ce ratios are 0.67 and 0.8 (Figure 1a, CeO@ZnO-0.67 and CeO@ZnO-0.8, respectively), the peaks corresponding to CeO_2 disappear, and those corresponding to ZnO increase in intensity. XPS analysis revealed that some CeO_2 was present in these two samples, and the lack of diffraction peaks is likely a result of the low quantity of CeO_2, which resulted in these peaks being X-ray invisible or obscured by those of ZnO. Finally, the pyrolysis product of Zn-MOF displays the characteristic diffraction peaks of ZnO. The phase evolution in the nanocomposites with respect to the Zn:Ce ratio is more distinct from $2\theta = 29°$ to $59°$ (Figure 1b), wherein the peaks corresponding to CeO_2 gradually disappear, whereas those related to ZnO gradually intensify. Notably, the positions of the peaks corresponding to CeO_2 and ZnO do not change with the Zn:Ce ratio, indicating that Zn did not enter the CeO_2 lattice and vice versa.

Figure 2 shows the morphologies of the CeO_2, ZnO, and $CeO_2@ZnO$ nanocomposites. As shown in Figure 2a, CeO_2 has a smooth nanosphere morphology and uniform size distribution (approximate diameter ~800 nm). With an increase in the Zn:Ce ratio, additional nanoparticles merge on the nanospheres (Figure 2b, $CeO_2@ZnO$-0.2), and the surfaces of the nanospheres become rough. For $CeO_2@ZnO$-0.45 and $CeO_2@ZnO$-0.67, the large nanospheres disappear, and only small nanoparticles are observed in the nanocomposites. With a further increase in the Zn:Ce ratio, the small nanoparticles aggregate, as shown in Figure 2e,f. Thus, adding Zn converts the large spheres into small nanoparticles, but excess Zn results in the aggregation of the small nanoparticles, which should decrease the specific surface area of the nanocomposites.

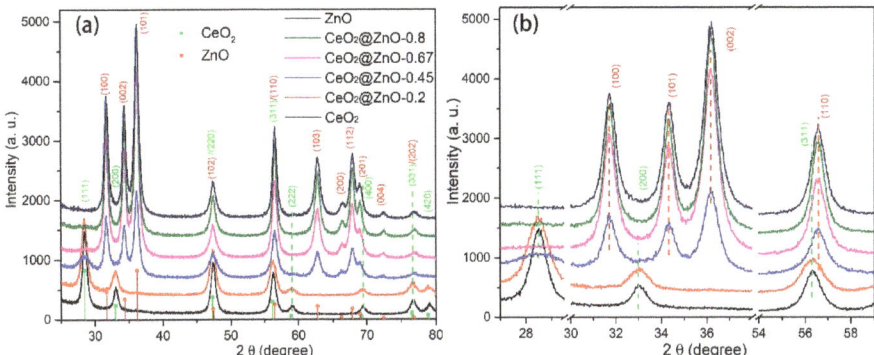

Figure 1. (a) XRD patterns for the CeO_2, ZnO, and CeO_2@ZnO nanocomposites (predicted peak positions for cubic CeO_2 and hexagonal ZnO are shown on the x-axis) and (b) an enlarged figure showing the most intense diffraction peaks in (a).

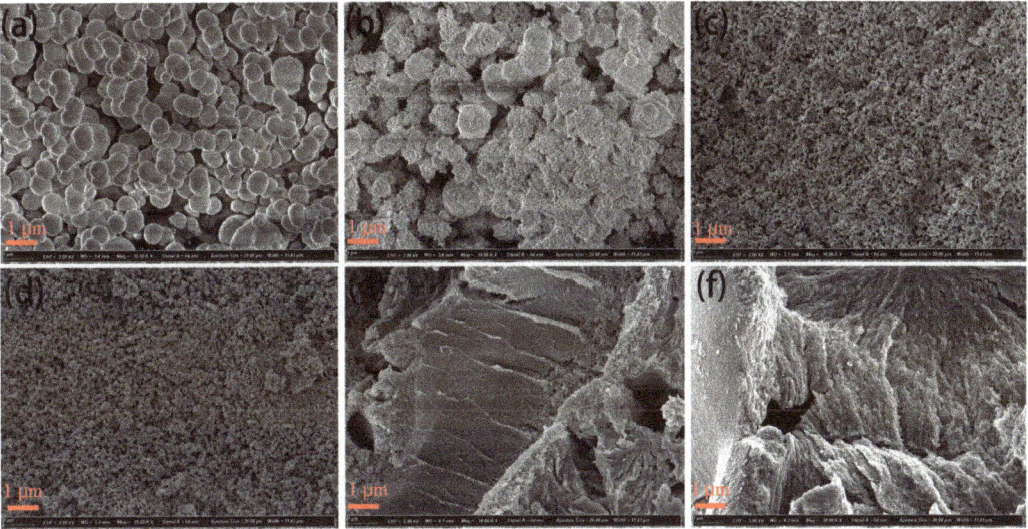

Figure 2. SEM images of (a) CeO_2, (b) CeO_2@ZnO-0.2, (c) CeO_2@ZnO-0.45, (d) CeO_2@ZnO-0.67, (e) CeO_2@ZnO-0.8, and (f) ZnO.

The microstructures of the CeO_2, ZnO, and CeO_2@ZnO nanocomposites were observed using TEM. The nanospheres in Figure 3a are approximately 758 nm in diameter. In addition, the high-resolution TEM (HRTEM) images reveal that the lattice fringes at the edges of the nanospheres have interplanar distances of 0.313 and 0.269 nm, which correspond to the (111) and (200) crystal planes of cubic CeO_2, respectively. Consistent with the SEM results, when the Zn:Ce ratio is 0.2, some nanoparticles are attached to the edges of the CeO_2 nanospheres, while the size of the CeO_2 nanospheres does not change significantly. These nanoparticles show lattice fringes corresponding to both cubic CeO_2 and hexagonal wurtzite ZnO; in particular, the 0.141 nm interplanar distance corresponds to the (200) plane of ZnO. With a further increase in the Zn:Ce ratio, the number of hexagonal wurtzite ZnO nanoparticles gradually increases, whereas the number of CeO_2 spheres in the cubic phase decreases. The diameter of the observed nanoparticles is approximately 10 nm. In addition, two sets of diffraction rings are observed in the selected area electron diffraction (SAED) pattern (Figure 3d-1). Importantly, an obvious boundary is observed between the ZnO and CeO$_2$ phases, indicating the formation of a ZnO@CeO$_2$ nanoheterojunction (Figure 3d-1). For the ZnO formed by the pyrolysis of the Zn-MOF precursor (Figure 3f-1), the size of the nanoparticles increases to approximately 25 nm, mainly because of the aggregation and growth of the nanoparticles at high temperatures. Its lattice fringes are detected at 0.247 and 0.2827 nm, corresponding to the (101) and (100) crystal phases of hexagonal wurtzite ZnO, respectively.

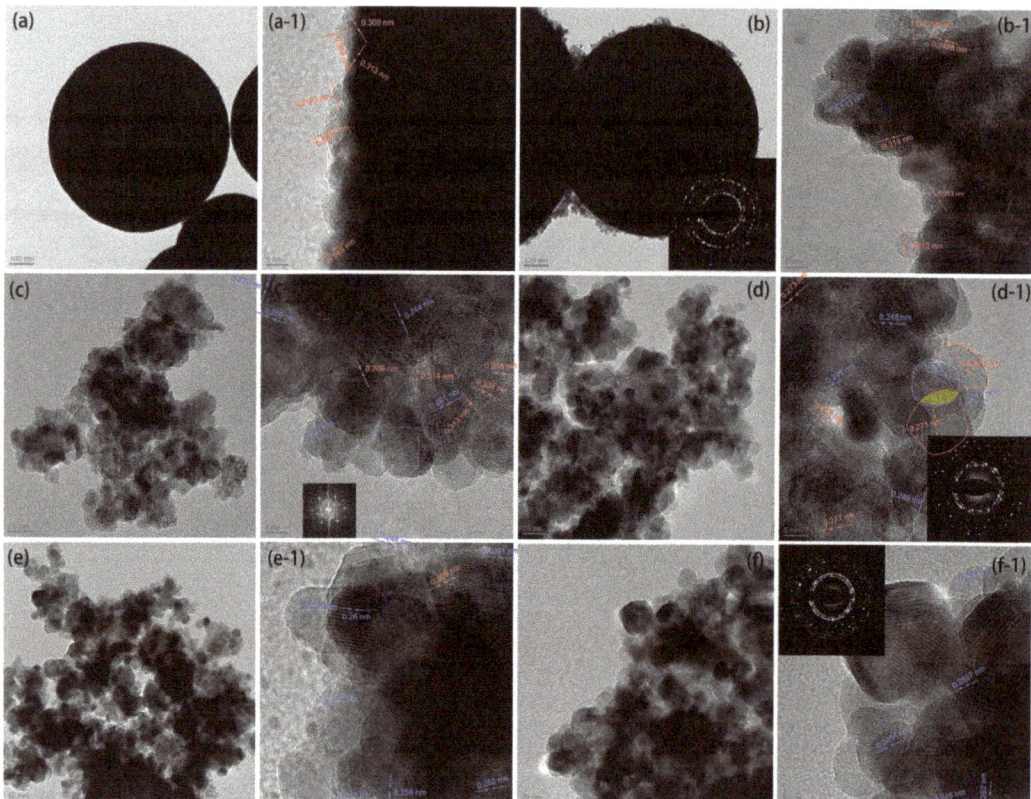

Figure 3. TEM images of (**a**) CeO$_2$, (**b**) CeO$_2$@ZnO-0.2, (**c**) CeO$_2$@ZnO-0.45, (**d**) CeO$_2$@ZnO-0.67, (**e**) CeO$_2$@ZnO-0.8, and (**f**) ZnO. The suffix "-1" indicates the high-resolution TEM images, and the insets show the corresponding selected area electron diffraction patterns. The lattice fringes in red and blue correspond to CeO$_2$ and ZnO, respectively.

The elemental distribution in CeO$_2$@ZnO-0.67 was also characterized via EDS mapping. As shown in Figure 4, Zn, Ce, and O are distributed uniformly in the nanocomposite. However, the distribution of Ce is sparser than those of Zn and O, which is consistent with the high Zn content in CeO$_2$@ZnO-0.67.

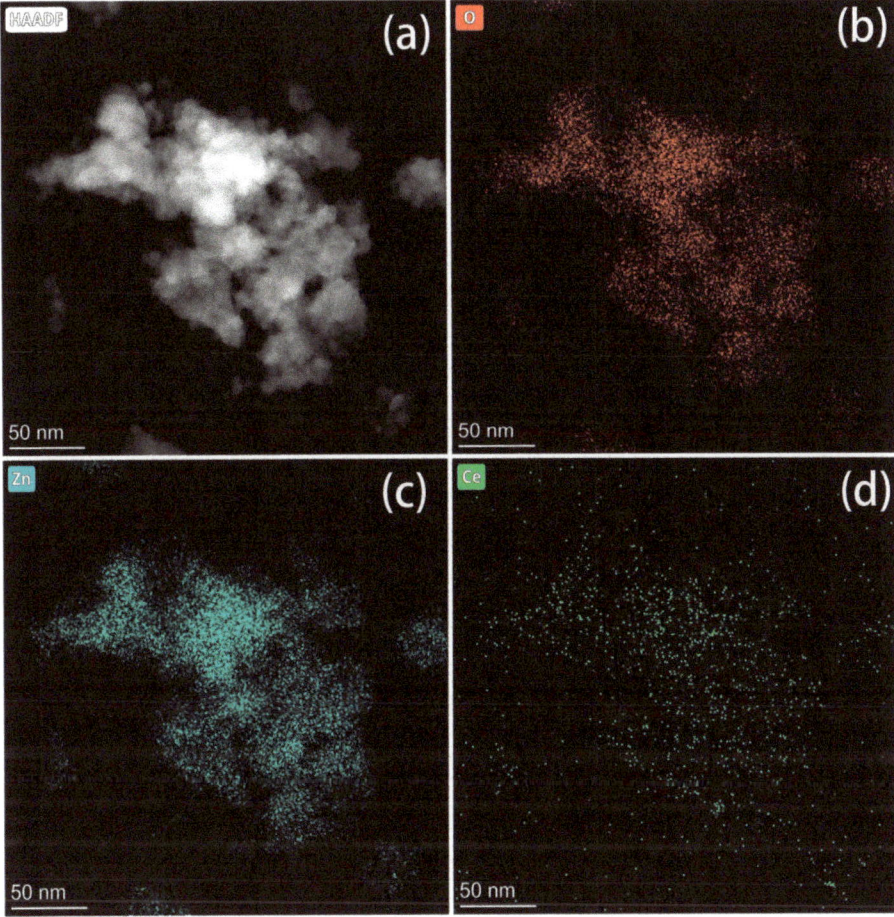

Figure 4. (a) Scanning tunneling electron microscopy high-angle annular dark field image of CeO$_2$@ZnO-0.67 and the corresponding (b) O, (c) Zn, and (d) Ce elemental maps.

XPS measurements were conducted to investigate the elemental composition and surface chemical valence states. The C 1s peak is related to the adventitious carbon introduced during pyrolysis. Therefore, the spectra were calibrated based on the C=C peak at a binding energy of 285.0 eV. As expected, the XPS survey spectra of the CeO$_2$@ZnO nanocomposites contain peaks corresponding to Ce, Zn, and O. In the Ce 3d high-resolution XPS spectra of CeO$_2$ (black curve in Figure 5b), the Ce^{3+} and Ce^{4+} peaks are detected [33–35], indicating the presence of CeO$_2$ and Ce$_2$O$_3$. The Ce$_2$O$_3$ phase was not detected via XRD, possibly because of its amorphous nature. In contrast, in the XPS spectra of the CeO$_2$@ZnO nanocomposites, the Ce^{3+} peaks are weakened until they disappear, indicating that Ce is present exclusively in the CeO$_2$ phase. The split peaks in the high-resolution Zn 2p spectra correspond to Zn 2p$_{3/2}$ and 2p$_{1/2}$, respectively, indicating that Zn is present as Zn^{2+} [36]. Interestingly, the Zn peaks in the CeO$_2$@ZnO nanocomposites are at lower binding energies than those in ZnO, suggesting the formation of an interface between CeO$_2$ and ZnO. In the high-resolution O 1s spectra (Figure 5d), three peaks at 529.1, 530.2, and 531.6 eV can be assigned to the oxygen bonded to Ce, Zn, and surface hydroxyl radicals [33,37,38], respectively. Overall, the XPS and HRTEM results indicate that a heterojunction interface between ZnO and CeO$_2$ nanoparticles is formed via pyrolysis. The interface is crucial for photocatalytic applications because photoelectrons can migrate across the interface and be effectively separated from the photogenerated holes.

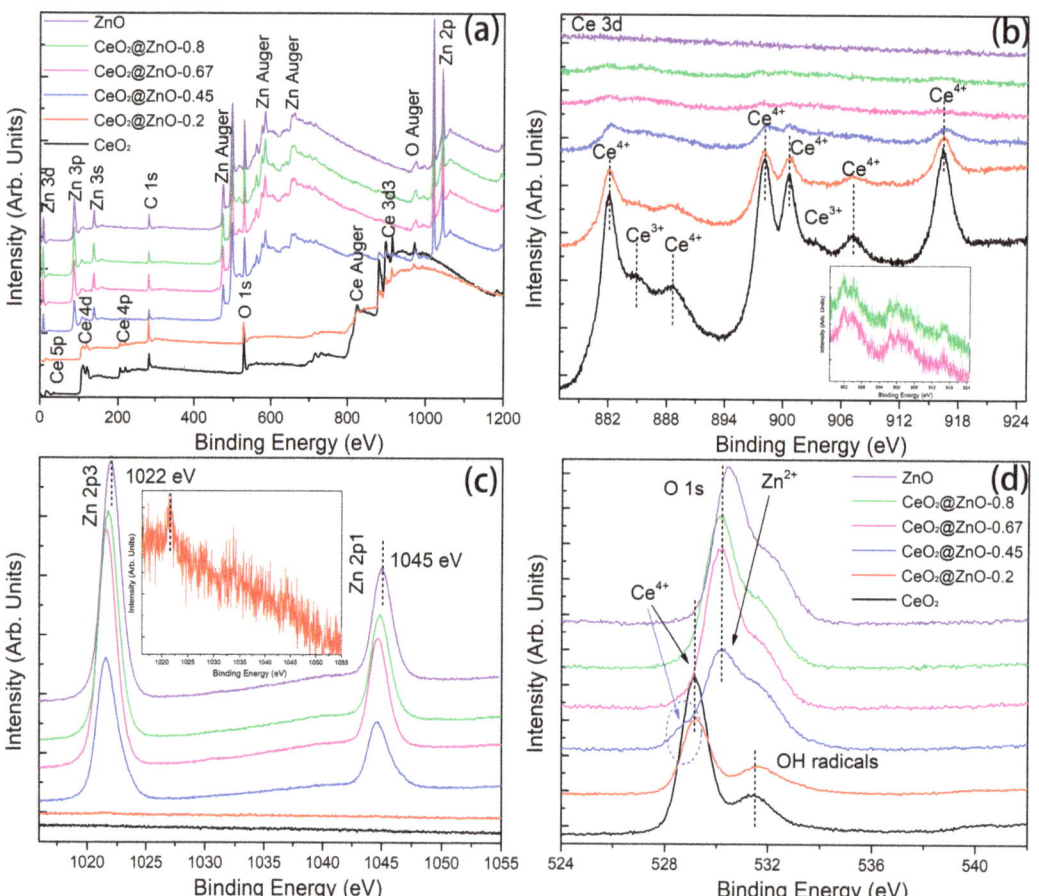

Figure 5. XPS spectra of the CeO$_2$, ZnO, and CeO$_2$@ZnO nanocomposites: (**a**) survey spectra and high-resolution (**b**) Ce 3d, (**c**) Zn 2p, and (**d**) O 1s spectra.

The optical properties of the CeO$_2$, ZnO, and CeO$_2$@ZnO nanocomposites were investigated using UV-vis absorption spectroscopy, and the results are shown in Figure 6a. With an increase in the Zn:Ce ratio, the optical absorption edge shows a progressive blue shift. The optical band gaps of the CeO$_2$, ZnO, and CeO$_2$@ZnO nanocomposites were obtained using the Tauc formula [39]: $\alpha h\nu = A(h\nu - E_g)^2$, where α, h, ν, A, and E_g are the absorption coefficient, Planck's constant, the frequency of the incident light, a constant, and the optical band gap, respectively. The fitting curves for $(\alpha h\nu)^2$ vs. $h\nu$ are shown in Figure 6b. The linear part of the curve is extrapolated, and the x intersection is the optical band gap. Thus, the optical band gaps of the CeO$_2$, CeO$_2$@ZnO-0.2, CeO$_2$@ZnO-0.45, CeO$_2$@ZnO-0.67, CeO$_2$@ZnO-0.8, and ZnO nanomaterials are 2.789, 3.06, 3.164, 3.195, 3.20, and 3.214 eV, respectively, showing an increasing trend with an increase in the Zn:Ce ratio. As the Zn:Ce ratio is increased, a transformation from CeO$_2$ to CeO$_2$@ZnO to ZnO occurs, and the size of the composite decreases from 800 nm to 10 nm. The quantum effects arising from particle size limitations cause a blue-shift in the absorption edge and an increase in the band gap.

Crucially, the band gap determines the range of light that can be absorbed and, thus, used during photocatalytic degradation.

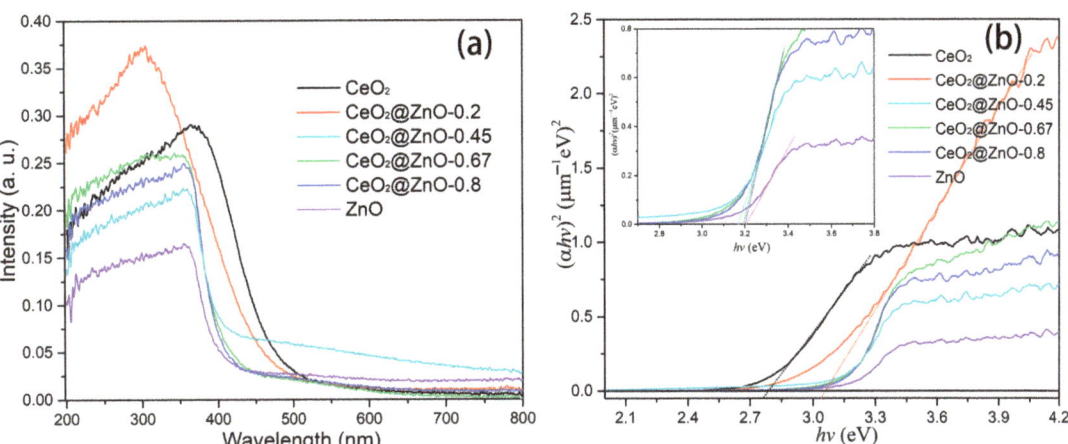

Figure 6. (**a**) UV-visible absorption spectra and (**b**) plots of $(\alpha h\nu)^2$ vs. $h\nu$ for the CeO_2, ZnO, and $CeO_2@ZnO$ nanocomposites.

The catalytic activity of the photocatalyst was evaluated by analyzing the degree of photodegradation of RhB, as calculated using Equation (1) [19]:

$$\text{Photodegradation efficiency (\%)} = (1 - C/C_0) \times 100\%, \tag{1}$$

where C_0 and C represent the UV-vis light absorption coefficients of RhB at adsorption equilibrium in the dark and upon light irradiation, respectively. Figure 7a shows the photodegradation efficiency for RhB with respect to irradiation time. The photocatalytic efficiency of CeO_2 for RhB is very low. The smooth and large nanospheres of CeO_2 have a low specific surface area, leading to a small contact area between CeO_2 and RhB. The photodegradation efficiency of $CeO_2@ZnO$-0.2 slightly increases, which can be attributed to the nanoparticles attached to the CeO_2 nanospheres. With a further increase in the Zn:Ce ratio, the $CeO_2@ZnO$-0.67 nanocomposite exhibits the best photocatalytic efficiency (approximately ~97% RhB degradation after 30 min of irradiation). However, the photocatalytic efficiencies of $CeO_2@ZnO$-0.8 and ZnO are lower than that of the $CeO_2@ZnO$-0.67.

The experimental data were fitted using the pseudo-first-order kinetic model shown in Equation (2) [30,34].

$$\ln\left(\frac{C}{C_0}\right) = -kt \tag{2}$$

Here, k (min^{-1}) is the kinetic degradation rate constant, and t (min) is the reaction time. As shown in Figure 7b, the plots of $-\ln(C/C_0)$ vs. t. approximately follow a linear relationship, indicating that this model can be used to analyze the photodegradation rate. The degradation rates over CeO_2 and $CeO_2@ZnO$-0.2 are very low and not reported here. The kinetic degradation rate constant (k) values for the photodegradation of RhB over $CeO_2@ZnO$-0.45, $CeO_2@ZnO$-0.67, $CeO_2@ZnO$-0.8, and ZnO were calculated as 0.0955, 0.124, 0.0749, and 0.0669, respectively. The $CeO_2@ZnO$-0.67 nanocomposite exhibits the best photodegradation rate constant (0.124), which is superior to the highest photodegradation rate constant observed in our previous study on $CeO_2@ZnO$ photocatalysts (0.1096). Therefore, the optimal Zn/Ce atomic ratio was 0.67 at the optimal pyrolysis temperature (450 °C). We have summarized recent reports on the photodegradation performance of $CeO_2@ZnO$, which are listed in Table 1. As can be seen from Table 1, $CeO_2@ZnO$-0.67, the nanocomposites prepared by the binary MOF pyrolysis method in this paper, show better performance.

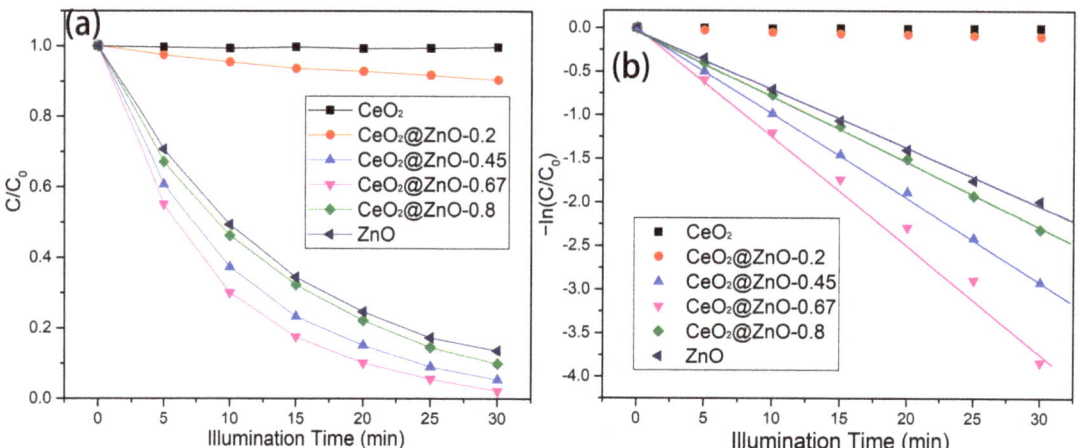

Figure 7. (a) Degradation efficiencies and (b) kinetic plots for the degradation of RhB over the CeO$_2$, ZnO, and CeO$_2$@ZnO nanocomposites during with respect to irradiation time.

Table 1. Summary of different synthetic methods of CeO$_2$, ZnO, and CeO$_2$@ZnO nanomaterials with regard to photodegradation.

Photocatalyst	Synthetic Method	Morphology	Light Source	Catalyst Amount	Degraded Object	Illumination Time	Photodegradation Efficiency	Reference
CeO$_2$@ZnO	Hydrothermal approach	Ordered mesoporous	380 nm < λ <780 nm	50 mg	MB	150 min	97.4%	[40]
CeO$_2$@ZnO	Electrospinning technique	Nanofibers	365 nm	10 mg	RhB	180 min	98%	[41]
CeO$_2$@ZnO	Sol–gel method	Nanocomposites	>420 nm	50 mg	RhB	250 min	50%	[42]
CeO$_2$/ZnO@Au	Hydrothermal method	Hierarchical heterojunction	Xe lamp	10 mg	RhB	20 min	99%	[43]
CuO/CeO$_2$/ZnO	Two-step sol–gel method	Nanoparticles	UV light	50 mg	RhB	30 min	98%	[44]
CeO$_2$/ZnO	In situ precipitation method	Nanocomposites	UV light	50 mg	RhB	80 mn	42%	[45]
CeO$_2$/ZnO	Pyrolyzing Ce@Zn metal–organic frameworks	Nanoheterojunction	UV light	50 mg	RhB	30 min	97%	This work

The production of free radicals was investigated under dark and light conditions using EPR spectroscopy. The peak intensities in the EPR spectra reflect the concentrations of free radicals. As shown in Figure 8, no free radicals are detected in the dark, whereas two types of free radicals are formed upon light irradiation. Furthermore, the concentration of free radicals produced by the CeO$_2$@ZnO-0.67 nanocomposite is the highest under light irradiation, confirming that these free radicals are responsible for enhancing the photodegradation efficiency.

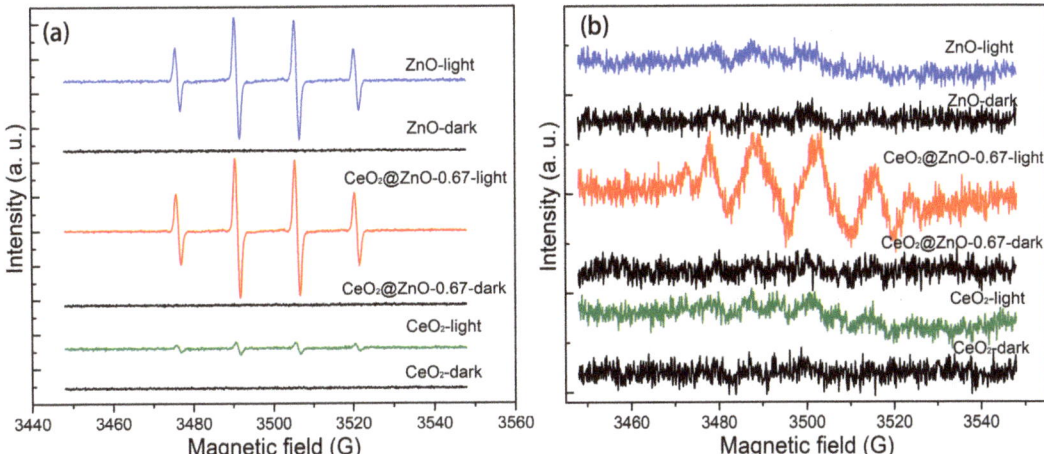

Figure 8. EPR spectra of the CeO$_2$, ZnO, and CeO$_2$@ZnO nanocomposites in the dark and under light irradiation: (**a**) DMPO-OH and (**b**) DMPO-O$_2^-$.

To investigate the separation ability of the photogenerated electrons and holes, the transient photocurrent characteristics of the CeO, ZnO, and CeO$_2$@ZnO-0.67 nanocomposites were measured. As shown in Figure 9a, the CeO$_2$@ZnO-0.67 nanocomposite produces a stronger photocurrent than CeO$_2$ and ZnO, indicating that has the highest number of photogenerated charge carriers under illumination. In addition, the PL spectra of the CeO$_2$, ZnO, and CeO$_2$@ZnO-0.67 were measured. As shown in Figure 9b, the intensity of the luminescent peak for CeO$_2$@ZnO-0.67 is significantly lower than that of CeO$_2$ and ZnO, which indicates that the electron–hole pairs generated by CeO$_2$@ZnO-0.67 have a low recombination rate. The efficiency of the direct electron transfer and separation of photogenerated electrons was evaluated using EIS (Figure 9c). In the EIS spectra, the arc radius determines the resistance of the interface layer, which affects the separation of electrons. A small arc radius means that electrons can be transported quickly. Among the three photocatalysts, CeO$_2$@ZnO-0.67 exhibits the smallest arc radius, indicating its excellent charge transfer ability. Cycling experiments were performed to evaluate the stability and recyclability of CeO$_2$@ZnO-0.67. Figure 9d shows that the high photodegradation efficiency of CeO$_2$@ZnO-0.67 is maintained after three cycles.

Based on the previously described analysis, the separation mechanism of the photogenerated electron–hole pairs is shown in Figure 10. Generally, the photoexcited electrons easily recombine with the holes in the VB. In CeO$_2$ or ZnO pure phase materials, the recombination of excited electrons in CB and holes in VB is dominant, which considerably reduces the efficiency of photodegradation. Therefore, the inhibition of charge carrier recombination is crucial for enhancing photocatalytic efficiency. Unlike CeO$_2$ and ZnO, the CeO$_2$@ZnO nanocomposite contains a heterojunction interface, which prevents the recombination of photogenerated charge carriers and ensures the production of free radicals for Z-scheme catalytic photodegradation. In detail, the excited electrons in ZnO preferentially recombine with the holes in CeO$_2$, which enables the electrons in the CB of CeO$_2$ and the holes in the VB of ZnO to interact fully with oxygen and water to generate free radicals for dye decomposition.

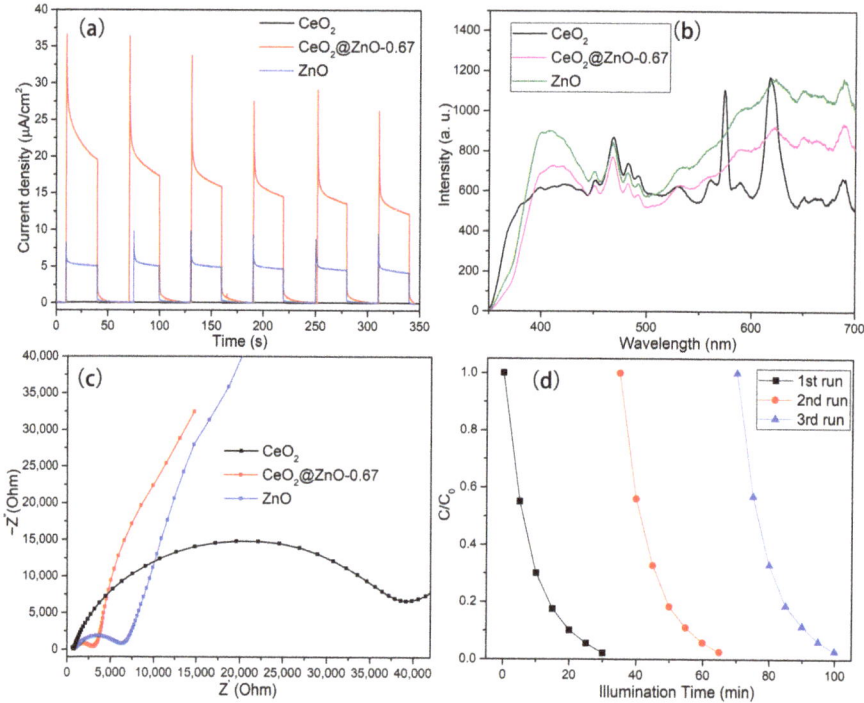

Figure 9. (**a**) Transient photocurrent curves, (**b**) PL spectra, (**c**) and EIS Nyquist plots for CeO, CeO$_2$@ZnO-0.67, and ZnO, respectively. (**d**) Cycling experiments of the photodegradation of RhB over CeO$_2$@ZnO-0.67.

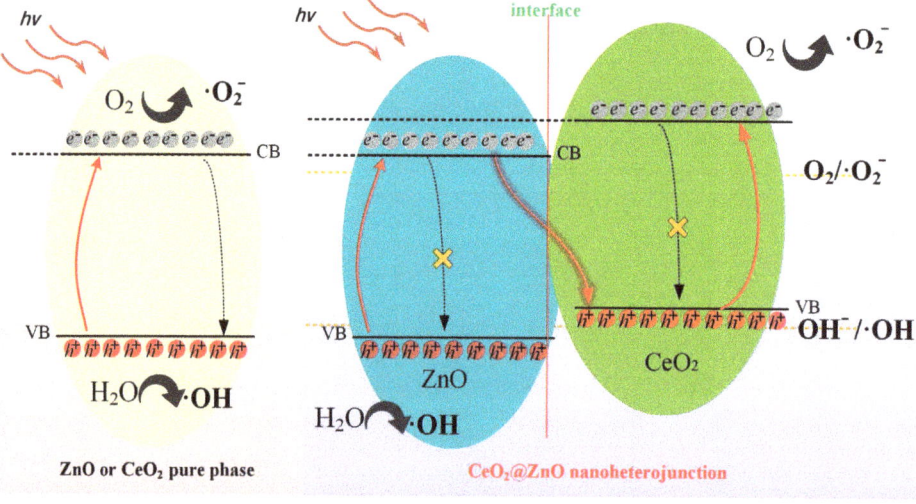

Figure 10. Separation mechanisms of photogenerated electron–hole pairs of the pure phase materials and the CeO$_2$@ZnO nanoheterojunction.

4. Conclusions

In this study, CeO_2@ZnO nanocomposites with various Zn:Ce ratios were prepared via the pyrolysis of Ce/Zn-MOFs precursors. As the Zn:Ce ratio increases from zero to one, pure CeO_2, CeO_2@ZnO nanocomposites, and pure ZnO are obtained. Pure CeO_2 exists as nanospheres with diameters of approximately 800 nm. With an increase in the Zn:Ce ratio, the CeO_2@ZnO nanocomposites gradually transform from nanospheres to nanoparticles of approximately 10 nm diameter, increasing the specific surface area. In addition, a heterojunction is formed, as evidenced by TEM and XPS analysis. The optical band gaps of the nanocomposites widen with an increase in the Zn:Ce ratio owing to the heterojunction interface and quantum size effects. Among the produced photocatalysts, the CeO_2@ZnO nanocomposite with a Zn:Ce ratio of 0.67 exhibits the best photocatalytic efficiency, which is higher than that of a CeO_2@ZnO nanocomposite with a Zn:Ce ratio of 1. In addition, this work can be extended to the preparation of other metal oxide nanocomposites, and excellent photocatalytic performance can be obtained.

Author Contributions: Methodology, X.A., C.L., Y.C. and L.M.; Writing-original draft, X.A., S.Y. and L.M.; Writing-review & editing, X.A.; Investigation, C.L., K.L. and L.M.; Formal analysis, C.L., Y.C. and K.L.; Supervision, L.M.; Project administration, L.M. All authors have read and agreed to the published version of the manuscript.

Funding: This work was jointly supported by the National Natural Science Foundation of China (No. 12204245), the Natural Science Foundation of the Jiangsu Higher Education Institutions of China (Nos. 21KJB140018, 21KJD430006, and 22KJB510030), Applied Fundamental Research Foundation of Nantong City, China (No. JC2021103), and Natural Science Foundation of Nanjing Xiaozhuang University (No. 2020NXY11).

Data Availability Statement: Not applicable.

Conflicts of Interest: The authors declare no conflict of interest.

References

1. Verma, A.K. Sustainable development and environmental ethics. *Int. J. Environ. Sci.* **2019**, *10*, 1–5.
2. Khan, S.A.R.; Sharif, A.; Golpîra, H.; Kumar, A. A green ideology in Asian emerging economies: From environmental policy and sustainable development. *Sustain. Dev.* **2019**, *27*, 1063–1075. [CrossRef]
3. Saleh, T.A. Global trends in technologies and nanomaterials for removal of sulfur organic compounds: Clean energy and green environment. *J. Mol. Liq.* **2022**, *359*, 119340. [CrossRef]
4. Gong, J.; Li, C.; Wasielewski, M.R. Advances in solar energy conversion. *Chem. Soc. Rev.* **2019**, *48*, 1862–1864. [CrossRef] [PubMed]
5. Hayat, M.B.; Ali, D.; Monyake, K.C.; Alagha, L.; Ahmed, N. Solar energy-A look into power generation, challenges, and a solar-powered future. *Int. J. Energy Res.* **2019**, *43*, 1049–1067. [CrossRef]
6. Fang, Z.; Hu, X.; Yu, D. Integrated Photo-Responsive Batteries for Solar Energy Harnessing: Recent Advances, Challenges, and Opportunities. *ChemPlusChem* **2020**, *85*, 600–612. [CrossRef] [PubMed]
7. Mevada, D.; Panchal, H.; ElDinBastawissi, H.A.; Elkelawy, M.; Sadashivuni, K.; Ponnamma, D.; Thakar, N.; Sharshir, S.W. Applications of evacuated tubes collector to harness the solar energy: A review. *Int. J. Ambient. Energy* **2022**, *43*, 344–361. [CrossRef]
8. Nagadurga, T.; Narasimham, P.V.R.L.; Vakula, V.S. Harness of maximum solar energy from solar PV strings using particle swarm optimisation technique. *Int. J. Ambient. Energy* **2021**, *42*, 1506–1515. [CrossRef]
9. Esswein, A.J.; Nocera, D.G. Hydrogen production by molecular photocatalysis. *Chem. Rev.* **2007**, *107*, 4022–4047. [CrossRef]
10. Preethi, V.; Kanmani, S. Photocatalytic hydrogen production. *Mater. Sci. Semicond. Process.* **2013**, *16*, 561–575. [CrossRef]
11. Guo, S.; Li, X.; Li, J.; Wei, B. Boosting photocatalytic hydrogen production from water by photothermally induced biphase systems. *Nat. Commun.* **2021**, *12*, 1343. [CrossRef] [PubMed]
12. Du, C.; Wang, X.; Chen, W.; Feng, S.; Wen, J.; Wu, Y.A. CO_2 transformation to multicarbon products by photocatalysis and electrocatalysis. *Mater. Today Adv.* **2020**, *6*, 100071. [CrossRef]
13. Ola, O.; Maroto-Valer, M.M. Review of material design and reactor engineering on TiO_2 photocatalysis for CO_2 reduction. *J. Photochem. Photobiol. C Photochem. Rev.* **2015**, *24*, 16–42. [CrossRef]
14. Tseng, I.-H.; Wu, J.C.; Chou, H.-Y. Effects of sol-gel procedures on the photocatalysis of Cu/TiO_2 in CO_2 photoreduction. *J. Catal.* **2004**, *221*, 432–440. [CrossRef]
15. Koe, W.S.; Lee, J.W.; Chong, W.C.; Pang, Y.L.; Sim, L.C. An overview of photocatalytic degradation: Photocatalysts, mechanisms, and development of photocatalytic membrane. *Environ. Sci. Pollut. Res.* **2020**, *27*, 2522–2565. [CrossRef] [PubMed]

16. Sakkas, V.A.; Islam, M.A.; Stalikas, C.; Albanis, T.A. Photocatalytic degradation using design of experiments: A review and example of the Congo red degradation. *J. Hazard. Mater.* **2010**, *175*, 33–44. [CrossRef]
17. Yang, L.; Liya, E.Y.; Ray, M.B. Degradation of paracetamol in aqueous solutions by TiO_2 photocatalysis. *Water Res.* **2008**, *42*, 3480–3488. [CrossRef]
18. Cheng, L.; Xiang, Q.; Liao, Y.; Zhang, H. CdS-based photocatalysts. *Energy Environ. Sci.* **2018**, *11*, 1362–1391. [CrossRef]
19. Ma, L.; Ai, X.; Yang, X.; Cao, X.; Han, D.; Song, X.; Jiang, H.; Yang, W.; Yan, S.; Wu, X. Cd (II)-based metal-organic framework-derived CdS photocatalysts for enhancement of photocatalytic activity. *J. Mater. Sci.* **2021**, *56*, 8643–8657. [CrossRef]
20. Ai, X.; Yan, S.; Ma, L. Morphologically Controllable Hierarchical ZnO Microspheres Catalyst and Its Photocatalytic Activity. *Nanomaterials* **2022**, *12*, 1124. [CrossRef]
21. Hezam, A.; Namratha, K.; Drmosh, Q.A.; Ponnamma, D.; Wang, J.; Prasad, S.; Ahamed, M.; Cheng, C.; Byrappa, K. CeO_2 nanostructures enriched with oxygen vacancies for photocatalytic CO_2 reduction. *ACS Appl. Nano Mater.* **2019**, *3*, 138–148. [CrossRef]
22. Nakata, K.; Fujishima, A. TiO_2 photocatalysis: Design and applications. *J. Photochem. Photobiol. C Photochem. Rev.* **2012**, *13*, 169–189. [CrossRef]
23. Peleyeju, M.G.; Viljoen, E.L. WO_3-based catalysts for photocatalytic and photoelectrocatalytic removal of organic pollutants from water–A review. *J. Water Process Eng.* **2021**, *40*, 101930. [CrossRef]
24. Wen, J.; Xie, J.; Chen, X.; Li, X. A review on g-C_3N_4-based photocatalysts. *Appl. Surf. Sci.* **2017**, *391*, 72–123. [CrossRef]
25. Abhilash, M.R.; Akshatha, G.; Srikantaswamy, S. Photocatalytic dye degradation and biological activities of the Fe_2O_3/Cu_2O nanocomposite. *RSC Adv.* **2019**, *9*, 8557. [CrossRef]
26. Liao, S.; Donggen, H.; Yu, D.; Su, Y.; Yuan, G. Preparation and characterization of ZnO/TiO_2, SO_4^{2-}/ZnO/TiO_2 photocatalyst and their photocatalysis. *J. Photochem. Photobiol. A Chem.* **2004**, *168*, 7–13. [CrossRef]
27. Srivastava, S.; Yadav, R.K.; Pande, P.P.; Singh, S.; Chaubey, S.; Singh, P.; Gupta, S.K.; Gupta, S.; Kim, T.W.; Tiwary, D. Dye degradation and sulfur oxidation of methyl orange and thiophenol using newly designed nanocomposite GQDs/NiSe-NiO photocatalyst under homemade LED light. In *Photochemistry and Photobiology*; Wiley: Hoboken, NJ, USA, 2022. [CrossRef]
28. Qamar, M.A.; Shahid, S.; Javed, M.; Iqbal, S.; Sher, M.; Bahadur, A.; AL-Anazy, M.M.; Laref, A.; Li, D. Designing of highly active g-C_3N_4/Ni-ZnO photocatalyst nanocomposite for the disinfection and degradation of the organic dye under sunlight radiations. *Colloids Surf. A Physicochem. Eng. Asp.* **2021**, *614*, 126176. [CrossRef]
29. He, H.; Lin, J.; Fu, W.; Wang, X.; Wang, H.; Zeng, Q.; Gu, Q.; Li, Y.; Yan, C.; Tay, B.K.; et al. MoS_2/TiO_2 edge-on heterostructure for efficient photocatalytic hydrogen evolution. *Adv. Energy Mater.* **2016**, *6*, 1600464. [CrossRef]
30. Ma, L.; Ai, X.; Jiang, W.; Liu, P.; Chen, Y.; Lu, K.; Song, X.; Wu, X. Zn/Ce metal-organic framework-derived ZnO@CeO_2 nano-heterojunction for enhanced photocatalytic activity. *Colloid Interface Sci. Commun.* **2022**, *49*, 100636. [CrossRef]
31. Kumar, P.; Kumar, P.; Kumar, A.; Meena, R.C.; Tomar, R.; Chand, F.; Asokan, K. Structural, morphological, electrical and dielectric properties of Mn doped CeO_2. *J. Alloys Compd.* **2016**, *672*, 543–548. [CrossRef]
32. Kaur, P.; Rani, S.; Lal, B. Excitation dependent photoluminescence properties of ZnO nanophosphor. *Optik* **2019**, *192*, 162929. [CrossRef]
33. Rajendran, S.; Khan, M.M.; Gracia, F.; Qin, J.; Gupta, V.K.; Arumainathan, S. Ce^{3+}-ion-induced visible-light photocatalytic degradation and electrochemical activity of ZnO/CeO_2 nanocomposite. *Sci. Rep.* **2016**, *6*, 31641. [CrossRef] [PubMed]
34. Zhu, L.; Li, H.; Xia, P.; Liu, Z.; Xiong, D. Hierarchical ZnO decorated with CeO_2 nanoparticles as the direct Z-scheme heterojunction for enhanced photocatalytic activity. *ACS Appl. Mater. Interfaces* **2018**, *10*, 39679–39687. [CrossRef] [PubMed]
35. Zeng, C.H.; Xie, S.; Yu, M.; Yang, Y.; Lu, X.; Tong, Y. Facile synthesis of large-area CeO_2/ZnO nanotube arrays for enhanced photocatalytic hydrogen evolution. *J. Power Sources* **2014**, *247*, 545–550. [CrossRef]
36. Zhu, L.; Li, H.; Liu, Z.; Xia, P.; Xie, Y.; Xiong, D. Synthesis of the 0D/3D CuO/ZnO heterojunction with enhanced photocatalytic activity. *J. Phys. Chem. C* **2018**, *122*, 9531–9539. [CrossRef]
37. Khan, M.M.; Ansari, S.A.; Pradhan, D.; Han, D.H.; Lee, J.; Cho, M.H. Defect-induced band gap narrowed CeO_2 nanostructures for visible light activities. *Ind. Eng. Chem. Res.* **2014**, *53*, 9754–9763. [CrossRef]
38. Mu, J.; Shao, C.; Guo, Z.; Zhang, Z.; Zhang, M.; Zhang, P.; Chen, B.; Liu, Y. High photocatalytic activity of ZnO-carbon nanofiber heteroarchitectures. *ACS Appl. Mater. Interfaces* **2011**, *3*, 590–596. [CrossRef]
39. Ma, L.; Ma, S.; Chen, H.; Ai, X.; Huang, X. Microstructures and optical properties of Cu-doped ZnO films prepared by radio frequency reactive magnetron sputtering. *Appl. Surf. Sci.* **2011**, *257*, 10036–10041. [CrossRef]
40. Xiao, Y.; Yu, H.; Dong, X.T. Ordered mesoporous CeO_2/ZnO composite with photodegradation concomitant photocatalytic hydrogen production performance. *J. Solid State Chem.* **2019**, *278*, 120893. [CrossRef]
41. Li, C.; Chen, R.; Zhang, X.; Shu, S.; Xiong, J.; Zheng, Y.; Dong, W. Electrospinning of CeO_2-ZnO composite nanofibers and their photocatalytic property. *Mater. Lett.* **2011**, *65*, 1327. [CrossRef]
42. Zhang, Q.; Zhao, X.; Duan, L.; Shen, H.; Liu, R. Controlling oxygen vacancies and enhanced visible light photocatalysis of CeO_2/ZnO nanocomposites. *J. Photochem. Photobiol. A Chem.* **2020**, *392*, 112156. [CrossRef]
43. Huang, L.; Bao, D.; Jiang, X.; Li, J.; Zhang, L.; Sun, X. Fabrication of stable high-performance urchin-like CeO_2/ZnO@Au hierarchical heterojunction photocatalyst for water remediation. *J. Colloid Interface Sci.* **2021**, *588*, 713. [CrossRef] [PubMed]

44. Luo, K.; Li, J.; Hu, W.; Li, H.; Zhang, Q.; Yuan, H.; Yu, F.; Xu, M.; Xu, S. Synthesizing CuO/CeO$_2$/ZnO ternary nano-photocatalyst with highly effective utilization of photo-excited carriers under sunlight. *Nanomaterials* **2020**, *10*, 1946. [CrossRef] [PubMed]
45. Shah, N.; Bhangaonkar, K.; Pinjari, D.V.; Mhaske, S.T. Ultrasound and conventional synthesis of CeO$_2$/ZnO nanocomposites and their application in the photocatalytic degradation of rhodamine B dye. *J. Adv. Nanomater.* **2017**, *2*, 133. [CrossRef]

Disclaimer/Publisher's Note: The statements, opinions and data contained in all publications are solely those of the individual author(s) and contributor(s) and not of MDPI and/or the editor(s). MDPI and/or the editor(s) disclaim responsibility for any injury to people or property resulting from any ideas, methods, instructions or products referred to in the content.

Article

Effect of Doping TiO$_2$ NPs with Lanthanides (La, Ce and Eu) on the Adsorption and Photodegradation of Cyanide—A Comparative Study

Ximena Jaramillo-Fierro [1,*] and Ricardo León [2]

[1] Departamento de Química, Facultad de Ciencias Exactas y Naturales, Universidad Técnica Particular de Loja, San Cayetano Alto, Loja 1101608, Ecuador
[2] Maestría en Química Aplicada, Facultad de Ciencias Exactas y Naturales, Universidad Técnica Particular de Loja, San Cayetano Alto, Loja 1101608, Ecuador
* Correspondence: xvjaramillo@utpl.edu.ec; Tel.: +593-7-3701444

Abstract: Free cyanide is a highly dangerous compound for health and the environment, so treatment of cyanide-contaminated water is extremely important. In the present study, TiO$_2$, La/TiO$_2$, Ce/TiO$_2$, and Eu/TiO$_2$ nanoparticles were synthesized to assess their ability to remove free cyanide from aqueous solutions. Nanoparticles synthesized through the sol–gel method were characterized by X-ray powder diffractometry (XRD), scanning electron microscopy (SEM), energy-dispersive X-ray spectroscopy (EDS), Fourier-transformed infrared spectroscopy (FTIR), diffuse reflectance spectroscopy (DRS), and specific surface area (SSA). Langmuir and Freundlich isotherm models were utilized to fit the adsorption equilibrium experimental data, and pseudo-first-order, pseudo-second-order, and intraparticle diffusion models were used to fit the adsorption kinetics experimental data. Cyanide photodegradation and the effect of reactive oxygen species (ROS) on the photocatalytic process were investigated under simulated solar light. Finally, reuse of the nanoparticles in five consecutive treatment cycles was determined. The results showed that La/TiO$_2$ has the highest percentage of cyanide removal (98%), followed by Ce/TiO$_2$ (92%), Eu/TiO$_2$ (90%), and TiO$_2$ (88%). From these results, it is suggested that La, Ce, and Eu dopants can improve the properties of TiO$_2$ as well as its ability to remove cyanide species from aqueous solutions.

Keywords: cyanide; adsorption; photocatalysis; anatase; lanthanide doping; nanoparticles

Citation: Jaramillo-Fierro, X.; León, R. Effect of Doping TiO$_2$ NPs with Lanthanides (La, Ce and Eu) on the Adsorption and Photodegradation of Cyanide—A Comparative Study. *Nanomaterials* 2023, 13, 1068. https://doi.org/10.3390/nano13061068

Academic Editors: Thomas Dippong and Marco Stoller

Received: 11 February 2023
Revised: 10 March 2023
Accepted: 11 March 2023
Published: 16 March 2023

Copyright: © 2023 by the authors. Licensee MDPI, Basel, Switzerland. This article is an open access article distributed under the terms and conditions of the Creative Commons Attribution (CC BY) license (https://creativecommons.org/licenses/by/4.0/).

1. Introduction

Cyanides are highly toxic chemical compounds that can cause severe harm to health and the environment [1]. The most common cyanide compounds in the environment are present both in their free form, which comprises the cyanide ion itself (CN$^-$) and hydrogen cyanide (HCN), and as water-soluble inorganic salts, including sodium cyanide (NaCN) and potassium cyanide (KCN) [2]. Some cyanide compounds are naturally produced by microorganisms, although they can also be found in plants and some foods, as well as in low concentrations in soil and water. On the other hand, generation of effluents by industries is the main source of release into the environment of cyanide complexes, which are formed by union of cyanide ions with metal ions [3].

All forms of cyanide can be lethal at high levels of exposure; however, free cyanide (HCN + CN$^-$) is the deadliest form of all [4]. This form of cyanide in a solution can be removed and/or transformed to fewer toxic forms through various physicochemical methods, such as acidification, photocatalysis, adsorption, coagulation, reverse osmosis, precipitation, filtration, and AVR (acidification, volatilization, and reneutralization) process [5,6]. Among these methods, adsorption and photocatalysis have shown to be two promising methodologies due to their high efficiency, simplicity of design, ease of handling, and low operating cost [7,8]. The adsorption process can be carried out by different mechanisms, including electrostatic interaction, ion exchange, and complexation [9]. Unfortunately, this process only allows for transfer of the contaminant from one medium to another without completely removing it from the environment. However, complete degradation of the adsorbed pollutant can be achieved through the series of oxidation–reduction reactions that take place in the photocatalytic process [10–13].

Photocatalysis applied to aqueous or gaseous systems involves formation of reactive oxygen species (ROS), such as superoxide ($^\bullet O_2^-$), hydroxyl radical ($^\bullet OH$), and singlet oxygen ($^1O_2$), which are produced by electron capture by oxygen or oxidation of water molecules [2]. ROS are very effective oxidizing agents, capable of degrading recalcitrant compounds, and can also be produced on the surface of semiconductor materials, such as titanium dioxide (TiO_2) [14]. TiO_2 is widely used as a photocatalyst because it is cheap, abundant, chemically inactive, and photostable to corrosion; it also has high thermal stability, intense photocatalytic activity, strong oxidizing power, and is friendly to the environment [15]. TiO_2 can produce ROS when it accepts a minimum quantity of energy (bandgap energy, E_g) that allows it to remove an electron from the valence band (VB) and transport it to the conduction band (CB), thus generating an electron/hollow (e^-/h^+) [16]. Unfortunately, the bandgap energy (E_g = 3.2 eV) required by TiO_2 for pair formation (e^-/h^+) restricts use of this semiconductor to wavelengths in the UV region ($\approx$ 390 nm), which represents only 5% of solar light [17]. Another important limitation of TiO_2 is its high recombination of photogenerated charges, which diminishes formation of ROS and, consequently, its photocatalytic efficiency [18].

In order to decrease recombination of the pairs (e^-/h^+) and shift the absorption wavelength to the wanted visible region (>λ = 400 nm), several specific surface modifications on TiO_2 have been suggested, including doping with different elements, including lanthanides [19–23]. In fact, lanthanide metal ions (Ln^{3+}) have been employed as TiO_2 dopants in numerous investigations since they can considerably modify the physicochemical properties of this photocatalyst [24]. Doping of TiO_2 with lanthanide elements can occur through two mechanisms. In one mechanism, elements can be incorporated in the TiO_2 lattice by direct bonding or substitution to produce a $\equiv$Ti–O–Ln–O–Ti$\equiv$ arrangement, causing lattice defects/distortions due to the difference in ionic radii of Ln^{3+} and Ti^{4+}. In the other mechanism, it is considered that there is not enough energy to promote ion substitution in the lattice, so the lanthanide elements can disperse on the TiO_2 surface, creating Ti–O–Ln bonds [25]. Presence of lanthanides, scattered as surface impurities in the TiO_2 structure, causes vacancies and surface defects that enable capture of the electrons produced on the surface of the photocatalyst, thus reducing recombination of photogenerated charge carriers [26]. However, an excess of lanthanide dopants on the TiO_2 surface can reduce its photocatalytic capacity, possibly due to generation of elevated density of defects and vacancies, which would behave as recombination centers and not as electron collectors [27].

Lanthanide elements have motivated numerous investigations due to their potential industrial applications, derived from their exceptional optical, magnetic, and/or redox properties. Lanthanides, in particular lanthanum (La), cerium (Ce), and europium (Eu), when applied as TiO_2 dopants, inhibit transition from the anatase phase to the rutile phase, cause lattice distortions in the surface layer, generate defects that reduce crystallite size, increase specific surface area, improve thermal stability, increase the number of oxygen vacancies, readily react with organic compounds to mineralize them, and can also provide luminescent properties to TiO_2 [28,29]. Lanthanum, cerium, and europium have been studied for their capacity to enhance the photocatalytic activity of TiO_2, probably due to the presence of two events: a decrease in their bandgap energy and generation of an imbalance in their surface charges [30,31]. Both events lead to states that contribute to capture of photogenerated electrons and reduce recombination of pairs (e^-/h^+), increasing the probability of formation of hydroxyl radicals ($^\bullet OH$) that are essential for photodegradation of organic molecules [32]. Presence of these elements on the TiO_2 surface leads to generation of Lewis acid sites that ensure catalytic stability of semiconductors in aqueous reaction media [25,33]. Likewise, Lewis acid sites enable increasing the adsorption of contaminant molecules on the TiO$_2$ surface, which improves transfer of electrons for direct degradation of the pollutant and increases the possibility of interaction between contaminant molecules and photogenerated radicals [34].

As can be seen, doping with lanthanide elements has demonstrated to be effective in adapting the TiO$_2$ surface to several applications [35]. However, almost surprisingly, no investigations have been found in the literature on use of TiO$_2$ doped with La, Ce, or Eu for adsorption and photocatalytic degradation of total cyanide in aqueous systems. Therefore, the objective of this comparative study is to evaluate the effect of La, Ce, and Eu ions on the adsorbent and photocatalytic capacity of TiO$_2$ for its application in removal of total cyanide from aqueous solutions. Thus, in the present study, adsorption and photocatalysis experiments were designed in batch reactors, where the amount of residual total cyanide (HCN + CN$^-$) in the solutions was determined by UV–visible spectrophotometry using the picrate alkaline method. The adsorption capacity of the synthesized nanoparticles was evaluated by modifying the composition of the doping element, the pH of the solutions, the initial concentration of the adsorbate, and the contact time, while the photocatalytic activity was determined by simulated solar irradiation (λ = 300–800 nm). The synthesized nanoparticles (TiO$_2$, La/TiO$_2$, Ce/TiO$_2$, and Eu/TiO$_2$) were characterized by X-ray powder diffractometry (XRD), Fourier-transformed infrared spectroscopy (FTIR), diffuse reflectance spectroscopy (DRS), scanning electron microscopy (SEM), energy-dispersive X-ray spectroscopy (EDS), and specific surface area (SSA) by the BET method.

The novelty of this comparative study lies in the fact that it was possible to demonstrate that doping of TiO$_2$ with lanthanide elements, particularly Lanthanum, enables efficient removal of cyanide species from aqueous systems at neutral pH thanks to the effective combination of the processes of adsorption and photocatalysis. The materials synthesized in this study could represent an important innovation in environmental technology due to their operability and reusability, thus providing a sustainable alternative with great application potential in effluent treatment.

2. Materials and Methods

2.1. Materials

The following reagents (analytical grade) were used in the present investigation without any further purification: Europium(III) nitrate pentahydrate (Eu(NO$_3$)$_3$·5H$_2$O, Sigma Aldrich, St. Louis, MO, USA, 99.9%), Cerium(III) nitrate hexahydrate (Ce(NO$_3$)$_3$·6H$_2$O, Sigma Aldrich, St. Louis, MO, USA, 99.9%), Lanthanum nitrate hexahydrate (La(NO$_3$)$_3$·6H$_2$O, Sigma Aldrich, St. Louis, MO, USA, 99.9%), Titanium (IV) isopropoxide (Ti(OC$_3$H$_7$)$_4$, Sigma Aldrich, St. Louis, MO, USA, 98.0%), Isopropyl alcohol (C$_3$H$_8$O, Sigma Aldrich, St. Louis, MO, USA, $\geq$99.5%), Hydrochloric acid (HCl, Sigma Aldrich, St. Louis, MO, USA, 37.0%), Sodium hydroxide (NaOH, Sigma Aldrich, St. Louis, MO, USA, $\geq$85.0%), Potassium cyanide (KCN, Sigma Aldrich, St. Louis, MO, USA, $\geq$97.0%), Picric acid ((O$_2$N)$_3$C$_6$H$_2$OH, Sigma Aldrich, St. Louis, MO, USA, $\geq$99.0%), Sodium carbonate (Na$_2$CO$_3$, Sigma Aldrich, St. Louis, MO, $\geq$99.0%), p-Benzoquinone (C$_6$H$_4$(=O)$_2$, Sigma Aldrich, St. Louis, MO, USA, $\geq$98.0%), Ethanol (CH$_3$CH$_2$OH, Sigma Aldrich, St. Louis, MO, USA, 95.0%).

2.2. Synthesis of the Nanoparticles

The TiO$_2$ (TO), La/TiO$_2$ (La/TO), Ce/TiO$_2$ (Ce/TO), and Eu/TiO$_2$ (Eu/TO) nanoparticles were synthesized following an adapted sol-gel method explained in previous studies [36]. To obtain the TiO$_2$ nanoparticles, a solution (Solution A) of titanium (IV) isopropoxide (TiPO) dissolved in isopropyl alcohol (iPrOH) at a TiPO/iPrOH ratio of 70% v/v was prepared at room temperature. A solution (Solution B) of isopropyl alcohol (iPrOH), iPrOH/water (50% v/v) was also prepared at room temperature. Solution B was gradually added dropwise to solution A, maintaining constant stirring and room temperature. After formation of a white precipitate, the reaction system was stirred for 60 min at room temperature. The precipitate was dried (60 °C) for 24 h and then calcined (500 °C) for 4 h. Finally, the obtained solids were cooled to room temperature. To achieve the La/TO,

Ce/TO, and Eu/TO doped nanoparticles, the procedure described above was repeated, adding the lanthanum, cerium, or europium salts to the aqueous solution (Solution B) to achieve a final concentration of the doping element of ~1% per gram of TiO_2.

2.3. Characterization of the Nanoparticles

Characterization of the nanoparticles was performed using the methodology described in our previous study [36]. For X-ray diffraction (XRD) measurements, a Bruker-AXS D8-Discover diffractometer (Bruker AXS, Karlsruhe, Germany) was used. For specific surface area (SSA) determination, ChemiSorb 2720 equipment (Micromeritics, Norcross, GA, USA) was used. Micrographs of the synthesized samples were obtained by field effect scanning electron microscopy (SEM) on a Zeiss Gemini ULTRA plus electron microscope (Carl Zeiss AG, Ober-kochen, Germany). A Phenom ProX (Phenom World BV, Eindhoven, The Netherlands) was used to achieve energy-dispersive X-ray (EDS) spectra. In the present study, quantification of the elements in the synthesized nanoparticles was also performed by inductively coupled plasma (ICP) optical emission spectrometry (OES) in an Optima 8000 ICP-OES Spectrometer (PerkinElmer, Inc., Waltham, MA, USA). FTIR spectra were recorded on a PerkinElmer GX2000-FTIR Spectrometer (PerkinElmer, Inc., Waltham, MA, USA). The diffuse UV–vis reflectance spectra (DRS) were obtained using a Phenom DesNicolet Evolution 201/220 Thermo UV–vis spectrophotometer (ThermoFisher, Waltham, MA, USA). To simulate solar light and evaluate the photoactivity of the nanoparticles, a solar box ATLAS, SUNTEST CPS+ (Atlas Material Testing Technology, Mount Prospect, IL, USA) was used. Finally, the amount of cyanide remaining in the solutions was quantified using a Jenway 7350 spectrophotometer (Cole-Parmer, Staffordshire, UK). IBM SPSS (version 25.0; statistic software for Windows; IBM Corp.; Armonk, NY, USA, 2017) was used for the ANOVA analysis. The crystalline phases were recognized using the ICDD database (International Center for Diffraction Data, version 2018). Finally, the Chemisoft TPx system (version 1.03; data analysis software; Micromeritics, Norcross, GA, USA, 2011) was used to calculate the SSA by the single-point method using the Brunauer–Emmet–Teller (BET) equation.

2.4. Adsorption Studies

Cyanide adsorption experiments from KCN aqueous solutions were designed to evaluate the following aspects: (a) the effect of pH on cyanide adsorption, (b) maximum cyanide adsorption capacity, (c) adsorption thermodynamics and (d) the kinetic behavior of cyanide adsorption. The data obtained from the adsorption experiments were fitted to isothermal and kinetic models using the least squares nonlinear regression method [37]. Experiments were performed with all nanoparticles using the methodology described in our previous study [36]. Cyanide species adsorption experiments were performed using a batch method at room temperature and keeping the pH of the solutions at 7.0 ± 0.1 by adding 0.1 M solutions of hydrochloric acid or sodium hydroxide. The amount of catalyst used in all the experiments was 0.2 g L^{-1}. The maximum cyanide adsorption capacity was investigated by varying the concentration of 500 mL of KCN solution from 0.20 to 40 mg L^{-1}. The adsorption thermodynamics and kinetic behavior of cyanide adsorption were investigated using 500 mL of water containing 20 mg L^{-1} of KCN [38]. The alkaline picrate analytical method was used to quantify total cyanide in aqueous solutions [39–41]. To accomplish this, first, the alkaline picrate solution (PAS) was prepared, consisting of 1 g of picric acid (($O_2N)_3C_6H_2OH$) and 5 g of sodium carbonate (Na_2CO_3) dissolved in 200 mL of HPLC water. A volume of 4 mL of this solution was added to 1 mL of cyanide or blank solution (HPLC water) contained in a test tube. This tube was incubated for 5 min in a 95 °C water bath. After this time, the absorbance at 490 nm was measured using a UV–vis spectrophotometer. The concentration of the solutions was determined based on the previously prepared calibration curve ($R^2 = 0.9996$) according to the Lambert–Beer Law. The tests were carried out in triplicate and the results were expressed as the average of three repetitions [38]. The amount of cyanide adsorbed (q_e) on the nanoparticles expressed in mg g^{-1} was estimated using the following equation [42]:

$$q_e = (C_0 - C_e) \times \frac{v}{w} \quad (1)$$

where C_0 and C_e are expressed in mg L^{-1} and correspond to the initial and equilibrium concentration, respectively. The mass (w) of the adsorbent is expressed in grams (g), and the volume (v) of the solution is expressed in liters (L).

The equilibrium total cyanide adsorption was assessed based on the Langmuir and Freundlich isotherm models. The expression of the Langmuir isotherm model can be represented using the following equation [43]:

$$\frac{C_e}{q_e} = \frac{1}{K_L q_{max}} + \frac{C_e}{q_{max}} \quad (2)$$

where q_{max} is expressed in mg g^{-1} and corresponds to the maximum monolayer adsorption, K_L is expressed in L mg^{-1} and corresponds to the equilibrium Langmuir constant related to the adsorption energy, and C_e is expressed in mg L^{-1} and corresponds to the concentration of solute at equilibrium. Additionally, the R_L separation factor values, which offer an idea of the adsorption characteristics, can be represented using the following equation [43]:

$$R_L = \frac{1}{(1 + K_L C_e)} \quad (3)$$

The suitability of the adsorption treatment based on the R_L value is given as follows: $0 < R_L < 1$; it denotes suitable adsorption, $R_L > 1$ unsuitable adsorption, $R_L = 0$ irreversible adsorption, and $R_L = 1$ linear adsorption [43].

Moreover, the Freundlich isotherm model can be represented by the following equation [43]:

$$q_e = K_F C_e^{1/n} \quad (4)$$

where K_F is expressed in L mg^{-1} and corresponds to the Freundlich constant, which specifies the adsorption affinity of the adsorbents, and $1/n$ is another constant that corresponds to the adsorption intensity. For favorable adsorption, the value of the constant n should be in the range of 1 to 10 [44].

For the thermodynamic studies, the experimental data were fitted according to the parameters of the thermodynamic laws described by Gibbs free energy (ΔG^0, kJ mol^{-1}), enthalpy (ΔH^0, kJ mol^{-1}), and entropy (ΔS^0, kJ mol^{-1} K^{-1}) conventionally used, represented by the following equation [45]

$$\Delta G^0 = -RT \ln k_C \quad (5)$$

The relationship between ΔG^0, ΔH^0, and ΔS^0, is obtained by the well-known van 't Hoff equation [45]

$$\ln k_C = \frac{-\Delta H^0}{R} \times \frac{1}{T} + \frac{\Delta S^0}{R} \quad (6)$$

where k_L (L mg^{-1}) is the Langmuir constant and could be achieved as a dimensionless parameter. T is the absolute temperature (K), and R is the universal gas constant (8.314 J mol^{-1} K^{-1}). k_C is achieved as a dimensionless parameter by multiplying k_L by a molecular weight of an adsorbate (M_w, g mol^{-1}) and then by factors 1000 and 55.5, which is the number of moles of pure water contained in a liter, described by the following equation [46].

$$k_C = k_L \times M_w \times 1000 \times 55.5 \quad (7)$$

The absorption kinetics were estimated by applying pseudo-first-order and pseudo-second-order models, as well as intraparticle diffusion, external-film diffusion, and internal-pore diffusion models. The pseudo-first-order kinetic model is represented by the following equation [44]:

$$\ln(q_e - q_t) = \ln(q_e) - k_1 t \quad (8)$$

where k_1 is the rate constant (min^{-1}) and q_e and q_t are the cyanide adsorbed per unit weight (mg g^{-1}) at equilibrium and at any time t, respectively.

The pseudo-second-order kinetic is represented by the following equation [44]:

$$\frac{t}{q_t} = \frac{1}{k_2 q_e^2} + \frac{1}{q_e} t \qquad (9)$$

where k_2 is the pseudo-second-order rate constant (g mg^{-1} min^{-1}).

Finally, to obtain an appropriate understanding of the cyanide adsorption mechanism on the nanoparticles surface, the rate-limiting step in the adsorption process was also estimated. The intraparticle diffusion model supposes that intraparticle diffusion is generally the rate-controlling step in well-mixed solutions. The intraparticle-diffusion model is represented by the following equation [44]:

$$q_t = k_3 t^{1/2} + A \qquad (10)$$

where k_3 (mg g^{-1} min$^{-1/2}$) is the intraparticle diffusion rate constant and A (mg g^{-1}) is a constant indicating the width of the boundary layer; that is, the larger the value of A, the greater the boundary layer effect. When the q_t plot against the square root of time shows multilinearity, it means that the diffusion occurs in several steps during the process.

The internal pore diffusion model was also utilized in this study to explain the kinetic adsorption data. When the adsorption rate is controlled by particle diffusion, the adsorption rate is represented using the following equation [44]:

$$-\ln\left(1 - \left(\frac{q_t}{q_e}\right)^2\right) = \frac{2\pi^2 D_p}{r^2} t \qquad (11)$$

On the other hand, regarding external-film-diffusion controls, the adsorption rate is expressed by the following equation [44]:

$$-\ln\left(1 - \left(\frac{q_t}{q_e}\right)\right) - \frac{D_f C_s}{h\, r\, C_z} t \qquad (12)$$

where q_e and q_t represent the amount of solute that is taken up by the adsorbent phase at equilibrium and at a specific time t (mg g^{-1}), respectively. The ion concentration in the solution and adsorbent are donated by C_s (mg L^{-1}) and C_z (mg kg^{-1}), respectively. The variable t refers to the contact time (min), while r represents the average radius of the adsorbent particles (1 × 10^{-7} m). The film thickness around the adsorbent particles is denoted by h and is assumed to be 10^{-6} m in poorly stirred solutions. Finally, the diffusion coefficient in the adsorbent phase is referred to as D_p (m^2 min^{-1}), while D_f (m^2 min^{-1}) represents the diffusion in the film phase surrounding the adsorbent particles.

2.5. Photodegradation Studies

Heterogeneous photocatalysis experiments were performed using the methodology described in our previous study [36]. These experiments were performed in an air-cooled solar box. The batch method was used, keeping the pH of the solution at 7.0 ± 0.1 by adding 0.1 M solutions of hydrochloric acid or sodium hydroxide. Typically, 0.2 g L^{-1} of nanoparticles were magnetically stirred in 500 mL of water containing 20 mg L^{-1} of KCN [38]. The photodegradation rate of cyanide species in the heterogeneous photocatalytic systems exposed to simulated solar light for the TO, La/TO, Ce/TO, and Eu/TO nanoparticles was followed by the Langmuir–Hinshelwood equation [47], which can be represented by the following equation [48]:

$$\ln\frac{C_o}{C_t} = k\, K\, t = k_{app} t \qquad (13)$$

where k (min^{-1}) is the actual rate constant, K is the adsorption constant of the substrate on the nanoparticles, C_0 (mg L^{-1}) represents the initial concentration of the substrate, and C_t (mg L^{-1}) represents the concentration at a specific time t (min), and k_{app} (min^{-1}) is the apparent rate constant. Plotting ln(C_0/C_t) against time t provides the apparent rate constant (k_{app}) for substrate degradation from the slope of the curve fit line, and the intercept is equal to zero.

To determine the influence of several reactive oxygen species (ROS) on the reaction system, the radical quenching experiment was performed using isopropanol (i-POH), ethanol (EtOH), and p-benzoquinone (p-BQ). These radical quenchers inhibit h$^+$, •OH, and •O_2^-, respectively. The experiments were carried out following the methodology described for the heterogeneous photocatalysis experiments but each time incorporating the radical quenchers (0.5 mM) in the reaction system [49].

2.6. Reuse of Nanoparticles

Finally, in order to verify the reusability of the nanoparticles in cyanide photodegradation, a recycling experiment was designed using the methodology described in our previous study [36]. The recycling experiment was performed for five consecutive cycles. At the end of each treatment cycle, the suspensions were precipitated by leaving them to stand for at least 1 h. After this time, the supernatant liquid was removed and the nanoparticles were carefully washed three times with HPLC water, avoiding loss of solid material. In each treatment cycle, 100 mL of fresh KCN solution (20 mg L^{-1}) was used. The amount of nanoparticles used in this experiment was 0.2 g L^{-1} [38].

3. Results

3.1. Characterization of the Nanoparticles

3.1.1. XRD and SSA Analysis

Figure 1 shows the diffraction patterns of pristine TiO_2 compared to TiO_2 doped with lanthanum (La/TO), cerium (Ce/TO), and europium (Eu/TO). These nanoparticles were synthesized at 500 °C. The diffraction peaks of anatase phase (TiO_2) at 2θ values of 25.30°, 36.95°, 37.79°, 38.57°, 48.04°, 53.89°, 55.06°, 62.11°, 62.69°, 68.76°, 70.29°, 75.04°, 76.03°, 82.67°, and 83.15° were assigned to planes (1 0 1), (1 0 3), (0 0 4), (1 1 2), (2 0 0), (1 0 5), (2 1 1), (2 1 3), (2 0 4), (1 1 6), (2 2 0), (2 1 5), (3 0 1), (2 2 4), and (3 1 2), respectively. The TiO_2 crystal structure was indexed to tetragonal phase with unit cell parameters a = b = 3.79 Å and c = 9.51 Å, α = β = γ = 90°, unit cell volume = 136.30 Å^3, and space group I41/amd(141) according to standard card JCPDS card No. 01-073-1764. Due to the low concentration of the doping elements, the maximum diffraction peaks reported in the literature are not observed for La_2O_3 (2θ = 29.96°) [50], Ce_2O_3 (2θ = 30.39°) [51], CeO_2 (2θ = 28.57°) [52], nor for europium oxide (2θ = 28.44°) [53].

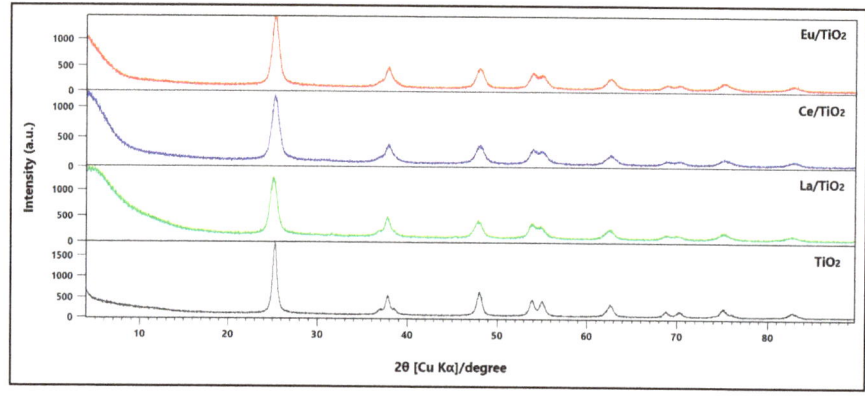

Figure 1. Comparison of XRD patterns of TiO_2, La/TiO_2, Ce/TiO_2, and Eu/TiO_2 nanoparticles.

The crystal sizes of the TO, La/TO, Ce/TO, and Eu/TO nanoparticles were calculated considering the most intense diffraction peak. The Scherrer equation (Equation (14)) was used for the respective calculation [54]

$$A = \frac{K\lambda}{\beta \cos\theta} \qquad (14)$$

where A is the size of the crystals expressed in nm, λ = 0.15406 nm and K = 0.89 represent the wavelength of the X-ray beam and the shape factor, respectively, θ represents the Bragg angle, while β represents the full width at half peak height maximum (FWHM) of the X-ray diffraction peak. The average crystal sizes of the TO, La/TO, Ce/TO, and Eu/TO nanomaterials were calculated at 28.54 ($\pm$0.98), 19.24 ($\pm$0.65), 20.68 ($\pm$0.87), and 21.94 ($\pm$1.07) nm, respectively.

On the other hand, the specific surface area (SSA) was evaluated by the single-point BET (Brunauer–Emmet–Teller) method. The SSA of the pristine TiO_2 (TO) was measured as 88 $m^2\ g^{-1}$, which was slightly lower than that of the La/TO, Ce/TO, and Eu/TO nanoparticles, with values of 126, 104, and 96 $m^2\ g^{-1}$, respectively. From these results, it is suggested that elements lanthanum (La), cerium (Ce), and europium (Eu) have a specific effect on inhibition of growth of TiO_2 crystallites, stabilization of this oxide in its anatase phase, as well as increasing specific surface area (SSA).

3.1.2. FTIR Analysis

FTIR spectroscopy was used for chemical characterization of TO, La/TO, Ce/TO, and Eu/TO nanoparticles. The FTIR spectra of all the synthesized nanoparticles are shown in Figure 2a–d, respectively.

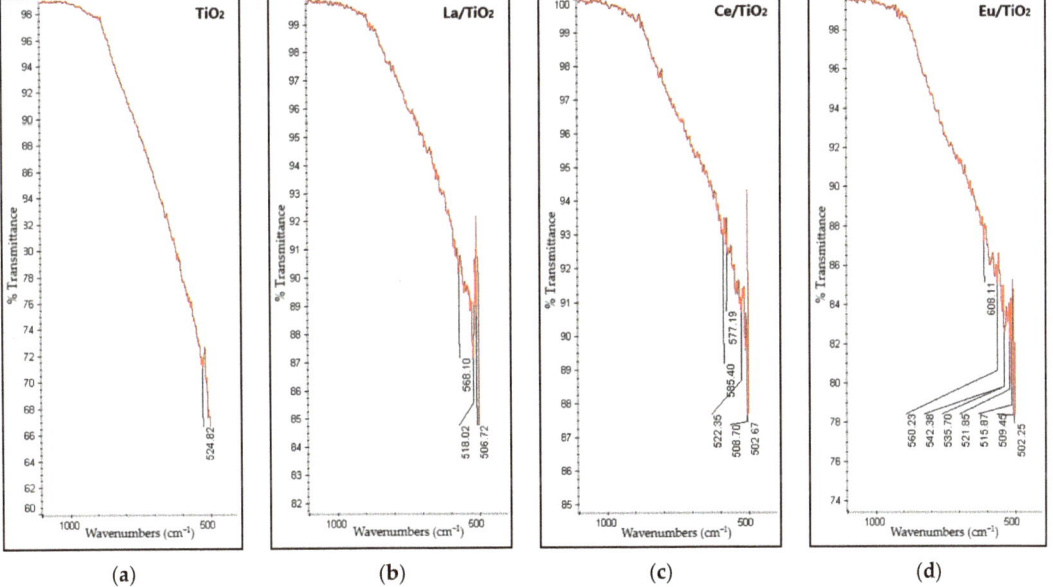

Figure 2. FTIR spectra of (a) TiO_2, (b) La/TiO_2, (c) Ce/TiO_2, and (d) Eu/TiO_2 nanoparticles.

The FTIR spectrum of TiO_2 reveals the existence of a broad band in the region of 540–900 cm^{-1}, which is assigned to Ti–O and Ti–O–Ti stretching vibrations [55,56]. After doping, the main change was identified in this region, presenting new unresolved bands around the 500–600 cm^{-1} region that could be attributed to Ln–O stretching modes of vibrations. The results obtained in this analysis agree with those reported in the literature [57,58].

On the other hand, the spectra in Figure 2 show that there is a tendency to increase the vibrational frequencies as the ionic radius of the doping elements decreases. Evidence from the literature indicates that Ln–O single-bond stretching frequencies and Ln–O single-bond stretching force constants show a linear correlation with inverse ionic radii [59].

3.1.3. SEM and EDS Analysis

Figure 3a–d displays the SEM photomicrographs of TO, La/TO, Ce/TO, and Eu/TO nanoparticles, respectively. In these figures, it is evident that all the nanoparticles appear almost spherical and are highly agglomerated. These results are in agreement with those informed by other authors [60]. Doped nanoparticles are smaller than pristine TO nanoparticles. The average size of the TO nanoparticles was 32 nm, in contrast to La/TO, Ce/TO, and Eu/TO nanoparticles, which had mean sizes of 26, 27, and 28 nm, respectively. The size of the synthesized nanoparticles was measured using ImageJ2, which is public domain software for scientific image processing and analysis [61,62]. The sizes of the nanoparticles are within the range of values reported by other authors [63].

The SEM images presented in this study support the results of the XRD analysis (Figure 1), demonstrating that doping elements lanthanum, cerium, and europium are effective in preventing crystallite growth and stabilizing the structure of the TiO_2 semiconductor.

Likewise, Figure 3a–d displays the EDS spectra of TO, La/TO, Ce/TO, and Eu/TO nanoparticles, respectively. These spectra support the results of the FTIR analysis (Figure 2), thus confirming incorporation of lanthanide elements La (1.32 wt%), Ce (1.34 wt%), and Eu (1.28 wt%) on the TiO_2 structure.

3.1.4. Optical and Photoelectric Properties

The optical absorption capacity of the synthesized nanoparticles was determined by UV–visible (UV–vis) DRS at room temperature in the range of 200 to 700 nm. Figure 4a shows the DRS UV–vis spectrum of TO, La/TO, Ce/TO, and Eu/TO nanoparticles. Comparatively, the absorption behavior in the visible light spectrum (around 400 nm) was slightly enhanced for La/TO, Ce/TO, and Eu/TO. Due to this bathochromic shift, it is suggested that, compared to TO, doped nanoparticles have improved response to visible light.

The Tauc formula was used to estimate the direct bandgap energy (E_g) for the nanoparticles using the following equation [64]:

$$\alpha h\nu = A\left(h\nu - E_g\right)^{n/2} \tag{15}$$

where α is the absorption coefficient, hv is photon energy, computed by hv = 1240/λ, A is a constant that is correlated with the absorption brink width parameter, and n is the optical transition phase. The value of n depends on the semiconductor transitions, that is, n = 1 for direct transitions and n = 4 for indirect transitions [64].

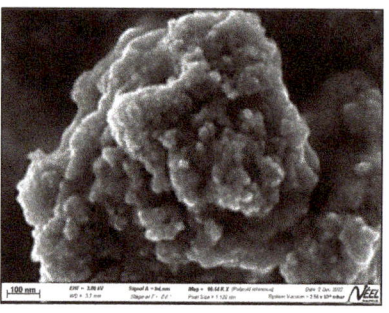 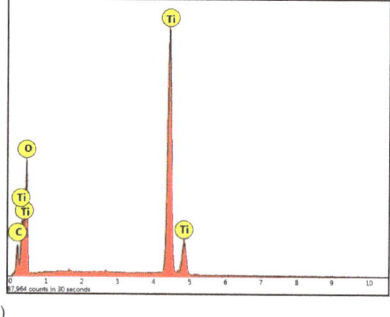

(a)

Figure 3. Cont.

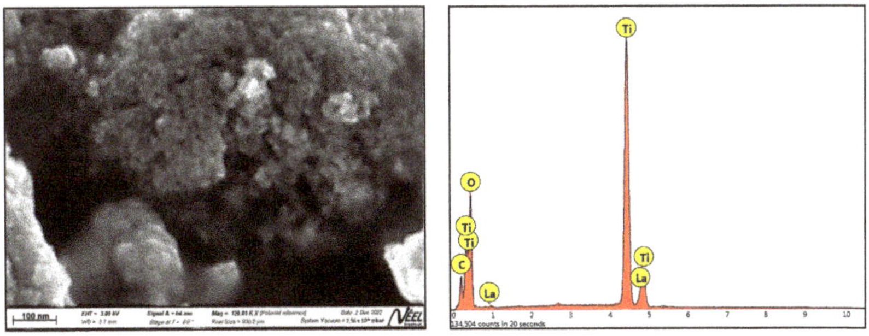

(b)

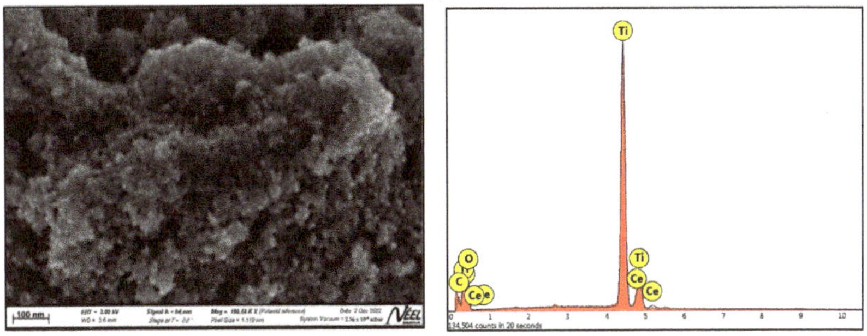

(c)

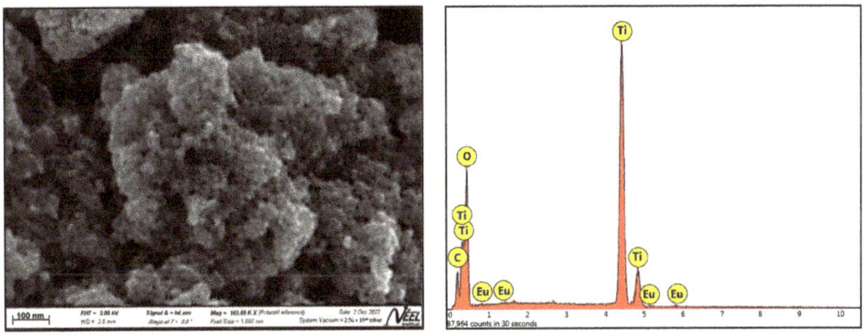

(d)

Figure 3. SEM images and EDS spectra of (**a**) TO, (**b**) La/TO, (**c**) Ce/TO, and (**d**) Eu/TO nanoparticles.

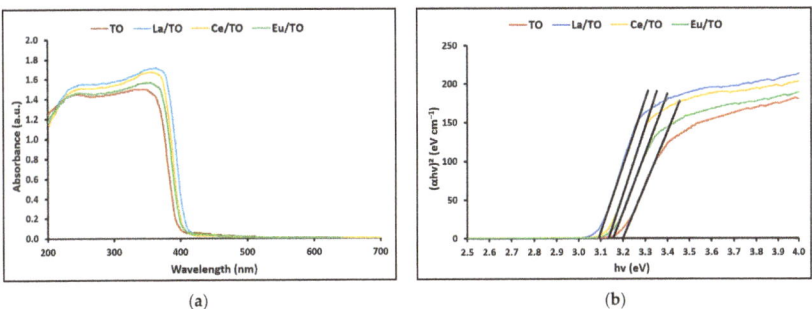

(a)

(b)

Figure 4. (**a**) UV–vis diffuse reflectance spectra and (**b**) plots of $(\alpha h\nu)^2$ vs. $h\nu$ of TO, La/TO, Ce/TO, and Eu/TO nanoparticles.

The relationship between $(\alpha h\nu)^2$ and photon energy (hν) shown in Figure 4b enabled estimating the direct bandgap energy (E_g) [65]. Based on the data presented in this figure, the direct E_g values achieved from the intersections of the straight line with the energy axis [66] were 3.20, 3.10, 3.14, and 3.16 eV for TO, La/TO, Ce/TO, and Eu/TO, respectively.

3.2. Effect of Nanoparticles Composition

The effect of the composition of TO, La/TO, Ce/TO, and Eu/TO nanoparticles on their total cyanide removal capacity q_e (mg g^{-1}) was evaluated in the absence of light (30 min) and then under simulated solar light. The results of this experiment are shown in Figure 5. For all the nanoparticles, it was found that, after 90 min of radiation exposure and under the test conditions, it is possible to achieve maximum removal. This time was denoted as the equilibrium time t_e (min).

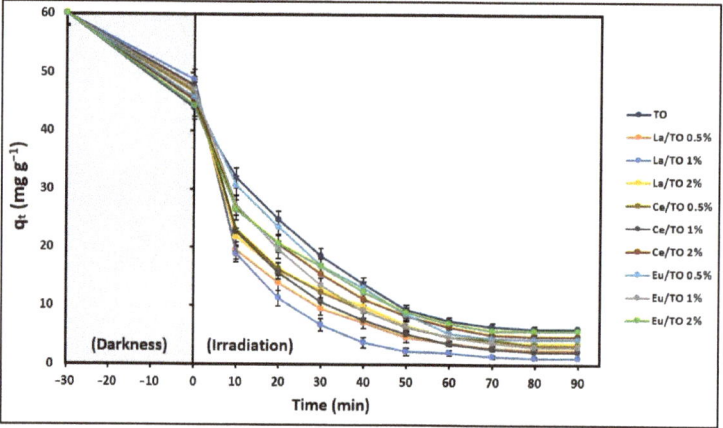

Figure 5. Cyanide removal capacity as a function of nanoparticles composition (catalyst concentration = 0.2 g L^{-1}; cyanide concentration = 20 mg L^{-1}; solution pH = 7.0 ± 0.1).

The findings of the analysis of variance (ANOVA) carried out at t_e (min) are summarized in Table 1. The values in the same column that have been assigned different letters (a–g) are significantly different from each other, with a high level of statistical significance ($p < 0.01$). Based on these results, it is suggested that, for each dopant, the best composition was 1.0 wt.%; therefore, the La/TO (1%), Ce/TO (1%), and Eu/TO (1%) nanoparticles were selected to perform cyanide removal tests and evaluate the effectiveness of doping in relation to non-doped TiO$_2$.

Table 1. Effect of nanoparticles compositions on their cyanide adsorption capacity.

Nanoparticles Composition	HSD Tukey *	Duncan *
	q_e (mg g^{-1})	q_e (mg g^{-1})
La/TO (1.0%)	1.18 ± 0.07 [a]	1.18 ± 0.07 [a]
Ce/TO (1.0%)	2.24 ± 0.11 [b]	2.24 ± 0.11 [b]
La/TO (0.5%)	2.43 ± 0.09 [b]	2.43 ± 0.09 [b]
Eu/TO (1.0%)	3.07 ± 0.02 [c]	3.07 ± 0.02 [c]
Ce/TO (0.5%)	3.22 ± 0.10 [c]	3.22 ± 0.10 [c]
La/TO (2.0%)	3.47 ± 0.23 [c]	3.47 ± 0.23 [c]
Eu/TO (0.5%)	4.28 ± 0.29 [d]	4.28 ± 0.29 [d]
Ce/TO (2.0%)	4.79 ± 0.31 [d]	4.79 ± 0.31 [e]
Eu/TO (2.0%)	5.73 ± 0.16 [e]	5.73 ± 0.16 [f]
TO	6.16 ± 0.07 [e]	6.16 ± 0.07 [g]
p-value	<0.001	<0.001

Different letters (a–g) indicate different groups with a high level of statistical significance ($p < 0.01$). The means for the groups in the homogeneous subsets are displayed. * Use the sample size of the harmonic mean = 3.0.

3.3. Adsorption Studies

3.3.1. Effect of pH on Cyanide Adsorption

Figure 6 shows the results of the cyanide adsorption experiment as a function of the pH of the solutions. Total cyanide adsorption capacity (HCN + CN$^-$) remained unchanged in the pH range 9 to 12 for pristine TO (pH$_{PZC}$ = 6.9) and for doped nanoparticles (pH$_{PZC}$ = 7.1). The point of zero charge (PZC) corresponds to the pH value at which the net surface charge of the solid is equal to zero under certain conditions of pressure, temperature, and composition of the aqueous solution [67]. The point of zero charge (PZC) for all nanoparticles was determined at room temperature (21 ± 2 °C) using the pH drift procedure (ΔpH = pH$_f$ − pH$_i$ = 0). The experiment was carried out in 50 mL tubes, in which 25 mL of a 0.1 M NaCl solution and 0.1 g of solid sample were placed. The pH values of the solutions contained in the tubes were adjusted to values between 3 and 12 using 0.1 M solutions of HCl or NaOH. These initial pH values were designated as pH$_i$. The tubes were shaken for 24 h at 230 rpm. After this time, the final pH of the supernatant liquid in each tube was measured and designated as pH$_f$. The PZC was determined from the graph of ΔpH (ΔpH = pH$_f$ − pH$_i$) vs. pH$_i$. The procedure was repeated for all nanoparticle samples using 0.01 and 0.05 M NaCl solutions. All experiments were carried out in triplicate, and, for each sample, the average pH$_{PZC}$ value was reported [68].

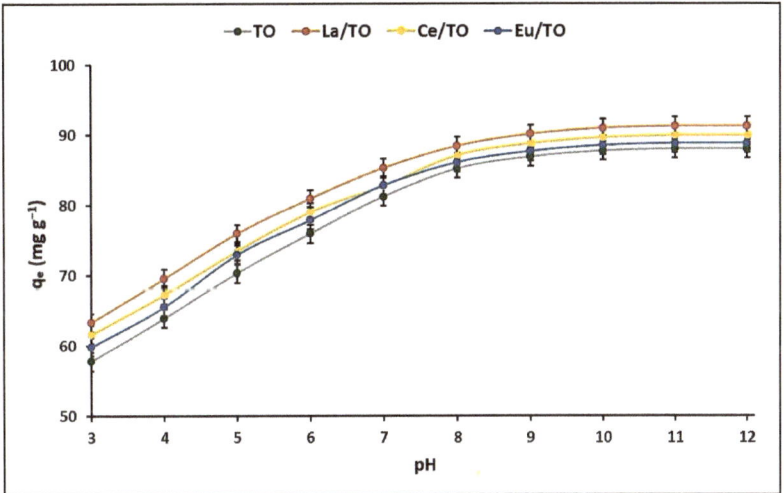

Figure 6. Solution pH effect experiment.

In this study, the pH$_{PZC}$ of all nanoparticles was on average 7.0. This implies that, at pH > pH$_{PZC}$ = 7.0, the surface of the nanoparticles has a negative charge and, at pH < pH$_{PZC}$ = 7.0, it has a positive charge.

The ability of ions to be adsorbed on a surface from a solution is greatly influenced by the pH of the solution. This is because pH affects both the surface charge of the ion adsorbent and the extent to which the ions ionize and exist in different forms. Therefore, to better understand the adsorption mechanisms of cyanide species on the surface of nanoparticles, it is important to consider the pH-dependent distribution of cyanide species in solution (Equation (16)) [7].

$$HCN \leftrightarrow CN^- + H^+ \quad pKa = 9.4 \quad (16)$$

Inorganic cyanides, such as KCN and NaCN, are weak acids and hydrolyze to form hydrocyanic acid (HCN), which has a pKa value of 9.4. At pH < 9.4, HCN is the predominant species in the aqueous solution, while, at pH > 9.4, HCN dissociates into H$^+$ and CN$^-$ ions [69]. CN$^-$ ions have nucleophilic characteristics and experience a strong electrostatic attraction in solutions at pH > 9.0, which is why several authors have informed higher cyanide adsorption at pH values between 9 and 11 [42]. However, other authors have also informed that inorganic cyanides can easily dissociate to form HCN and CN^- species at a neutral pH of 7.0 [70]. In fact, several investigations related to cyanide adsorption have been carried out at pH values around 7.0 (pH < pKa), suggesting that the high adsorption of HCN species is based on interactions with the existing Lewis acids sites on the adsorbent surface [71,72]. Consequently, the adsorption experiments were performed at a pH of 7.0.

3.3.2. Maximum Cyanide Adsorption Capacity

In this study, the Langmuir and Freundlich isotherms were examined as equilibrium models that rely on the initial concentration. The Langmuir isotherm represents a credible theoretical framework for the process of adsorption occurring on a uniform and completely homogeneous surface, whereby a finite quantity of identical and specific sites are available for adsorption. In this model, there is minimal interaction among the molecules. The Freundlich equation is a non-theoretical formula that does not assume uniformity in the energy of surface sites, allows for unlimited adsorption capacity, and corresponds to an exponential distribution of active sites that reflects a heterogeneous surface. Figure 7 shows the cyanide adsorption isotherms of TO, La/TO, Ce/TO, and Eu/TO nanoparticles. This figure clearly shows that, for all nanoparticles, the Langmuir model is a more suitable description of the adsorption process compared to the Freundlich model.

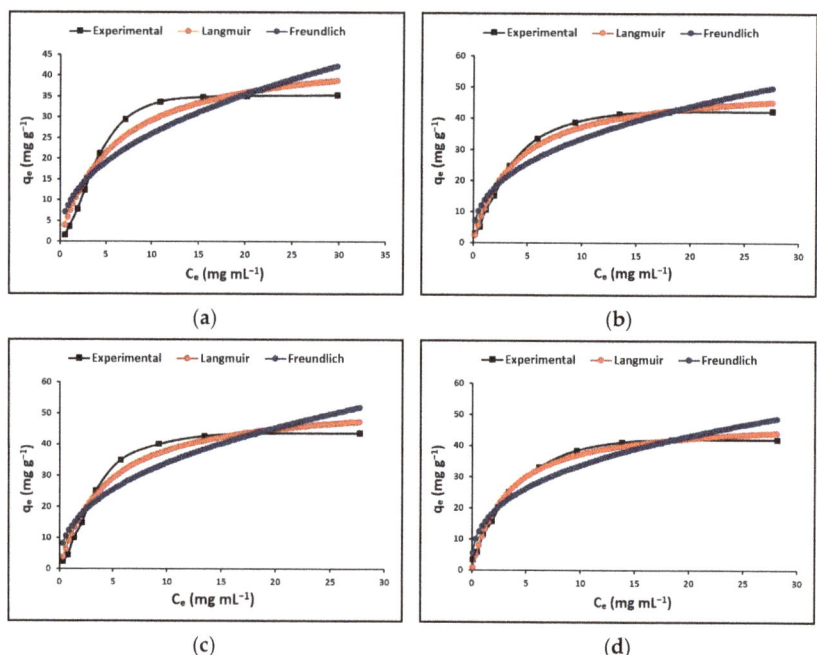

Figure 7. Adsorption isotherms of (**a**) TO, (**b**) La/TO, (**c**) Ce/TO, and (**d**) Eu/TO.

The values estimated at different temperatures for the constants of the Langmuir and Freundlich models are given in Table 2. In this table, it is evident that the values of the R_L separation factor or equilibrium parameter were in the range of 0 to 1, while the values of the coefficient n, which represents the intensity of adsorption, were in the range of 1 to 10. Consequently, it is suggested that adsorption of cyanide species on the surface of all materials was satisfactory.

Table 2. Isotherm parameters for cyanide sorption on nanoparticles at different temperatures.

	Isotherm Parameters		293.15 K				298.15 K				303.15 K			
			TO	La/TO	Ce/TO	Eu/TO	TO	La/TO	Ce/TO	Eu/TO	TO	La/TO	Ce/TO	Eu/TO
Langmuir	q_{max} (mg g^{-1})		46.48 (±4.10)	54.96 (±3.52)	51.39 (±2.11)	49.25 (±1.52)	51.13 (±2.56)	60.46 (±2.42)	56.53 (±3.18)	54.17 (±1.24)	55.78 (±2.32)	68.95 (±2.76)	61.67 (±3.01)	59.10 (±1.43)
	K_L (L mg^{-1})		0.17 (±0.04)	0.22 (±0.04)	0.26 (±0.04)	0.31 (±0.03)	0.18 (±0.04)	0.24 (±0.03)	0.28 (±0.04)	0.33 (±0.04)	0.19 (±0.04)	0.27 (±0.03)	0.31 (±0.04)	0.37 (±0.03)
	R_L		0.23	0.18	0.16	0.14	0.22	0.17	0.15	0.13	0.21	0.16	0.14	0.12
	χ^2		2.89	3.21	3.97	2.45	2.67	3.30	3.78	2.36	3.64	2.41	2.81	3.65
	R^2		0.95	0.97	0.98	0.99	0.97	0.96	0.99	0.98	0.96	0.98	0.97	0.99
Freundlich	K_F (L mg^{-1})		9.35 (±2.24)	13.02 (±2.59)	13.65 (±1.94)	14.76 (±1.94)	10.28 (±2.62)	14.32 (±2.63)	15.02 (±2.41)	16.24 (±1.78)	11.22 (±2.12)	15.62 (±2.85)	16.38 (±2.34)	17.71 (±1.36)
	n		2.25 (±0.44)	2.40 (±0.44)	2.56 (±0.41)	2.80 (±0.39)	2.48 (±0.47)	2.64 (±0.41)	2.82 (±0.44)	3.07 (±0.37)	2.70 (±0.41)	2.88 (±0.40)	3.08 (±0.39)	3.35 (±0.35)
	1/n		0.44	0.42	0.39	0.36	0.40	0.38	0.35	0.33	0.37	0.35	0.33	0.30
	χ^2		3.95	3.15	2.96	2.70	4.20	3.02	3.12	2.81	3.56	3.42	3.15	2.15
	R^2		0.84	0.86	0.89	0.92	0.86	0.89	0.90	0.94	0.87	0.90	0.88	0.92

3.3.3. Adsorption Thermodynamics

The thermodynamic parameters provide information about the spontaneity and possibility of a process. To determine these parameters, namely Gibbs free energy change ($\Delta G°$), enthalpy change ($\Delta H°$), and surface entropy change ($\Delta S°$), the equilibrium constant was measured at different temperatures, as shown in Figure 8.

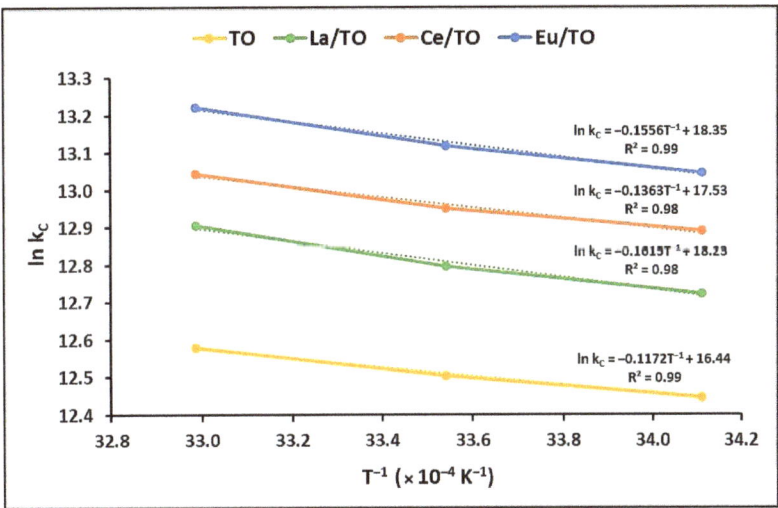

Figure 8. Thermodynamic study of cyanide adsorption onto nanoparticles.

The results of the thermodynamic parameters are shown in Table 3. $\Delta G°$ indicates the degree of spontaneity of the process; negative values reflect higher adsorption favorability. Likewise, negative $\Delta H°$ values reflect that the process is exothermic, while negative $\Delta S°$ values indicate a decrease in randomness at the solution–solid interface during adsorption.

3.3.4. Kinetic Behavior of Cyanide Adsorption

To design and evaluate adsorbents for adsorption, it is important to determine the rate of the time-dependent process. Two models, the pseudo-first-order (Lagergren) and pseudo-second-order (Ho) models, were used to describe the adsorption kinetics, which are commonly used as simplified models. Both models, as shown in Figure 9, exhibit a rapid initial adsorption stage followed by a plateau stage. Table 4 indicates that the correlation coefficient of the pseudo-second-order model is higher than that of the pseudo-first-order model, suggesting a chemisorption process, according to previous literature [9].

Table 3. Thermodynamic parameters of cyanide adsorption onto nanoparticles.

NPs	Temperature (K)	ln k_C	$\Delta G°$ (kJ mol^{-1})	$\Delta H°$ (kJ mol^{-1})	$\Delta S°$ (kJ mol^{-1} K^{-1})
TO	293.15	12.45	−30.34		
	298.15	12.50	−30.99	9.74	0.14
	303.15	12.58	−31.70		
La/TO	293.15	12.72	−31.01		
	298.15	12.80	−31.72	13.43	0.15
	303.15	12.91	−32.53		
Ce/TO	293.15	12.89	−31.42		
	298.15	12.95	−32.10	11.33	0.15
	303.15	13.04	−32.87		
Eu/TO	293.15	13.05	−31.80		
	298.15	13.12	−35.52	12.94	0.15
	303.15	13.22	−33.32		

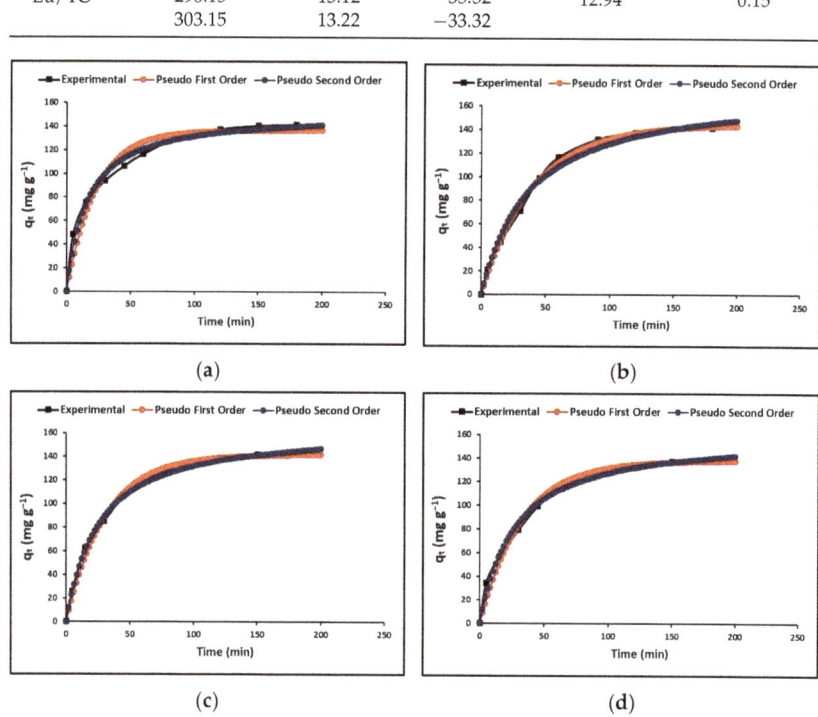

Figure 9. Adsorption kinetics of (a) TO, (b) La/TO, (c) Ce/TO, and (d) Eu/TO.

On the other hand, the intraparticle diffusion model was utilized to explain the adsorption rate, which is dependent on the rate at which the cyanide species transfer from the aqueous solution to the adsorption sites on the nanoparticles. Figure 10 shows the variation in the q_t (mg g^{-1}) curves as a function of time ($t^{1/2}$) for the TO, La/TO, Ce/TO, and Eu/ZTO nanoparticles.

Table 4 shows the total cyanide adsorption kinetic parameters estimated in this study for all nanoparticles.

3.4. Photodegradation Studies

3.4.1. Kinetics of Cyanide Photodegradation

TiO_2 is a widely used photocatalyst to efficiently degrade organic compounds due to its strong oxidizing capacity, which is generated when subjected to the action of light.

In this study, the photoactivity of pristine TiO_2 nanoparticles and La/TiO_2, Ce/TiO_2, and Eu/TiO_2 nanoparticles was tested through model cyanide photodegradation reaction in aqueous solutions under simulated solar radiation.

Table 4. Kinetic parameters for cyanide sorption on nanoparticles.

Kinetic Parameters		TO	La/TO	Ce/TO	Eu/TO
Pseudo-first-order	q_{max} (mg g^{-1})	136.53 (±2.51)	143.85 (±1.69)	141.83 (±1.81)	137.82 (±2.94)
	k_1 (L mg^{-1})	0.04 (±2.29 × 10^{-3})	0.03 (±1.02 × 10^{-3})	0.03 (±1.56 × 10^{-3})	0.03 (±2.45 × 10^{-3})
	χ^2	8.94	8.48	7.69	7.13
	R^2	0.95	0.99	0.99	0.99
Pseudo-second-order	q_{max} (mg g^{-1})	152.22 (±3.67)	175.08 (±6.36)	165.40 (±2.45)	160.69 (±3.99)
	k_2 (L mg^{-1})	3.19 × 10^{-4} (±2.75 × 10^{-5})	1.55 × 10^{-4} (±2.41 × 10^{-5})	2.37 × 10^{-4} (±1.70 × 10^{-5})	2.37 × 10^{-4} (±2.81 × 10^{-5})
	χ^2	2.82	3.03	3.67	2.64
	R^2	0.99	1.00	1.00	0.99
Intraparticle diffusion	k_3 (mg g^{-1} min$^{-1/2}$)	9.94 (±0.27)	10.80 (±0.17)	10.40 (±0.21)	9.89 (±0.16)
	A	8.09 (±1.25)	9.99 (±1.05)	8.37 (±1.15)	9.12 (±1.17)
	R^2	0.91	0.92	0.90	0.92
External-film diffusion	Df (m^2 min^{-1})	1.14 × 10^{-11}	1.14 × 10^{-11}	1.25 × 10^{-11}	1.10 × 10^{-11}
	R^2	0.98	0.99	0.98	0.99
Internal-pore diffusion	Dp (m^2 min^{-1})	1.40 × 10^{-17}	1.42 × 10^{-17}	1.60 × 10^{-17}	1.40 × 10^{-17}
	R^2	0.97	0.97	0.96	0.97

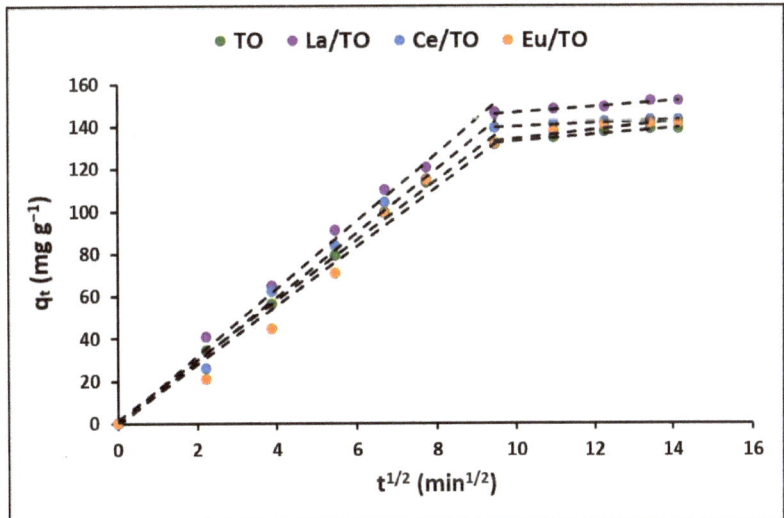

Figure 10. Intraparticle diffusion plots for cyanide removal by nanoparticles.

The Langmuir–Hinshelwood equation showed a linear correlation between $\ln(C_0/C_t)$ and t, confirming that the photocatalytic degradation reaction proceeds via a pseudo-first-order reaction. The calculated apparent rate constants (k_{app}) were calculated to be 0.016, 0.030, 0.018, and 0.017 min^{-1} for TO, La/TO, Ce/TO, and Eu/TO nanoparticles, respectively. These results are in agreement with those reported by other authors [48,73]. The results obtained in the photocatalytic degradation test are shown in Figure 11. From this figure, the maximum percentage of cyanide degradation for all nanoparticles is reached around the first 90 min, after which photodegradation becomes almost constant. Evidence from the literature suggests that there is a limit to the efficacy of photocatalysis for complete degradation of some pollutants [74]. In this study, maximum efficiency was reached by La/TO (96.8%), followed by Ce/TO (88.9%), Eu/TO (86.5%), and TO (83.9%).

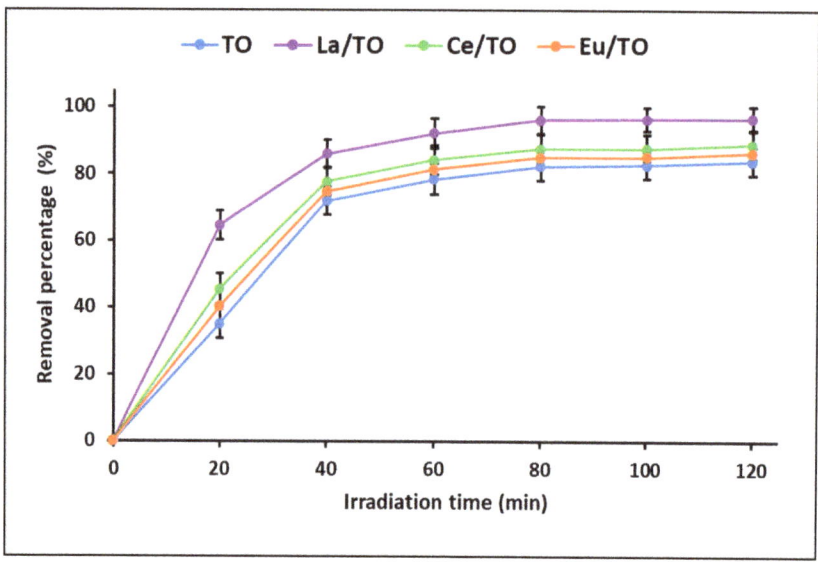

Figure 11. Photocatalytic cyanide degradation by nanoparticles.

3.4.2. Effect of the Photogenerated Radicals

It is widely known that photogenerated electrons (e^-), photogenerated holes (h+), hydroxyl radicals ($^\bullet OH$), and superoxide radicals ($^\bullet O_2^-$) are the main reactive oxygen species (ROS) in photocatalytic processes. Therefore, to establish the impact of various ROS on the reaction system, a radical quenching experiment was performed incorporating different radical quenchers [75]. Figure 12 shows the results of this experiment. From this figure, it is evident that, for all nanoparticles, the efficiency of photocatalytic degradation of total cyanide decreased with introduction of isopropanol (i-PrOH) and ethanol (EtOH), which are quenchers of h+ and $^\bullet OH$ radicals, respectively. On average, the nanoparticles decreased their cyanide removal efficiency by 9.0% and 5.0% when introducing ethanol and isopropanol, respectively. On the other hand, introduction of p-Benzoquinone (p-BQ) as a quencher of radials $^\bullet O_2^-$ did not affect the efficiency of the nanoparticles for cyanide removal.

3.5. Total Efficiency and Reuse of Nanoparticles

The percentage of cyanide adsorbed and photodegraded by the synthesized nanoparticles is shown comparatively in Figure 13. In this figure, for all the nanoparticles, the photocatalysis process was more efficient for removal of cyanide species than the adsorption process. From this figure, it can be inferred that, under the conditions tested in this study, the highest adsorption and photodegradation capacity of cyanide species was achieved by La/TO nanoparticles, followed by Ce/TO, Eu/TO, and finally by TO nanoparticles.

Finally, knowing that the stability and recyclability of materials with adsorbent and photocatalytic applications are considered crucial aspects for their large-scale application, in this study, reuse experiments were performed for five consecutive cyanide removal cycles. Figure 14 shows the removal efficiency of the synthesized nanoparticles during the five cycles.

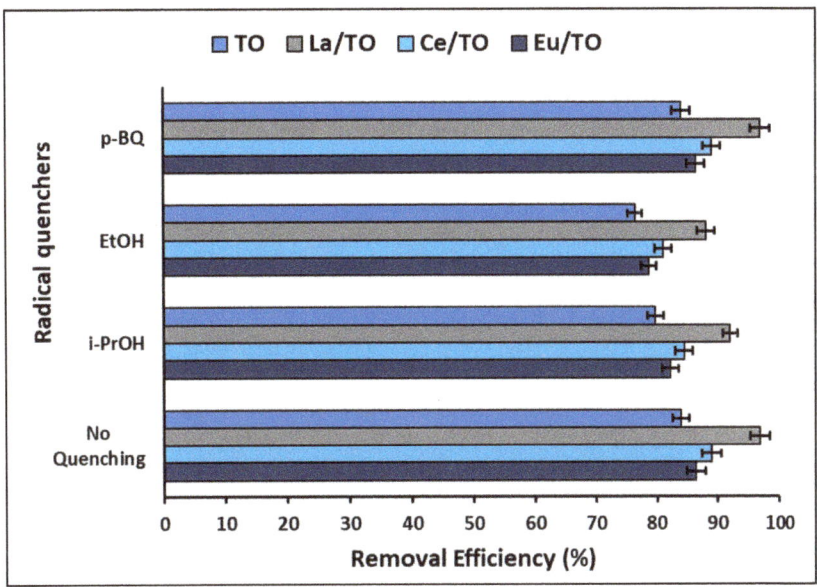

Figure 12. Photogenerated radical quenching experiments.

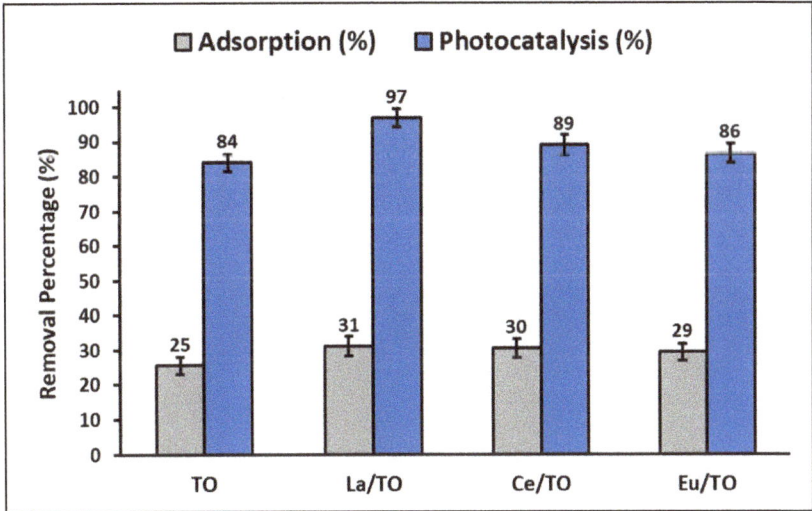

Figure 13. Percentage of cyanide adsorbed and photodegraded by nanoparticles.

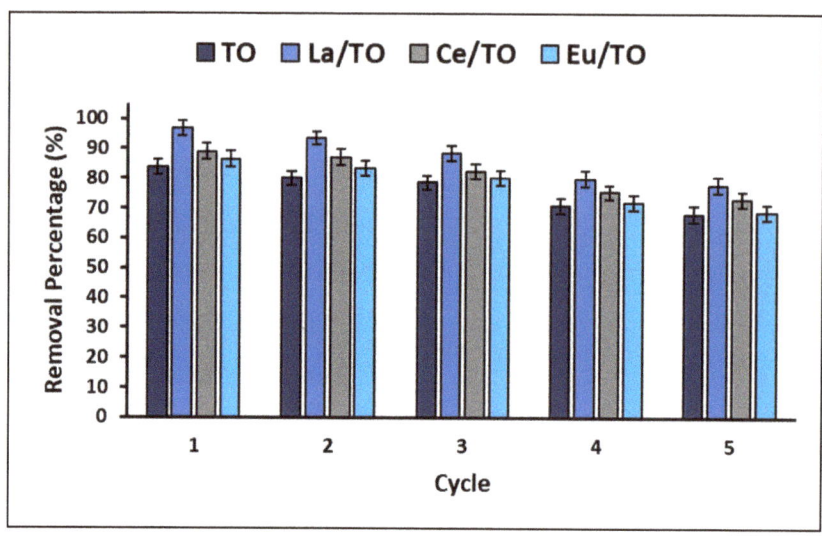

Figure 14. Nanoparticle reuse experiment.

4. Discussion

4.1. Characterization of Nanoparticles

4.1.1. XRD and SSA Analysis

The results depicted in Figure 1 indicate that doping with lanthanide (Ln) can impact the crystal structure of pure TiO_2 (anatase), causing slight broadening and reduction in intensity of its characteristic diffraction peaks. This suggests that the adsorbent and photocatalytic properties of this semiconductor oxide can be affected by the presence of Ln dopants [76]. According to the literature, it is very difficult for lanthanide ions to substitute Ti^{4+} ions in the TiO_2 crystal lattice. This is primarily due to the significant difference in ionic radii between Ti^{4+} (0.068 nm) and the lanthanide ions, such as La^{3+} (0.116 nm), Ce^{3+} (0.114 nm), Ce^{4+} (0.101 nm), and Eu^{3+} (0.107 nm). Therefore, La^{3+}, $Ce^{3+/4+}$, and Eu^{3+} could rather be uniformly dispersed on the TiO_2 surface in the form of La_2O_3, Ce_2O_3/CeO_2, and Eu_2O_3 particles, respectively [77]. It should be noted that, even though the doping elements were present in low concentration (1 wt%) and were evenly dispersed on the TiO_2 surface, Figure 1 does not exhibit identifiable diffraction peaks of the corresponding lanthanide oxides [78]. Nevertheless, the reduction in crystallite size may be attributed to the existence of La–O–Ti, Ce–O–Ti, or Eu–O–Ti bonds on the surface of TiO_2. These findings are consistent with those reported by other researchers, who have demonstrated a decrease in crystallite size caused by formation of Ln–O–Ti bonds [79].

The results of the BET analysis would corroborate this appreciation since the doped nanoparticles (La/TO, Ce/TO, and Eu/TO) show higher specific surface area related to pristine TiO_2. From the results of the BET analysis, it is suggested that La/TO, Ce/TO, and Eu/TO nanoparticles could have better cyanide species removal capacity than TO due to the greater accessibility of active sites on the increased surface area. Furthermore, the high surface area of the doped nanoparticles could facilitate diffusion of cyanide species and photogenerated ROS. In this way, photodegradation of cyanide could be improved by allowing more energetic photons to be adsorbed on the surface of the photocatalytic nanoparticles. Finally, the high surface area also facilitates contact between the nanoparticles and the cyanide species present in the aqueous medium, which benefits the subsequent photocatalytic reaction [48].

4.1.2. FTIR Analysis

The FTIR spectra shown in Figure 2b–d allowed to demonstrate that, due to doping, the broad band of the TiO_2 semiconductor, shown in the region of 540–900 cm^{-1} in Figure 2a, changes drastically. According to the literature, this change can be attributed to the presence of Ln—O stretching modes of vibrations [24]. Unresolved bands suggest the existence of uniform stresses in the lattice, which is generally associated with defects in the material structure. These defects, probably derived from doping, can cause shortening of the Ti–O bond, promoting vibrational mode changes observed in doped nanoparticles relative to pristine TiO_2 [80]. The results obtained in this analysis agree with those reported in the literature [57,58].

4.1.3. SEM and EDS Analysis

Figure 3 shows a comparison of the SEM micrographs of the TO, La/TO, Ce/TO, and Eu/TO nanoparticles. As evidenced in this figure, lanthanide elements La, Ce, and Eu, by acting as TiO_2 dopants, could promote reduction in size of semiconductor crystallites. This is probably due to the fact that dispersion of dopant ions La^{3+}, $Ce^{3+/4+}$, and Eu^{3+} on the TiO_2 surface could restrict direct contact between neighboring crystallites of this semiconductor, thus inhibiting their growth [34]. According to the literature, restriction of direct contact between neighboring crystallites is due to generation of large defects in the crystal lattice of the material. These defects can change the interatomic distance and, therefore, affect the stability of the bonds between TiO_2 atoms. As a result, TiO_2 microcrystals with different sizes and random orientations can be formed [81]. The results of this study agree with those reported by other authors, who have shown that doping enables stabilization of small particles [76]. Regarding EDS analysis, the results obtained allowed us to confirm the presence of La (1.32 wt%), Ce (1.34 wt%), and Eu (1.28 wt%) on the TiO_2 structure. These results support the results obtained in the XRD and FTIR assays.

4.1.4. Optical and Photoelectric Properties

Figure 4a displays the UV–vis absorption spectra of TO, La/TO, Ce/TO, and Eu/TO nanoparticles. In this study, the absorption threshold of the pristine TiO_2 (TO) semiconductor was around 350 nm. The strong absorption band observed in this region for TO would be associated with electronic transition from the valence band (VB) dominated by the 2p orbital of O to the conduction band (CB) dominated by the 3d orbital of Ti [35]. In contrast to the TO spectrum, the spectra of the nanoparticles that have been doped exhibit a red shift. This phenomenon has been described in the literature as being caused by a charge transfer transition between the f-electrons of the lanthanide ions and either the conduction band or valence band of TiO_2 [82]. As can be seen in Figure 4a, the shift of the absorption edge towards a slightly longer wavelength depends on the lanthanide ions used as dopants; however, it is evident that doping with these ions could improve the response of TO to visible light.

On the other hand, to estimate the bandgap energy (E_g) of the synthesized nanoparticles, the (UV–vis) DRS method was used, and the absorption data were fitted to equations for direct bandgap transitions. Figure 4b shows a reduction in the E_g value for the doped nanoparticles relative to the value calculated for the pristine TiO_2. According to the literature, the reduction in the value of E_g is due to formation of impure energy levels (impure orbitals) between the VB and the CB of the doped semiconductor [83]. Furthermore, due to the presence of these impure orbitals, less energy is required to achieve the transfer of electrons from the valence band to the impure orbital and/or from the latter to the conduction band. From these results, it is suggested that doping of the TiO_2 semiconductor with lanthanide elements La, Ce, and Eu allows to reduce the bandgap energy and, consequently, to optimize the capacity of TiO_2 to absorb wavelengths with less energy (including solar radiation), as well as reduce recombination of photogenerated charges and modify the adsorption capacity of the surface of this semiconductor [13].

4.2. Effect of Nanoparticles Composition

In this study, a preliminary experiment was developed in order to evaluate the best concentration of elements La, Ce, and Eu as TiO_2 dopants for effective removal of total cyanide from aqueous systems. Figure 5 shows that the total cyanide removal capacity of the doped nanoparticles increases when the percentage by weight of the dopant in the semiconductor increases from 0.5 to 1%. However, the total cyanide removal capacity of the doped nanoparticles decreases when the percentage by weight of the dopant in the semiconductor increases from 1 to 2%. According to the literature, the reason for this is that the presence of high concentrations of doping elements can lead to an increase in the number of oxygen vacancies in the material, resulting in an increase in the number of recombination centers for photoinduced charges [84]. Similarly, when the dopant particles agglomerate and form clusters, they can obstruct the active sites on the surface of the semiconductor, leading to a reduction in its photocatalytic efficiency. This has been reported in the literature as well [85]. These findings align with previous studies conducted by other researchers, who have shown that incorporating lanthanide elements at levels of 1–2 wt% can effectively enhance photoactivity of semiconductors across a broad range of the electromagnetic spectrum [78].

4.3. Adsorption Studies

4.3.1. Effect of pH on Cyanide Adsorption

Due to the protonation–deprotonation balance of the surface hydroxyl groups, the surface of titanium oxide (TiO_2) has a pH-dependent charge, so this oxide presents a positive charge under acidic conditions and a negative charge under basic conditions [67]. In the middle of these two regions is the point of zero charge (PZC), that is, the pH where the total charge on the TiO_2 surface is zero. As reported in the literature, adsorption of ionic doping species, including lanthanides La, Ce, and Eu, could change the surface charge density and position of PZC of TiO_2 [86,87].

Several studies have reported that TiO_2 behaves as a weak Brönsted acid and that, when hydrated, it can form surface hydroxyl groups (Ti–OH) [88]. The hydroxyl groups present can participate in chemical reactions involving association and dissociation of protons, thus generating a pH-dependent surface charge:

$$TiOH + H^+ \leftrightarrow TiOH_2^+ \quad \text{at pH} < pH_{PZC} \quad (17)$$

and

$$TiOH \leftrightarrow TiO^- + H^+ + H_2O \quad \text{at pH} > pH_{PZC} \quad (18)$$

where the positive, neutral, and negative surface hydroxyl groups are represented by $TiOH_2^+$, $TiOH$, and TiO^-, respectively [87].

From Figure 6, it can be inferred that, as the pH of the solution enhances, the amount of total cyanide adsorbed also increases until reaching an adsorption maximum at pH = 9. Above this value, cyanide adsorption remains constant for all nanoparticles. In this figure, it is also observed that the doped nanoparticles have a higher total cyanide removal capacity compared to pristine TiO_2. This is probably because doping with lanthanum (La), cerium (Ce), and europium (Eu) enables greater total number of available active sites on the surface of TiO_2 nanoparticles, thus driving favorable kinetics [34].

According to the literature, the pH level of a solution has an impact on the surface charge of the adsorbents and the ionization and speciation of the adsorbate. Pristine and doped TiO_2 nanoparticles have a pH_{PZC} of 6.9 and 7.1, respectively. This implies that, at higher pH levels, there is a preference for an increase in negatively charged groups, while at lower pH levels, there is a preference for an increase in positively charged groups. Regarding cyanide (pKa = 9.4), at pH values < pKa, this molecule in solution associates in its molecular form (HCN), while, at pH > pKa, it dissociates mainly in its ionic form ($CN^- + H^+$).

Evidence from several studies indicates that, at pH < pH_{PZC}, the specific adsorption of HCN species on the OH groups present on the surface of hydrated nanoparticles is driven through formation of N–H polar covalent bonds, as indicated below [69]

$$TiOH + HCN \leftrightarrow TiOH \cdots NCH \quad (19)$$

or

$$TiOH_2^+ + HCN \leftrightarrow TiOH_2^+ \cdots NCH \quad (20)$$

For pH < 7.0, the surface of the nanoparticles exhibits a positive charge and the degree of HCN dissociation is negligible. Therefore, electrostatic forces of attraction between a charged surface and a neutral molecule are unlikely to develop, especially at very low pH values. At pH > 7.0, the negative charge on the surface of the nanoparticles increases, but the dissociation of HCN into CN^- ions is negligible until pH = pKa = 9.4 is reached. However, in the pH range between 7.0 and 9.4, an increase in the forces of electrostatic attraction is expected and, therefore, an increase in the adsorption of HCN species on the negative surface of the nanoparticles. At pH > pKa = 9.4, the proportion of CN^- species in the solution increases, but, in parallel, the surface of the nanoparticles becomes more negative since pH > pH_{PZC}. This generates an augment in the repulsive forces between the negative surface sites of the nanoparticles and the negative CN^- ions, which causes a reduction in the adsorption capacity of these ions at very high pH values, as shown in Figure 6.

The strong dependence between the pH of the solution and the adsorption capacity of the nanoparticles is a clear indication that, in this study, the total cyanide adsorption mechanism was mainly driven by electrostatic interactions, although other types of interactions cannot be excluded (van der Waals and/or specific interactions). In fact, it is suggested that adsorption of CN^- species on La/TO, Ce/TO, and Eu/TO nanoparticles could occur predominantly by chemical adsorption instead of physical adsorption (outer sphere complex), although the presence of physical adsorption is not ruled out completely. The results of this study are consistent with those reported in the literature [44]. When CN^- is in contact with active cationic sites on the adsorbent surface, it forms a bond with them. According to the literature, lanthanum, cerium, and europium ions show strong electron withdrawal effect [25,33], so the presence of these lanthanides on the surface of TiO_2 nanoparticles contributes to generation of Lewis acids sites [28,29,89]. It should be mentioned that these active cationic sites could certainly improve cyanide adsorption on the TiO_2 surface [34], in addition to providing this semiconductor with catalytic stability in an aqueous reaction system [90].

4.3.2. Maximum Cyanide Adsorption Capacity

Figure 7 shows the adsorption isotherms obtained for the synthesized nanoparticles. This figure displays that, at low concentrations of cyanide species (HCN + CN^-), the adsorption rate increases; instead, at high concentrations of these species, the adsorption rate reaches a maximum where it stabilizes. One possible reason for this phenomenon is the excessive presence of cyanide compounds that compete for the available active sites on the surface of the adsorbent nanoparticles. Evidence from these results suggests that the initial concentration of cyanide species in solution could generate a significant driving force to allow these species to migrate from the liquid phase to the nanoparticle surface [91].

The adsorption data obtained in this study for TO, La/TO, Ce/TO, and Eu/TO nanoparticles were fitted to the Langmuir and Freundlich isotherm models. Table 2 describes the parameters corresponding to adjustment of the experimental data. As shown in this table, given the values of the correlation coefficient (R^2), it is concluded that the Langmuir model is the most suitable to describe the equilibrium adsorption behavior. Consequently, it is suggested that monolayer adsorption of cyanide species on the synthesized nanoparticles proceeds as a phenomenon of electrostatic attraction. This attraction takes place in regions with homogeneous surfaces, where the strongest binding sites are initially filled, and, as the saturation level increases, the binding strength decreases [44]. Finally, the values of the Langmuir and Freundlich constants (R_L and n) shown in Table 2 confirm the favorable adsorption of cyanide species on the nanoparticles synthesized in this study.

4.3.3. Adsorption Thermodynamics

As mentioned above, thermodynamic parameters offer the most reliable indications for effective implementation of a process in practice. Table 3 shows the thermodynamic parameters obtained in this study for removal of cyanide species. Negative $\Delta G°$ values suggest the feasibility and thermodynamic spontaneity of total cyanide adsorption on TO, La/TO, Ce/TO, and Eu/TO nanoparticles. Furthermore, the decrease in negative values of $\Delta G°$ with increasing temperature reveals an increase in the effectiveness of the adsorption process at high temperatures. On the other hand, the positive values of $\Delta H°$ and $\Delta S°$ shown in Table 3 suggest that adsorption of cyanide species on the nanoparticles occurred as an endothermic process. Finally, the positive $\Delta S°$ also suggests an increase in randomness at the solution/solid interface, with several structural alterations in the active sites of the nanoparticles during cyanide fixation. These results are in agreement with those informed by other authors [43].

4.3.4. Kinetic Behavior of Cyanide Adsorption

The kinetic models of adsorption allow to determine the contact time necessary for the whole adsorption of the chemical species. Figure 9 shows the fit of the experimental data obtained in this investigation with the pseudo-first-order and pseudo-second-order kinetic models. In this figure, for all nanoparticles (TO, La/TO, Ce/TO, and Eu/TO), the concentration of cyanide species decreases very fast at the beginning of the process and tends to be constant after ~90 min. The rapid adsorption that occurs in the initial stage could be due to the high concentration gradient, as well as the presence of vacant adsorption sites. Table 4 shows the kinetic parameters of adsorption obtained for all the nanoparticles. From the values shown in this table for the correlation coefficient (R^2), it can be concluded that the experimental data fit the pseudo-second-order model more than the pseudo-first-order model, suggesting chemical adsorption of cyanide species on the surface of nanoparticles [43].

On the other hand, Figure 10 shows the fit of the experimental data to the intraparticle diffusion model. From this figure, the adsorption of total cyanide on the nanoparticles occurs conceptually in two stages (linear regions), after which intrinsic adsorption occurs, either by chemical binding on the active sites on the nanoparticles or by physical processes. The initial stage of fast speed that is observed in Figure 10 could be described as a process of diffusion of cyanide particles through the stationary film that surrounds each adsorbent nanoparticle; therefore, it corresponds to the mass transfer from the aqueous solution to the surface of the adsorbent. In contrast, the slow-rate second stage corresponds to intraparticle mass transfer and describes the process of diffusion of cyanide particles through the pores of adsorbent nanoparticles. The linear regression analysis for the diffusion kinetic models is displayed in Table 4. From the relatively high values of A reported in this table, it is suggested that surface adsorption could be the rate-limiting step for all synthesized nanoparticles [44].

4.4. Photodegradation Studies

4.4.1. Kinetics of Cyanide Photodegradation

Regarding the photocatalytic behavior of TO, La/TO, Ce/TO, and Eu/TO, in this study, it was possible to demonstrate that these nanoparticles achieved effective removal of total cyanide in the aqueous solution. Possibly, this occurred because the doping of TiO_2 with lanthanide elements La, Ce, and Eu not only provided greater quantity of active centers for adsorption of cyanide species but also the presence of these lanthanide elements on the surface of TiO_2 promoted decrease in bandgap energy, which definitely enhanced the photoactivity of the doped nanoparticles. It is widely known that magnitude of bandgap energy is fundamental in photoactivity of semiconductors since it determines the recombination rate of the electron/hole pairs (e^-/h^+). Moreover, the recombination rate may depend on other factors, such as carrier concentration, carrier mobility, and semiconductor structure [81]. Therefore, the evidence from this study suggests that the presence of La, Ce, and Eu ions on the TiO_2 surface enhanced the photoactivity of this semiconductor, probably by modifying the recombination rate of the pair (e^-/h^+). In addition, in this study, it was shown that the nanoparticles doped with La/TO, Ce/TO, and Eu/TO had lower bandgap energy than TO, so it is suggested that, given the smaller separation between the valence bands (VB) and conduction (CB), doped nanoparticles could easily transfer photoinduced electrons from the bulk to the surface and be more active than TO nanoparticles under simulated solar light [92].

The photocatalytic route involving the doped nanoparticles (La/TO, Ce/TO, and Eu/TO) begins with electronic excitation under simulated solar light. The photoexcited electrons (e^-) are transferred from the VB (e^-_{VB}) to the CB (e^-_{CB}), leaving holes (h^+) in the VB (h^+_{VB}) of the photocatalyst (Equation (21)). Due to this electronic transfer, electron/hole pairs (e^-/h^+) are photogenerated that migrate to the surface of the photocatalyst to react directly with the adsorbed species, such as H_2O (($H_2O)_{ads}$), OH^- (OH^-_{ads}), O_2 (($O_2)_{ads}$), and other molecules. These (e^-/h^+) pairs can also recombine immediately after their formation (Equation (22)). On the other hand, the photogenerated holes (h^+) in the VB of TiO_2 promote oxidation of both adsorbed water molecules and hydroxyl ions to produce highly reactive hydroxyl radicals (Equations (23) and (24)). Furthermore, the holes (h^+) can migrate towards the surface of the photocatalyst to create more reactive radicals and oxidize the molecules adsorbed on the surface. Lanthanide ions (Ln^{n+}) have empty 5d orbitals, so they can trap photoexcited electrons in the CB of TiO_2 (Equation (25)); however, these electrons are very unstable, so they are quickly transferred to oxygenated molecules adsorbed on the surface of the photocatalyst. Due to this transfer, radical superoxide anions ($^\bullet O_2^-$) and hydroxyl radicals ($^\bullet OH$) are generated through a series of sequential reactions (Equations (26)–(29)). The following reactions suggest the likely pathway for ROS photogeneration on the surface of $Ln^{(n+)}/TiO_2$ ($Ln^{(n+)} = La^{(3+)}$, $Ce^{(3+/4+)}$, or $Eu^{(3+)}$) [73,89]:

$$Ln^{n+}/TiO_2 \xrightarrow{h\nu} Ln^{n+}/TiO_2 + e^-_{CB} + h^+_{VB} \quad (21)$$

$$e^-_{CB} + h^+_{VB} \rightarrow heat \quad (22)$$

$$(H_2O)_{ads} + h^+_{VB} \rightleftharpoons (H^+ + OH^-)_{ads} + h^+_{VB} \rightarrow OH^\bullet_{ads} \quad (23)$$

$$OH^-_{ads} + h^+_{VB} \rightarrow OH^\bullet_{ads} \quad (24)$$

$$Ln^{(n+)} + e^-_{CB} \rightarrow Ln^{(n+)-1} \quad (25)$$

$$Ln^{(n+)-1} + (O_2)_{ads} \rightarrow Ln^{(n+)} + O_2^{\bullet -} \quad (26)$$

$$O_2^{\bullet -} + H^+ \rightarrow HO_2^\bullet \quad (27)$$

$$2HO_2^\bullet \rightarrow H_2O_2 + O_2 \quad (28)$$

$$H_2O_2 + e^-_{CB} \rightarrow OH^\bullet + OH^- \quad (29)$$

Cyanide photodegradation occurs at the surface of the photocatalyst as a complex process involving transfer of multiple electrons and protons from the catalyst to cyanide species. Therefore, one of the key steps in photocatalytic degradation of cyanide is adsorption of cyanide ions on the surface of the photocatalyst [93].

Photocatalytic degradation of cyanide involves various reactive oxygen species (ROS), such as hydroxyl radicals ($^\bullet OH$), superoxide radicals ($^\bullet O_2^-$), and singlet oxygen ($^1O_2$), among others. As is well-known, the hydroxyl radical ($^\bullet OH$) is the most reactive species and can oxidize a wide range of contaminants, including cyanide. Oxidation of cyanide by hydroxyl radicals ($^\bullet OH$) can proceed through several pathways. In the biocatalytic pathway, the hydroxyl radical attacks the carbon atom of the cyanide ion to form formamide (NH_2CHO), which can then react with the hydroxyl radicals to form formic acid (HCOOH) and ammonia (NH_3) [94].

In the photocatalytic pathway, the hydroxyl radical attacks the nitrogen atom of the cyanide ion to form cyanate (CNO$^-$) (Equation (30)), which can be oxidized to produce nitrogen gas, carbon dioxide gas, nitrate, or nitrite, and bicarbonate (Equations (31)–(34)) [95–98]. It is worth mentioning that hydrolysis of cyanate to bicarbonate and ammonium ions (Equation (35)) is more favored under acidic conditions (pH < 7) [99]. On the other hand, direct oxidation of cyanide can also occur if the molecule reacts directly with the photogenerated holes [79].

$$CN^- + 2\,OH^\bullet_{ads} \rightarrow CNO^- + H_2O \tag{30}$$

$$OCN^- + OH^\bullet_{ads} \rightarrow CO_2(g) + 1/2\,N_2(g) + H^+ \tag{31}$$

$$OCN^- + 3\,OH^\bullet_{ads} \rightarrow HCO_3^- + 1/2\,N_2(g) + H_2O \tag{32}$$

$$OCN^- + 6\,OH^\bullet_{ads} \rightarrow HCO_3^- + NO_2^- + H^+ + 2H_2O \tag{33}$$

$$OCN^- + 8\,OH^\bullet_{ads} \rightarrow HCO_3^- + NO_3^- + H^+ + 3H_2O \tag{34}$$

$$OCN^- + H^+ + 2\,H_2O \rightarrow HCO_3^- + NH_4^+ \tag{35}$$

In addition to oxidation of cyanide by hydroxyl radicals, electrons in the conduction band of the photocatalyst may also participate in the degradation process. Electrons can reduce oxidized species (CNO$^-$, CO$_2$, N$_2$) to less harmful and more stable forms, such as formate (HCOO$^-$) or bicarbonate (HCO$_3^-$). The reduction process is driven by transfer of electrons from the photocatalyst to the oxidized species [94].

In summary, photocatalytic cyanide degradation is a process that can reduce cyanide toxicity in wastewater by generating reactive species through light absorption by a photocatalyst. The photodegradation mechanism of cyanide in an aqueous solution involves a series of reactions that depend on the type of photocatalyst used and the reaction conditions, including the initial cyanide concentration, light intensity, and the presence of other contaminants in the solution [91]. Although there are many factors that can affect the effectiveness of the cyanide photocatalysis process, the process has great potential as a treatment strategy for cyanide-contaminated wastewater.

4.4.2. Effect of the Photogenerated Radicals

As shown in Figure 12, the efficiency of nanoparticles for photodegradation of cyanide species decreased with incorporation of i-POH and EtOH. This probably occurs due to quenching of h$^+$ and $^\bullet$OH, suggesting that these play a crucial role in photocatalytic degradation of cyanide species. However, Figure 12 did not show a significant effect on efficiency of photodegradation using p-BQ, probably because the presence of the radical $^\bullet$O$_2^-$ is not essential for photocatalytic degradation of cyanide species under the conditions evaluated in this study. These results are in agreement with those informed by other authors [100].

4.5. Total Efficiency and Reuse of Nanoparticles

In this study, it was shown that photodegradation of cyanide species is more efficient than adsorption of these species under the test conditions (Figure 13). As is known, adsorption is a process that can be affected by a series of parameters, such as sorbent properties, sorbate properties, solution conditions, among others. In fact, in this study, it was shown that the pH of the solution had an important effect on adsorption capacity of cyanide species on the surface of nanoparticles. Thus, Figure 6 shows that the maximum adsorption capacity of cyanide species occurs at pH values > 9. However, since the adsorption tests were performed at pH = 7, it is possible that the relatively low adsorption percentages shown in Figure 13 were due to the limited electrostatic attraction between cyanide species (pKa = 9.4) and nanoparticle surface (pH$_{ZPC}$ = 7). Although adsorption of cyanide species was limited by the pH of the solution, in this study, it was also possible to demonstrate that incorporation of lanthanide elements, particularly lanthanum, in the TiO$_2$ structure is a good alternative to improve adsorption of these species on the semiconductor surface. Possibly, this was because incorporation of lanthanide elements as dopants enables generation of Lewis acid sites on the TiO_2 surface, which undoubtedly contributes to improving the adsorption capacity of a semiconductor.

Likewise, in this study, it was demonstrated that the combination of the adsorption and photocatalysis processes enables improvement in the efficiency of removal of cyanide species from aqueous solutions. This is probably because the cyanide species that first adsorbed and accumulated on the nanoparticle surface at the beginning of photocatalytic degradation are the ones that first degraded under simulated solar light. Consequently, the constant migration and successive photocatalytic oxidation on the surface of the nanoparticles certainly contribute to improving the removal efficiency at the solid–liquid interface. This is due to generation of a concentration gradient, which acts as the main driving force of the removal process of cyanide species from an aqueous solution. From these results, it is suggested that, in this study, removal of cyanide species is due to the cooperative effect between both adsorption and photocatalysis processes. Thus, the synthesized nanoparticles are effective for removal of cyanide from aqueous systems because they have coupled "adsorption–photodegradation" performances [88].

On the other hand, it is well-known that both the useful life of a material and its potential applications are closely related to its structural and chemical stability. In this study, an experiment was designed to evaluate reuse of synthesized nanoparticles in order to estimate their effectiveness after five cycles of consecutive use. The results of this experiment are shown in Figure 14. From this figure, it could be inferred that percentage of cyanide removal decreases with each treatment cycle. However, after five consecutive cycles, loss of cyanide removal capacity of the nanoparticles did not exceed 20% on average. In fact, in this study, it was found that La/TO nanoparticles had a lower loss of effectiveness (15.3%) at the end of the fifth reuse cycle compared to Ce/TO (16.6%) and Eu/TO (17.8%) nanoparticles. The loss in effectiveness was probably due to chemical adsorption of cyanide species on the surface of the nanoparticles since the possible formation of covalent bonds and complexes could decrease the availability of active sites on the surface. At the end of the fifth cycle, the chemical composition of the nanoparticles was verified by ICP analysis, confirming that there was no lanthanide leakage into the solution. Consequently, the synthesized nanoparticles (TO, La/TO, Ce/TO, and Eu/TO) are stable and maintain adequate activity until the fifth treatment cycle, being able to effectively remove cyanide species in aqueous solution.

Finally, Table 5 compares the maximum adsorption capacity (mg g^{-1}) of the present nanoparticles and some adsorbents used for cyanide removal from aqueous solutions.

Table 5. Comparison of adsorption capacity (mg g^{-1}) of various materials for cyanide removal.

Adsorbent	q_{max} (mg g^{-1})	Isotherm Model	Kinetic Model	Reference
ZnO	275.00	Langmuir	Pseudo-second-order	[43]
NiO	185.00	Langmuir	Pseudo-first-order	[43]
ZnO-NiO	320.00	Langmuir	Pseudo-second-order	[43]
LTA zeolite modified with HDMTMAB	24.09	Langmuir	-	[44]
Activated Carbon (AC)	78.10	Redlich-Peterson	Pseudo-second-order	[48]
Fe$_2$O$_3$/AC	86.20	Redlich-Peterson	Pseudo-second-order	[48]
TiO$_2$/AC	90.90	Redlich-Peterson	Pseudo-second-order	[48]
ZnO/AC	91.70	Redlich-Peterson	Pseudo-second-order	[48]
TiO$_2$/Fe$_2$O$_3$/AC	96.20	Langmuir	Pseudo-second-order	[48]
ZnO/Fe$_2$O$_3$/AC	101.00	Langmuir	Pseudo-second-order	[48]
Clay-K	253.98	-	Pseudo-second-order	[91]
TiO$_2$/Fe$_2$O$_3$	124.87	-	Pseudo-second-order	[91]
Fe-MFI zeolite	33.98	Langmuir	Pseudo-second-order	[101]
Activated Periwinkle Shell Carbon (APSC)	2.85	Langmuir	Pseudo-second-order	[102]

Table 5. Cont.

Adsorbent	q_{max} (mg g^{-1})	Isotherm Model	Kinetic Model	Reference
SiO$_2$/TiO$_2$	39.79	Temkin	Pseudo-second-order	[103]
Activated Carbon (AC)	1.66	Freundlich	Pseudo-second-order	[104]
LDH loaded MMB	80	Langmuir	-	[105]
Corncob biochar	2.57	Langmuir	-	[106]
TiO$_2$	46.48	Langmuir	Pseudo-second-order	In this study
La/TiO$_2$	54.96	Langmuir	Pseudo-second-order	In this study
Ce/TiO$_2$	51.39	Langmuir	Pseudo-second-order	In this study
Eu/TiO$_2$	49.25	Langmuir	Pseudo-second-order	In this study

Likewise, Table 6 compares the photodegradation efficiency (%) of the present nanoparticles and some photocatalysts used for cyanide removal from aqueous solutions.

Table 6. Comparison of photodegradation efficiency (%) of various materials for cyanide removal.

Material	[CN] (mg L^{-1})	[Catalyst] (g L^{-1})	Time (min)	Efficiency (%)	Reference
TiO$_2$/Fe$_2$O$_3$/zeolite	200	1.4	160	89	[48]
TiO$_2$/Fe$_2$O$_3$/PAC	300	1.4	170	97	[48]
Blast furnace sludge (BFS)	750	2.0	120	97	[91]
Cts-Ag	71.6	2.5	180	98	[100]
SiO$_2$/TiO$_2$	61.54	3.5	30	96	[103]
Fe^{2+}	10	0.14	30	86	[107]
TiO$_2$	30	0.05	60	72	[108]
Co/TiO$_2$/SiO$_2$	100	2.0	60	55	[109]
TiO$_2$/SiO$_2$	100	1.7	180	93	[110]
Ce/ZnO	250	4.0	180	84	[111]
Degussa P25 TiO$_2$	13.2	0.1	60	73	[112]
Cu(II)-cryptate, complex 1	6.07	0.5–1.0	180	77	[113]
S–TiO$_2$@rGO-FeTCPP	100	1.6	120	75	[114]
N-rGO-ZnO-CoPc(COOH)$_8$	25	2.0	120	91	[115]
TiO$_2$/ZSM-5	71.55	2.5	240	94	[116]
Me (Fe, Mn)-N/TiO$_2$/SiO$_2$	75	2.5	120	97	[117]
Carbon/nano-TiO$_2$	61.53	3.0	180	98	[118]
rGO/TiO$_2$ P25	50	1.0	180	100	[119]
TiO$_2$	20	0.2	90	84	In this study
La/TiO$_2$	20	0.2	90	97	In this study
Ce/TiO$_2$	20	0.2	90	89	In this study
Eu/TiO$_2$	20	0.2	90	86	In this study

As can be seen in Tables 5 and 6, among the nanoparticles synthesized in this study, La/TiO$_2$ showed the highest capacity for adsorption and photodegradation of cyanide species in aqueous solutions. However, compared to other materials based on metal oxides (ZnO, NiO, Fe$_2$O$_3$) reported in the literature, La/TiO$_2$ (q_{max} = 54.96 mg g^{-1}) has a lower adsorption capacity for cyanide species under the tested conditions, although it turned out to be a better adsorbent material for cyanide species than certain zeolitic materials and biochar. On the other hand, La/TiO$_2$ also proved to be more efficient for cyanide photodegradation than the other nanoparticles synthesized in this study. Likewise, La/TiO$_2$ proved to be the same and even more efficient for cyanide photodegradation than other TiO$_2$-based photocatalytic materials recently reported in the literature, reaching an efficiency of 97% after 90 min of reaction, with a catalyst load of 0.2 g L^{-1} and a cyanide concentration in the solution of 20 mg L^{-1}.

Based on the adsorption and photodegradation results obtained in this study and comparing them with those reported in the literature cited above, it is suggested that La/TO, Ce/TO, Eu/TO, and TO nanoparticles, in that order, are effective for removal of cyanide species from aqueous solutions. Furthermore, the evidence from this study suggests that the removal of cyanide species was governed by a combination of electrostatic interactions, complex formation by coordinate covalent bonds, and photo-oxidation on the surface of the synthesized nanoparticles when the reaction systems were exposed to simulated solar light.

5. Conclusions

The evidence from this study indicates that the adsorption and photodegradation capacity of the nanoparticles synthesized in the present study (TiO_2, La/TiO_2, Ce/TiO_2, and Eu/TiO_2) were influenced by several factors. Some of these were dependent on material properties (chemical composition, bandgap energy, crystal structure, morphology, particle size, specific surface area, and pH_{PZC}), as well as the operating conditions (initial concentration of adsorbate, contact time, system temperature, and pH of the solution) of the removal processes used. These results are in agreement with those reported by other authors [120]. Investigation of these parameters enabled us to obtain valuable information on the mechanisms of adsorption and photocatalysis, which was fundamental to determining the best operational conditions for efficient removal of cyanide species from aqueous solutions at neutral pH. Likewise, synergistic coupling of the adsorption and photocatalysis processes enabled significant improvement in the capacity of the TiO_2 semiconductor for effective removal of cyanide species from aqueous solutions.

Our results showed that doping of the TiO_2 semiconductor with lanthanide elements La, Ce, and Eu represents a promising alternative to improve the adsorption capacity of the TiO_2 semiconductor and extend its photoresponse to the visible light region. Among the doped nanoparticles, La/TO was more efficient than Ce/TO and Eu/TO for total cyanide removal, probably because the La/TO nanoparticles had a higher specific surface area (126 $m^2\ g^{-1}$) in comparison with the specific surface area of Ce/TO (104 $m^2\ g^{-1}$) and Eu/TO (96 $m^2\ g^{-1}$). Regarding cyanide photodegradation, it was also influenced by type of doping lanthanide ion. La/TO nanoparticles were slightly more effective under simulated solar light than Ce/TO and Eu/TO nanoparticles, possibly because the La/TO bandgap energy (E_g = 3.10 eV) was lower than that of Ce/TO (E_g = 3.14 eV) and Eu/TO (E_g = 3.16 eV), which is essential for absorption of light and generation of pairs of electrons and voids for photodegradation of pollutants. On the other hand, reuse of doped nanoparticles was also affected by type of lanthanide element. In this study, it was found that La/TO nanoparticles had lower loss of effectiveness (15.3%) at the end of the fifth reuse cycle compared to Ce/TO (16.6%) and Eu/TO (17.8%) nanoparticles, probably due to the smaller particle size of La/TO compared to Ce/TO and Eu/TO, which could provide greater structural stability to the photocatalyst in aqueous reaction systems. Therefore, the evidence from this comparative study suggests that each doping element has an intrinsic effect on the properties of TiO_2, making it essential to understand these effects in the design of effective materials for wastewater treatment.

In summary, this study provides information on the potential capacity of doping TiO_2 nanoparticles with lanthanides (La, Ce, and Eu) as an innovative and effective approach to improve the adsorption and photocatalysis properties of a semiconductor. In addition, it highlights the synergistic effect of combining both techniques, adsorption and photocatalysis, to achieve complete and efficient removal of cyanide in wastewater, which is valuable for development of new technologies that contribute to treatment and recovery of hydric resources. Finally, it supports use of solar energy as a lighting source during removal of cyanide in aqueous systems, contributing to generation of beneficial effects in terms of environmental impact, energy efficiency, and remediation costs, thus promoting sustainable practices.

Author Contributions: Conceptualization, X.J.-F.; methodology, X.J.-F. and R.L.; software, X.J.-F.; validation, X.J.-F.; formal analysis, X.J.-F.; investigation, X.J.-F. and R.L.; resources, X.J.-F.; data curation, X.J.-F.; writing—original draft preparation, X.J.-F.; writing—review and editing, X.J.-F.; supervision, X.J.-F. All authors have read and agreed to the published version of the manuscript.

Funding: This research received no external funding.

Institutional Review Board Statement: Not applicable.

Informed Consent Statement: Not applicable.

Data Availability Statement: Data are contained within the article.

Acknowledgments: The authors would like to thank Xavier Cattoen from the Institut NEEL-CNRS (Grenoble-France) for the SEM measurements. This work was financially supported by Universidad Técnica Particular de Loja (Ecuador).

Conflicts of Interest: The authors declare no conflict of interest.

References

1. González-Valoys, A.C.; Arrocha, J.; Monteza-Destro, T.; Vargas-Lombardo, M.; Esbrí, J.M.; Garcia-Ordiales, E.; Jiménez-Ballesta, R.; García-Navarro, F.J.; Higueras, P. Environmental challenges related to cyanidation in Central American gold mining; the Remance mine (Panama). *J. Environ. Manag.* **2022**, *302*, 113979. [CrossRef] [PubMed]
2. Alvillo-Rivera, A.; Garrido-Hoyos, S.; Buitrón, G.; Thangarasu-Sarasvathi, P.; Rosano-Ortega, G. Biological treatment for the degradation of cyanide: A review. *J. Mater. Res. Technol.* **2021**, *12*, 1418–1433. [CrossRef]
3. Zhang, Y.; Zhang, Y.; Huang, Y.; Chen, X.; Cui, H.; Wang, M. Enhanced photocatalytic reaction and mechanism for treating cyanide-containing wastewater by silicon-based nano-titania. *Hydrometallurgy* **2020**, *198*, 105512. [CrossRef]
4. Cosmos, A.; Erdenekhuyag, B.O.; Yao, G.; Li, H.; Zhao, J.; Laijun, W.; Lyu, X. Principles and methods of bio detoxification of cyanide contaminants. *J. Mater. Cycles Waste Manag.* **2020**, *22*, 939–954. [CrossRef]
5. Das, P.P.; Anweshan, A.; Mondal, P.; Sinha, A.; Biswas, P.; Sarkar, S.; Purkait, M.K. Integrated ozonation assisted electrocoagulation process for the removal of cyanide from steel industry wastewater. *Chemosphere* **2021**, *263*, 128370. [CrossRef]
6. Li, X.-M.; Wang, Q.-W.; Zhan, P.-Y.; Pan, Y.-R. Synthesis, Crystal Structure and Theoretical Calculations of a Cadmium(II) Coordination Polymer Assembled by 4,4′-Oxydibenzoic Acid and 1,3-Bis(imidazol-1-ylmethyl)-Benzene Ligands. *J. Chem. Crystallogr.* **2016**, *46*, 163–169. [CrossRef]
7. Eletta, O.; Ajayi, O.; Ogunleye, O.; Akpan, I. Adsorption of cyanide from aqueous solution using calcinated eggshells: Equilibrium and optimisation studies. *J. Environ. Chem. Eng.* **2016**, *4*, 1367–1375. [CrossRef]
8. Núñez-Salas, R.E.; Hernández-Ramírez, A.; Hinojosa-Reyes, L.; Guzmán-Mar, J.L.; Villanueva-Rodríguez, M.; Maya-Treviño, M.D.L. Cyanide degradation in aqueous solution by heterogeneous photocatalysis using boron-doped zinc oxide. *Catal. Today* **2019**, *328*, 202–209. [CrossRef]
9. Maulana, I.; Takahashi, F. Cyanide removal study by raw and iron-modified synthetic zeolites in batch adsorption experiments. *J. Water Process. Eng.* **2018**, *22*, 80–86. [CrossRef]
10. Upadhyay, G.K.; Rajput, J.K.; Pathak, T.K.; Kumar, V.; Purohit, L. Synthesis of ZnO:TiO$_2$ nanocomposites for photocatalyst application in visible light. *Vacuum* **2019**, *160*, 154–163. [CrossRef]
11. Siwińska-Stefańska, K.; Kubiak, A.; Piasecki, A.; Goscianska, J.; Nowaczyk, G.; Jurga, S.; Jesionowski, T. TiO$_2$-ZnO Binary Oxide Systems: Comprehensive Characterization and Tests of Photocatalytic Activity. *Materials* **2018**, *11*, 841. [CrossRef] [PubMed]
12. Belver, C.; Hinojosa, M.; Bedia, J.; Tobajas, M.; Alvarez, M.A.; Rodríguez-González, V.; Rodriguez, J.J. Ag-Coated Heterostructures of ZnO-TiO$_2$/Delaminated Montmorillonite as Solar Photocatalysts. *Materials* **2017**, *10*, 960. [CrossRef] [PubMed]
13. Sridevi, A.; Ramji, B.; Venkatesan, G.P.; Sugumaran, V.; Selvakumar, P. A facile synthesis of TiO$_2$/BiOCl and TiO$_2$/BiOCl/La2O3 heterostructure photocatalyst for enhanced charge separation efficiency with improved UV-light catalytic activity towards Rhodamine B and Reactive Yellow 86. *Inorg. Chem. Commun.* **2021**, *130*, 108715. [CrossRef]
14. Zhang, L.; Djellabi, R.; Su, P.; Wang, Y.; Zhao, J. Through converting the surface complex on TiO$_2$ nanorods to generate superoxide and singlet oxygen to remove CN. *J. Environ. Sci.* **2023**, *124*, 300–309. [CrossRef]
15. Coronel, S.; Endara, D.; Lozada, A.; Manangón-Perugachi, L.; de la Torre, E. Photocatalytic Study of Cyanide Oxidation Using Titanium Dioxide (TiO$_2$)-Activated Carbon Composites in a Continuous Flow Photo-Reactor. *Catalysts* **2021**, *11*, 924. [CrossRef]
16. Kanakaraju, D.; Chandrasekaran, A. Recent advances in TiO$_2$/ZnS-based binary and ternary photocatalysts for the degradation of organic pollutants. *Sci. Total Environ.* **2023**, *868*, 161525. [CrossRef] [PubMed]
17. Dubsok, A.; Khamdahsag, P.; Kittipongvises, S. Life cycle environmental impact assessment of cyanate removal in mine tailings wastewater by nano-TiO$_2$/FeCl3 photocatalysis. *J. Clean. Prod.* **2022**, *366*, 132928. [CrossRef]
18. Lincho, J.; Zaleska-Medynska, A.; Martins, R.C.; Gomes, J. Nanostructured photocatalysts for the abatement of contaminants by photocatalysis and photocatalytic ozonation: An overview. *Sci. Total Environ.* **2022**, *837*, 155776. [CrossRef]
19. Fang, W.; Xing, M.; Zhang, J. Modifications on reduced titanium dioxide photocatalysts: A review. *J. Photochem. Photobiol. C Photochem. Rev.* **2017**, *32*, 21–39. [CrossRef]
20. Shoueir, K.; Kandil, S.; El-Hosainy, H.; El-Kemary, M. Tailoring the surface reactivity of plasmonic Au@TiO$_2$ photocatalyst bio-based chitosan fiber towards cleaner of harmful water pollutants under visible-light irradiation. *J. Clean. Prod.* **2019**, *230*, 383–393. [CrossRef]

21. Elleuch, L.; Messaoud, M.; Djebali, K.; Attafi, M.; Cherni, Y.; Kasmi, M.; Elaoud, A.; Trabelsi, I.; Chatti, A. A new insight into highly contaminated landfill leachate treatment using Kefir grains pre-treatment combined with Ag-doped TiO_2 photocatalytic process. *J. Hazard. Mater.* **2019**, *382*, 121119. [CrossRef] [PubMed]
22. Kadi, M.W.; Hameed, A.; Mohamed, R.; Ismail, I.M.; Alangari, Y.; Cheng, H.-M. The effect of Pt nanoparticles distribution on the removal of cyanide by TiO_2 coated Al-MCM-41 in blue light exposure. *Arab. J. Chem.* **2019**, *12*, 957–965. [CrossRef]
23. Basavarajappa, P.S.; Patil, S.B.; Ganganagappa, N.; Reddy, K.R.; Raghu, A.V.; Reddy, C.V. Recent progress in metal-doped TiO_2, non-metal doped/codoped TiO_2 and TiO_2 nanostructured hybrids for enhanced photocatalysis. *Int. J. Hydrogen Energy* **2020**, *45*, 7764–7778. [CrossRef]
24. Ndabankulu, V.O.; Maddila, S.; Jonnalagadda, S.B. Synthesis of lanthanide-doped TiO_2 nanoparticles and their photocatalytic activity under visible light. *Can. J. Chem.* **2019**, *97*, 672–681. [CrossRef]
25. Mazierski, P.; Mikolajczyk, A.; Bajorowicz, B.; Malankowska, A.; Zaleska-Medynska, A.; Nadolna, J. The role of lanthanides in TiO_2-based photocatalysis: A review. *Appl. Catal. B Environ.* **2018**, *233*, 301–317. [CrossRef]
26. Han, M.; Dong, Z.; Liu, J.; Ren, G.; Ling, M.; Yang, X.; Zhang, L.; Xue, B.; Li, F. The role of lanthanum in improving the visible-light photocatalytic activity of TiO_2 nanoparticles prepared by hydrothermal method. *Appl. Phys. A* **2020**, *126*, 1–10. [CrossRef]
27. Tang, X.; Xue, Q.; Qi, X.; Cheng, C.; Yang, M.; Yang, T.; Chen, F.; Qiu, F.; Quan, X. DFT and experimental study on visible-light driven photocatalysis of rare-earth-doped TiO_2. *Vacuum* **2022**, *200*, 110972. [CrossRef]
28. Priyanka, K.; Revathy, V.; Rosmin, P.; Thrivedu, B.; Elsa, K.; Nimmymol, J.; Balakrishna, K.; Varghese, T. Influence of La doping on structural and optical properties of TiO_2 nanocrystals. *Mater. Charact.* **2016**, *113*, 144–151. [CrossRef]
29. Khan, S.; Cho, H.; Kim, D.; Han, S.S.; Lee, K.H.; Cho, S.-H.; Song, T.; Choi, H. Defect engineering toward strong photocatalysis of Nb-doped anatase TiO_2: Computational predictions and experimental verifications. *Appl. Catal. B Environ.* **2017**, *206*, 520–530. [CrossRef]
30. Wang, R.; An, S.; Zhang, J.; Song, J.; Wang, F. Existence form of lathanum and its improving mechanism of visible-light-driven La-F co-doped TiO_2. *J. Rare Earths* **2020**, *38*, 39–45. [CrossRef]
31. Lopes Colpani, G.; Zanetti, M.; Carla Frezza Zeferino, R.; Luiz Silva, L.; Maria Muneron de Mello, J.; Gracher Riella, H.; Padoin, N.; Antônio Fiori, M.; Soares, C. Lanthanides effects on TiO_2 photocatalysts. In *Photocatalysts–Applications and Attributes*; IntechOpen: London, UK, 2019; ISBN 978-1-78985-476-3.
32. Islam, S.; Alshoaibi, A.; Bakhtiar, H.; Alamer, K.; Mazher, J.; Alhashem, Z.H.; Hatshan, M.R.; Aleithan, S.H. Mesoporous CdTe supported SiO_2-TiO_2 nanocomposite: Structural, optical, and photocatalytic applications. *Mater. Res. Bull.* **2023**, *161*, 112172. [CrossRef]
33. Kumar, V.V.; Naresh, G.; Sudhakar, M.; Tardio, J.; Bhargava, S.K.; Venugopal, A. Role of Brønsted and Lewis acid sites on Ni/TiO_2 catalyst for vapour phase hydrogenation of levulinic acid: Kinetic and mechanistic study. *Appl. Catal. A Gen.* **2015**, *505*, 217–223. [CrossRef]
34. Prakash, J.; Samriti; Kumar, A.; Dai, H.; Janegitz, B.C.; Krishnan, V.; Swart, H.C.; Sun, S. Novel rare earth metal–doped one-dimensional TiO_2 nanostructures: Fundamentals and multifunctional applications. *Mater. Today Sustain.* **2021**, *13*, 100066. [CrossRef]
35. Jaramillo-Fierro, X.; Gaona, S.; Valarezo, E. La^{3+}'s Effect on the Surface (101) of Anatase for Methylene Blue Dye Removal, a DFT Study. *Molecules* **2022**, *27*, 6370. [CrossRef] [PubMed]
36. Jaramillo-Fierro, X.; González, S.; Medina, F. La-Doped $ZnTiO_3$/TiO_2 Nanocomposite Supported on Ecuadorian Diatomaceous Earth as a Highly Efficient Photocatalyst Driven by Solar Light. *Molecules* **2021**, *26*, 6232. [CrossRef]
37. Benmessaoud, A.; Nibou, D.; Mekatel, E.H.; Amokrane, S. A Comparative Study of the Linear and Non-Linear Methods for Determination of the Optimum Equilibrium Isotherm for Adsorption of Pb^{2+} Ions onto Algerian Treated Clay. *Iran. J. Chem. Chem. Eng.* **2020**, *39*, 153–171. [CrossRef]
38. Jaramillo-Fierro, X.; Cuenca, G.; Ramón, J. Comparative Study of the Effect of Doping $ZnTiO_3$ with Rare Earths (La and Ce) on the Adsorption and Photodegradation of Cyanide in Aqueous Systems. *Int. J. Mol. Sci.* **2023**, *24*, 3780. [CrossRef] [PubMed]
39. Wang, Z.; Yuan, T.; Yao, J.; Li, J.; Jin, Y.; Cheng, J.; Shen, Z. Development of an unmanned device with picric acid strip for on-site rapid detections of sodium cyanide in marine water. *IOP Conf. Ser. Earth Environ. Sci.* **2021**, *734*, 012026. [CrossRef]
40. Pramitha, A.R.; Harijono, H.; Wulan, S.N. Comparison of cyanide content in arbila beans (Phaseolus lunatus l) of East Nusa Tenggara using picrate and acid hydrolysis methods. *IOP Conf. Ser. Earth Environ. Sci.* **2021**, *924*, 012031. [CrossRef]
41. Castada, H.Z.; Liu, J.; Barringer, S.A.; Huang, X. Cyanogenesis in Macadamia and Direct Analysis of Hydrogen Cyanide in Macadamia Flowers, Leaves, Husks, and Nuts Using Selected Ion Flow Tube–Mass Spectrometry. *Foods* **2020**, *9*, 174. [CrossRef]
42. Tsunatu, D.; Taura, U.; Jirah, E. Kinetic Studies of Bio-sorption of Cyanide Ions from Aqueous Solution Using Carbon Black Developed from Shea Butter Seed Husk as an Adsorbent. *Am. Chem. Sci. J.* **2015**, *8*, 1–12. [CrossRef]
43. Pirmoradi, M.; Hashemian, S.; Shayesteh, M.R. Kinetics and thermodynamics of cyanide removal by ZnO@NiO nanocrystals. *Trans. Nonferrous Met. Soc. China* **2017**, *27*, 1394–1403. [CrossRef]
44. Noroozi, R.; Al-Musawi, T.J.; Kazemian, H.; Kalhori, E.M.; Zarrabi, M. Removal of cyanide using surface-modified Linde Type-A zeolite nanoparticles as an efficient and eco-friendly material. *J. Water Process. Eng.* **2018**, *21*, 44–51. [CrossRef]
45. Tran, H.N.; You, S.-J.; Hosseini-Bandegharaei, A.; Chao, H.-P. Mistakes and inconsistencies regarding adsorption of contaminants from aqueous solutions: A critical review. *Water Res.* **2017**, *120*, 88–116. [CrossRef]

46. Zhou, X.; Zhou, X. The unit problem in the thermodynamic calculation of adsorption using the Langmuir equation. *Chem. Eng. Commun.* **2014**, *201*, 1459–1467. [CrossRef]
47. Bettoni, M.; Falcinelli, S.; Rol, C.; Rosi, M.; Sebastiani, G. Gas-Phase TiO$_2$ Photosensitized Mineralization of Some VOCs: Mechanistic Suggestions through a Langmuir–Hinshelwood Kinetic Approach. *Catalysts* **2021**, *11*, 20. [CrossRef]
48. Eskandari, P.; Farhadian, M.; Nazar, A.R.S.; Goshadrou, A. Cyanide adsorption on activated carbon impregnated with ZnO, Fe2O3, TiO$_2$ nanometal oxides: A comparative study. *Int. J. Environ. Sci. Technol.* **2021**, *18*, 297–316. [CrossRef]
49. Wang, W.; Huang, G.; Yu, J.C.; Wong, P.K. Advances in photocatalytic disinfection of bacteria: Development of photocatalysts and mechanisms. *J. Environ. Sci.* **2015**, *34*, 232–247. [CrossRef] [PubMed]
50. Razali, N.A.; Conte, M.; McGregor, J. The role of impurities in the La2O3 catalysed carboxylation of crude glycerol. *Catal. Lett.* **2019**, *149*, 1403–1414. [CrossRef]
51. Ma, R.; Islam, M.J.; Reddy, D.A.; Kim, T.K. Transformation of CeO$_2$ into a mixed phase CeO$_2$/Ce$_2$O$_3$ nanohybrid by liquid phase pulsed laser ablation for enhanced photocatalytic activity through Z-scheme pattern. *Ceram. Int.* **2016**, *42*, 18495–18502. [CrossRef]
52. Hernández-Castillo, Y.; García-Hernández, M.; López-Marure, A.; Luna-Domínguez, J.H.; López-Camacho, P.Y.; Ramirez, A.M. Antioxidant activity of cerium oxide as a function of europium doped content. *Ceram. Int.* **2018**, *45*, 2303–2308. [CrossRef]
53. Stanković, D.M.; Kukuruzar, A.; Savić, S.; Ognjanović, M.; Janković-Častvan, I.M.; Roglić, G.; Antić, B.; Manojlović, D.; Dojčinović, B. Sponge-like europium oxide from hollow carbon sphere as a template for an anode material for Reactive Blue 52 electrochemical degradation. *Mater. Chem. Phys.* **2021**, *273*, 125154. [CrossRef]
54. Holzwarth, U.; Gibson, N. The Scherrer equation versus the 'Debye-Scherrer equation'. *Nat. Nanotechnol.* **2011**, *6*, 534. [CrossRef] [PubMed]
55. Juan, J.L.X.; Maldonado, C.S.; Sánchez, R.A.L.; Díaz, O.J.E.; Ronquillo, M.R.R.; Sandoval-Rangel, L.; Aguilar, N.P.; Delgado, N.A.R.; Martínez-Vargas, D.X. TiO$_2$ doped with europium (Eu): Synthesis, characterization and catalytic performance on pesticide degradation under solar irradiation. *Catal. Today* **2021**, *394-396*, 304–313. [CrossRef]
56. Anilkumar, P.; Kalaivani, T.; Deepak, S.; Jasmin, J.; El-Rehim, A.A.; Kumar, E.R. Evaluation of structural, optical and morphological properties of La doped TiO$_2$ nanoparticles. *Ceram. Int.* **2023**. [CrossRef]
57. Oliveira, L.D.S.; Barbosa, E.F.; Martins, F.C.B.; Silva, G.D.F.; Rezende, T.K.d.L.; Filho, J.C.S.; Barbosa, H.P.; Andrade, A.A.; de Oliveira, L.F.C.; Góes, M.S.; et al. Emission of TiO$_2$: Y^{3+} and Eu^{3+} in water medium, under UV excitation and band gap theoretical calculus. *J. Lumin.* **2023**, *257*, 119639. [CrossRef]
58. Yan, Z.; Yang, X.; Gao, G.; Gao, R.; Zhang, T.; Tian, M.; Su, H.; Wang, S. Understanding of photocatalytic partial oxidation of methanol to methyl formate on surface doped La(Ce) TiO$_2$: Experiment and DFT calculation. *J. Catal.* **2022**, *411*, 31–40. [CrossRef]
59. Mink, J.; Skripkin, M.; Hajba, L.; Németh, C.; Abbasi, A.; Sandström, M. Infrared and Raman spectroscopic and theoretical studies of nonaaqua complexes of trivalent rare earth metal ions. *Spectrochim. Acta Part A Mol. Biomol. Spectrosc.* **2005**, *61*, 1639–1645. [CrossRef]
60. Ramakrishnan, V.M.; Natarajan, M.; Santhanam, A.; Asokan, V.; Velauthapillai, D. Size controlled synthesis of TiO$_2$ nanoparticles by modified solvothermal method towards effective photo catalytic and photovoltaic applications. *Mater. Res. Bull.* **2018**, *97*, 351–360. [CrossRef]
61. Rueden, C.T.; Schindelin, J.; Hiner, M.C.; Dezonia, B.E.; Walter, A.E.; Arena, E.T.; Eliceiri, K.W. ImageJ2: ImageJ for the next generation of scientific image data. *BMC Bioinform.* **2017**, *18*, 529. [CrossRef]
62. Schneider, C.A.; Rasband, W.S.; Eliceiri, K.W. NIH Image to ImageJ: 25 Years of image analysis. *Nat. Methods* **2012**, *9*, 671–675. [CrossRef] [PubMed]
63. Vorontsov, A.V.; Tsybulya, S.V. Influence of Nanoparticles Size on XRD Patterns for Small Monodisperse Nanoparticles of Cu^0 and TiO$_2$ Anatase. *Ind. Eng. Chem. Res.* **2018**, *57*, 2526–2536. [CrossRef]
64. Nada, A.A.; Nasr, M.; Viter, R.; Miele, P.; Roualdes, S.; Bechelany, M. Mesoporous ZnFe$_2$O$_4$@TiO$_2$ nanofibers prepared by electrospinning coupled to PECVD as highly performing photocatalytic materials. *J. Phys. Chem. C* **2017**, *121*. [CrossRef]
65. Orellana, W. D-π-A dye attached on TiO$_2$(101) and TiO$_2$(001) surfaces: Electron transfer properties from ab initio calculations. *Sol. Energy* **2021**, *216*, 266–273. [CrossRef]
66. Mehrabi, M.; Javanbakht, V. Photocatalytic degradation of cationic and anionic dyes by a novel nanophotocatalyst of TiO$_2$/ZnTiO$_3$/αFe$_2$O$_3$ by ultraviolet light irradiation. *J. Mater. Sci. Mater. Electron.* **2018**, *29*, 9908–9919. [CrossRef]
67. Kosmulski, M. The pH dependent surface charging and points of zero charge. VII. Update. *Adv. Colloid Interface Sci.* **2018**, *251*, 115–138. [CrossRef]
68. Bakatula, E.N.; Richard, D.; Neculita, C.M.; Zagury, G.J. Determination of point of zero charge of natural organic materials. *Environ. Sci. Pollut. Res.* **2018**, *25*, 7823–7833. [CrossRef]
69. Stavropoulos, G.; Skodras, G.; Papadimitriou, K. Effect of solution chemistry on cyanide adsorption in activated carbon. *Appl. Therm. Eng.* **2015**, *74*, 182–185. [CrossRef]
70. Dash, R.R.; Gaur, A.; Balomajumder, C. Cyanide in industrial wastewaters and its removal: A review on biotreatment. *J. Hazard. Mater.* **2009**, *163*, 1–11. [CrossRef]
71. Gupta, N.; Balomajumder, C.; Agarwal, V.K. Adsorption of cyanide ion on pressmud surface: A modeling approach. *Chem. Eng. J.* **2012**, *191*, 548–556. [CrossRef]

72. Saxena, S.; Prasad, M.; Amritphale, S.; Chandra, N. Adsorption of cyanide from aqueous solutions at pyrophyllite surface. *Sep. Purif. Technol.* **2001**, *24*, 263–270. [CrossRef]
73. Biswas, P.; Bhunia, P.; Saha, P.; Sarkar, S.; Chandel, H.; De, S. In situ photodecyanation of steel industry wastewater in a pilot scale. *Environ. Sci. Pollut. Res.* **2020**, *27*, 33226–33233. [CrossRef]
74. Li, R.; Li, T.; Zhou, Q. Impact of Titanium Dioxide (TiO_2) Modification on Its Application to Pollution Treatment—A Review. *Catalysts* **2020**, *10*, 804. [CrossRef]
75. Nada, E.A.; El-Maghrabi, H.H.; Raynaud, P.; Ali, H.R.; El-Wahab, S.A.; Sabry, D.Y.; Moustafa, Y.M.; Nada, A.A. Enhanced Photocatalytic Activity of WS_2/TiO_2 Nanofibers for Degradation of Phenol under Visible Light Irradiation. *Inorganics* **2022**, *10*, 54. [CrossRef]
76. He, L.; Meng, J.; Feng, J.; Yao, F.; Zhang, L.; Zhang, Z.; Liu, X.; Zhang, H. Investigation of $4f$-Related Electronic Transitions of Rare-Earth Doped ZnO Luminescent Materials: Insights from First-Principles Calculations. *ChemPhysChem* **2019**, *21*, 51–58. [CrossRef] [PubMed]
77. Tanyi, A.R.; Rafieh, A.I.; Ekaneyaka, P.; Tan, A.L.; Young, D.J.; Zheng, Z.; Chellappan, V.; Subramanian, G.S.; Chandrakanthi, R. Enhanced efficiency of dye-sensitized solar cells based on Mg and La co-doped TiO_2 photoanodes. *Electrochim. Acta* **2015**, *178*, 240–248. [CrossRef]
78. Armaković, S.J.; Grujić-Brojčin, M.; Šćepanović, M.; Armaković, S.; Golubović, A.; Babić, B.; Abramović, B.F. Efficiency of La-doped TiO_2 calcined at different temperatures in photocatalytic degradation of β-blockers. *Arab. J. Chem.* **2019**, *12*, 5355–5369. [CrossRef]
79. Wang, B.; Zhang, G.; Sun, Z.; Zheng, S.; Frost, R.L. A comparative study about the influence of metal ions (Ce, La and V) doping on the solar-light-induced photodegradation toward rhodamine B. *J. Environ. Chem. Eng.* **2015**, *3*, 1444–1451. [CrossRef]
80. Bispo, A.G.; Ceccato, D.A.; Lima, S.A.M.; Pires, A.M. Red phosphor based on Eu^{3+}-isoelectronically doped Ba_2SiO_4 obtained via sol-gel route for solid state lightning. *RSC Adv.* **2017**, *7*, 53752–53762. [CrossRef]
81. Orđević, V.; Milićević, B.; Dramićanin, M.D. Rare earth-doped anatase TiO_2 nanoparticles. In *Titanium Dioxide*; Janus, M., Ed.; IntechOpen: Rijeka, Croatia, 2017.
82. Štengl, V.; Bakardjieva, S.; Murafa, N. Preparation and photocatalytic activity of rare earth doped TiO_2 nanoparticles. *Mater. Chem. Phys.* **2009**, *114*, 217–226. [CrossRef]
83. Gharaei, S.K.; Abbasnejad, M.; Maezono, R. Bandgap reduction of photocatalytic TiO_2 nanotube by Cu doping. *Sci. Rep.* **2018**, *8*, 14192. [CrossRef]
84. Chahal, S.; Singh, S.; Kumar, A.; Kumar, P. Oxygen-deficient lanthanum doped cerium oxide nanoparticles for potential applications in spintronics and photocatalysis. *Vacuum* **2020**, *177*, 109395. [CrossRef]
85. Elsellami, L.; Lachheb, H.; Houas, A. Synthesis, characterization and photocatalytic activity of Li-, Cd-, and La-doped TiO_2. *Mater. Sci. Semicond. Process.* **2015**, *36*, 103–114. [CrossRef]
86. Rouster, P.; Pavlovic, M.; Szilagyi, I. Destabilization of Titania Nanosheet Suspensions by Inorganic Salts: Hofmeister Series and Schulze-Hardy Rule. *J. Phys. Chem. B* **2017**, *121*, 6749–6758. [CrossRef] [PubMed]
87. Muráth, S.; Sáringer, S.; Somosi, Z.; Szilágyi, I. Effect of Ionic Compounds of Different Valences on the Stability of Titanium Oxide Colloids. *Colloids Interfaces* **2018**, *2*, 32. [CrossRef]
88. Muthirulan, P.; Devi, C.N.; Sundaram, M.M. Synchronous role of coupled adsorption and photocatalytic degradation on CAC–TiO_2 composite generating excellent mineralization of alizarin cyanine green dye in aqueous solution. *Arab. J. Chem.* **2017**, *10*, S1477–S1483. [CrossRef]
89. Eskandarloo, H.; Badiei, A.; Behnajady, M.A.; Tavakoli, A.; Ziarani, G.M. Ultrasonic-assisted synthesis of Ce doped cubic–hexagonal $ZnTiO_3$ with highly efficient sonocatalytic activity. *Ultrason. Sonochemistry* **2016**, *29*, 258–269. [CrossRef]
90. Shu, Q.; Liu, X.; Huo, Y.; Tan, Y.; Zhang, C.; Zou, L. Construction of a Brönsted-Lewis solid acid catalyst La-PW-SiO_2/SWCNTs based on electron withdrawing effect of La(III) on π bond of SWCNTs for biodiesel synthesis from esterification of oleic acid and methanol. *Chin. J. Chem. Eng.* **2022**, *44*, 351–362. [CrossRef]
91. Amaro-Medina, B.M.; Martinez-Luevanos, A.; Soria-Aguilar, M.D.J.; Sanchez-Castillo, M.A.; Estrada-Flores, S.; Carrillo-Pedroza, F.R. Efficiency of Adsorption and Photodegradation of Composite TiO_2/Fe_2O_3 and Industrial Wastes in Cyanide Removal. *Water* **2022**, *14*, 3502. [CrossRef]
92. Chaker, H.; Ameur, N.; Saidi-Bendahou, K.; Djennas, M.; Fourmentin, S. Modeling and Box-Behnken design optimization of photocatalytic parameters for efficient removal of dye by lanthanum-doped mesoporous TiO_2. *J. Environ. Chem. Eng.* **2021**, *9*, 104584. [CrossRef]
93. Meng, Y.-J.; Chen, S.-S.; Luo, C.-B.; Song, Y.-J.; Xiong, Z.-W.; Li, J.; Li, D.-Q. Competitive coordination strategy for preparing TiO_2/C nanocomposite with adsorption-photocatalytic synergistic effect. *Appl. Surf. Sci.* **2022**, *603*, 154395. [CrossRef]
94. Martínková, L.; Bojarová, P.; Sedova, A.; Křen, V. Recent trends in the treatment of cyanide-containing effluents: Comparison of different approaches. *Crit. Rev. Environ. Sci. Technol.* **2023**, *53*, 416–434. [CrossRef]
95. Mediavilla, J.J.V.; Perez, B.F.; de Cordoba, M.C.F.; Espina, J.A.; Ania, C.O. Photochemical Degradation of Cyanides and Thiocyanates from an Industrial Wastewater. *Molecules* **2019**, *24*, 1373. [CrossRef]
96. Betancourt-Buitrago, L.A.; Hernandez-Ramirez, A.; Colina-Marquez, J.A.; Bustillo-Lecompte, C.F.; Rehmann, L.; Machuca-Martinez, F. Recent Developments in the Photocatalytic Treatment of Cyanide Wastewater: An Approach to Remediation and Recovery of Metals. *Processes* **2019**, *7*, 225. [CrossRef]

97. Dagaut, P.; Glarborg, P.; Alzueta, M.U. The oxidation of hydrogen cyanide and related chemistry. *Prog. Energy Combust. Sci.* **2008**, *34*, 1–46. [CrossRef]
98. Luque-Almagro, V.M.; Cabello, P.; Sáez, L.P.; Olaya-Abril, A.; Moreno-Vivián, C.; Roldán, M.D. Exploring anaerobic environments for cyanide and cyano-derivatives microbial degradation. *Appl. Microbiol. Biotechnol.* **2018**, *102*, 1067–1074. [CrossRef] [PubMed]
99. Eskandari, P.; Farhadian, M.; Solaimany Nazar, A.R.; Jeon, B.H. Adsorption and Photodegradation Efficiency of $TiO_2/Fe_2O_3/PAC$ and TiO_2/Fe_2O_3/Zeolite Nanophotocatalysts for the Removal of Cyanide. *Ind. Eng. Chem. Res.* **2019**, *58*, 2099–2112. [CrossRef]
100. Pan, Y.; Zhang, Y.; Huang, Y.; Jia, Y.; Chen, L.; Cui, H. Functional Ag-doped coralloid titanosilicate zeolite (CTS-Ag) for efficiently catalytic and photodegradative removal of free cyanides and copper/zinc-cyanide complexes in real wastewater. *J. Alloys Compd.* **2022**, *926*, 166848. [CrossRef]
101. Amamou, S.; Cheniti-Belcadhi, L. Tutoring in Project-Based Learning. *Proc. Procedia Comput. Sci.* **2018**, *126*, 176–185. [CrossRef]
102. Eke-Emezie, N.; Etuk, B.R.; Akpan, O.P.; Chinweoke, O.C. Cyanide removal from cassava wastewater onto H_3PO_4 activated periwinkle shell carbon. *Appl. Water Sci.* **2022**, *12*, 1–12. [CrossRef]
103. Wei, P.; Zhang, Y.; Huang, Y.; Chen, L. Structural design of SiO_2/TiO_2 materials and their adsorption-photocatalytic activities and mechanism of treating cyanide wastewater. *J. Mol. Liq.* **2023**, *377*, 121519. [CrossRef]
104. Aliprandini, P.; Veiga, M.M.; Marshall, B.G.; Scarazzato, T.; Espinosa, D.C. Investigation of mercury cyanide adsorption from synthetic wastewater aqueous solution on granular activated carbon. *J. Water Process. Eng.* **2020**, *34*, 101154. [CrossRef]
105. Ravuru, S.S.; Jana, A.; De, S. Cyanide removal from blast furnace effluent using layered double hydroxide based mixed matrix beads: Batch and fixed bed study. *J. Clean. Prod.* **2022**, *371*, 133634. [CrossRef]
106. Xiong, Q.; Jiang, S.; Fang, R.; Chen, L.; Liu, S.; Liu, Y.; Yin, S.; Hou, H.; Wu, X. An environmental-friendly approach to remove cyanide in gold smelting pulp by chlorination aided and corncob biochar: Performance and mechanisms. *J. Hazard. Mater.* **2021**, *408*, 124465. [CrossRef]
107. Golbaz, S.; Jafari, A.J.; Kalantari, R.R. The study of Fenton oxidation process efficiency in the simultaneous removal of phenol, cyanide, and chromium (VI) from synthetic wastewater. *Desalination Water Treat.* **2013**, *51*, 5761–5767. [CrossRef]
108. Kim, S.H.; Lee, S.W.; Lee, G.M.; Lee, B.-T.; Yun, S.-T.; Kim, S.-O. Monitoring of TiO_2-catalytic UV-LED photo-oxidation of cyanide contained in mine wastewater and leachate. *Chemosphere* **2016**, *143*, 106–114. [CrossRef]
109. Baeissa, E.S. Photocatalytic removal of cyanide by cobalt metal doped on TiO_2–SiO_2 nanoparticles by photo-assisted deposition and impregnation methods. *J. Ind. Eng. Chem.* **2014**, *20*, 3761–3766. [CrossRef]
110. Faisal, M.; Jalalah, M.; Harraz, F.A.; El-Toni, A.M.; Labis, J.P.; Al-Assiri, M. A novel $Ag/PANI/ZnTiO_3$ ternary nanocomposite as a highly efficient visible-light-driven photocatalyst. *Sep. Purif. Technol.* **2021**, *256*, 117847. [CrossRef]
111. Karunakaran, C.; Gomathisankar, P.; Manikandan, G. Preparation and characterization of antimicrobial Ce-doped ZnO nanoparticles for photocatalytic detoxification of cyanide. *Mater. Chem. Phys.* **2010**, *123*, 585–594. [CrossRef]
112. Andrade, L.R.S.; Cruz, I.A.; de Melo, L.; Vilar, D.D.S.; Fuess, L.T.; de Silva, G.R.; Manhães, V.M.S.; Torres, N.H.; Soriano, R.N.; Bharagava, R.N.; et al. Oyster shell-based alkalinization and photocatalytic removal of cyanide as low-cost stabilization approaches for enhanced biogas production from cassava starch wastewater. *Process. Saf. Environ. Prot.* **2020**, *139*, 47–59. [CrossRef]
113. Ghosh, T.K.; Ghosh, R.; Chakraborty, S.; Saha, P.; Sarkar, S.; Ghosh, P. Cyanide contaminated water treatment by di-nuclear Cu(II)-cryptate: A supramolecular approach. *J. Water Process. Eng.* **2020**, *37*, 101364. [CrossRef]
114. Pattanayak, D.S.; Mishra, J.; Nanda, J.; Sahoo, P.K.; Kumar, R.; Sahoo, N.K. Photocatalytic degradation of cyanide using polyurethane foam immobilized Fe-TCPP-S-TiO_2-rGO nano-composite. *J. Environ. Manag.* **2021**, *297*, 113312. [CrossRef] [PubMed]
115. Bhattacharya, S.; Das, A.A.; Dhal, G.C.; Sahoo, P.K.; Tripathi, A.; Sahoo, N.K. Evaluation of N doped rGO-ZnO-CoPc$(COOH)_8$ nanocomposite in cyanide degradation and its bactericidal activities. *J. Environ. Manag.* **2022**, *302*, 114022. [CrossRef] [PubMed]
116. Pan, Y.; Zhang, Y.; Huang, Y.; Jia, Y.; Chen, L.; Cui, H. Synergistic effect of adsorptive photocatalytic oxidation and degradation mechanism of cyanides and Cu/Zn complexes over TiO_2/ZSM-5 in real wastewater. *J. Hazard. Mater.* **2021**, *416*, 125802. [CrossRef] [PubMed]
117. Zhang, Y.; Zhang, Y.; Huang, Y.; Jia, Y.; Chen, L.; Pan, Y.; Wang, M. Adsorptive-photocatalytic performance and mechanism of Me (Mn, Fe)-N co-doped TiO_2/SiO_2 in cyanide wastewater. *J. Alloys Compd.* **2021**, *867*, 159020. [CrossRef]
118. Li, P.; Zhang, Y.; Huang, Y.; Chen, L. Activity and mechanism of macroporous carbon/nano-TiO_2 composite photocatalyst for treatment of cyanide wastewater. *Colloids Surf. A Physicochem. Eng. Asp.* **2023**, *658*, 130728. [CrossRef]
119. Albiter, E.; Barrera-Andrade, J.M.; Calzada, L.A.; García-Valdés, J.; Valenzuela, M.A.; Rojas-García, E. Enhancing Free Cyanide Photocatalytic Oxidation by rGO/TiO_2 P25 Composites. *Materials* **2022**, *15*, 5284. [CrossRef]
120. Daou, I.; Zegaoui, O.; Elghazouani, A. Physicochemical and photocatalytic properties of the ZnO particles synthesized by two different methods using three different precursors. *Comptes Rendus Chim.* **2017**, *20*, 47–54. [CrossRef]

Disclaimer/Publisher's Note: The statements, opinions and data contained in all publications are solely those of the individual author(s) and contributor(s) and not of MDPI and/or the editor(s). MDPI and/or the editor(s) disclaim responsibility for any injury to people or property resulting from any ideas, methods, instructions or products referred to in the content.

Article

Scanning Photocurrent Microscopy in Single Crystal Multidimensional Hybrid Lead Bromide Perovskites

Elena Segura-Sanchis, Rocío García-Aboal ᵈ, Roberto Fenollosa, Fernando Ramiro-Manzano * and Pedro Atienzar *

Instituto de Tecnología Química, Consejo Superior de Investigaciones Científicas, Universitat Politècnica de València, Avenida de los Naranjos s/n, 46022 Valencia, Spain; elsesan@itq.upv.es (E.S.-S.); rociogarciaaboal@gmail.com (R.G.-A.); rfenollo@ter.upv.es (R.F.)
* Correspondence: ferraman@fis.upv.es (F.R.-M.); pedatcor@itq.upv.es (P.A.)

Citation: Segura-Sanchis, E.; García-Aboal, R.; Fenollosa, R.; Ramiro-Manzano, F.; Atienzar, P. Scanning Photocurrent Microscopy in Single Crystal Multidimensional Hybrid Lead Bromide Perovskites. *Nanomaterials* **2023**, *13*, 2570. https://doi.org/10.3390/nano13182570

Academic Editor: Thomas Dippong

Received: 7 August 2023
Revised: 8 September 2023
Accepted: 12 September 2023
Published: 16 September 2023

Copyright: © 2023 by the authors. Licensee MDPI, Basel, Switzerland. This article is an open access article distributed under the terms and conditions of the Creative Commons Attribution (CC BY) license (https://creativecommons.org/licenses/by/4.0/).

Abstract: We investigated solution-grown single crystals of multidimensional 2D–3D hybrid lead bromide perovskites using spatially resolved photocurrent and photoluminescence. Scanning photocurrent microscopy (SPCM) measurements where the electrodes consisted of a dip probe contact and a back contact. The crystals revealed significant differences between 3D and multidimensional 2D–3D perovskites under biased detection, not only in terms of photocarrier decay length values but also in the spatial dynamics across the crystal. In general, the photocurrent maps indicate that the closer the border proximity, the shorter the effective decay length, thus suggesting a determinant role of the border recombination centers in monocrystalline samples. In this case, multidimensional 2D–3D perovskites exhibited a simple fitting model consisting of a single exponential, while 3D perovskites demonstrated two distinct charge carrier migration dynamics within the crystal: fast and slow. Although the first one matches that of the 2D–3D perovskite, the long decay of the 3D sample exhibits a value two orders of magnitude larger. This difference could be attributed to the presence of interlayer screening and a larger exciton binding energy of the multidimensional 2D–3D perovskites with respect to their 3D counterparts.

Keywords: perovskite; diffusion length; multidimensional; dynamic; SPCM; single crystal

1. Introduction

Hybrid halide perovskites are highly promising materials, not only for the next-generation photovoltaics but also in many other fields where optoelectronics plays a key role. These applications include light-emitting diodes (LEDs), photodetectors, lasers, sensors, and field-effect transistors [1]. Their tunable bandgap, defect tolerance, long carrier diffusion length and high carrier mobility are noteworthy. In addition, they can be easily processed with a low fabrication cost. On the other hand, the huge amount of possibilities that these materials offer in terms of composition [2], morphology and synthetic methods makes a deep understanding of their intrinsic properties mandatory in order to know their potential advantages in commercial applications.

Typically, hybrid perovskites are based on a structural base such as ABX_3 in the three dimensions, where A and B correspond to monovalent and divalent cations, respectively, with different sizes, and X are halides. However, these perovskites have certain limitations, mainly related to the stability to moisture. The reduction of the perovskite dimension in the form of a 2D layered structure, analogous to a conventional van der Waals material, has been shown to attenuate such an undesirable effect by providing hydrophobic shielding. This is achieved, for instance, by introducing a large organic cation (M) between $[PbX_6]^{4-}$ layers as a spacer. In addition, 2D structures yield other effects related to quantum confinement, band alignment, passivation of trap states and inhibition of ion movement [3–5]. On the other hand, the quantum confinement confers a strong exciton binding energy, which hinders their suitability for use in solar cells because the free carrier generation is significantly suppressed. In this regard, multidimensional perovskites that combine 2D–3D structures have proven to be a promising alternative, because they represent a tradeoff between stability and performance. The general formula of these perovskites is $M_{2(n)}A_{n-1}B_nX_{3n+1}$, where n represents the number of metal halide interlayers, estimated from the stichometry of the precursors [6,7]. Therefore, the n value defines the bulk-like block and thus the 3D-like behavior inside the van der Waals stacking.

One of the most important parameters employed in the evaluation of various perovskite structures and compositions in terms of charge transport is the diffusion length [8]. This parameter corresponds to the average distance an excited charge carrier can travel towards a collecting electrode before recombining through mechanisms such as radiative recombination or trap-assisted recombination [9]. It has already been demonstrated that large single crystals of perovskites containing iodine in their structure can achieve diffusion lengths ranging from several to even thousands of micrometers [10,11]. Furthermore, hybrid perovskite materials containing iodide ions in their structure have been observed to exhibit longer electron and hole diffusion lengths compared to bromide perovskites [12]. However, it is worth noting that $MAPbBr_3$ perovskite materials are significantly more stable under ambient conditions than $MAPbI_3$ perovskite materials [13], making this composition preferable for studying intrinsic electronic properties. The introduction of a large cation tends to reduce the diffusion length of the photogenerated carriers, although it remains within the same order of magnitude as that of 3D perovskites [10]. However, different values have been reported in the literature and depend on the experimental tool used for evaluating it. For instance, optical probing methods based on photoluminescence or transient absorption do not monitor all the possible charges produced, such as those thermally emitted from traps states [14]. Furthermore, various types of charge carrier transport have been studied. For example, Guo et al. [14] proposed different transport regimes, including quasiballistic transport where charge carriers can travel long distances without significant scattering or collisions, nonequilibrium transport depending on conditions and energy input, and diffusive transport characterized by random scattering events, leading to a random walk-like behavior [14,15]. Keeping in mind that the transport is strongly influenced by defects and the grain boundaries the interfaces present in polycrystalline films could be a determinant in their photocarrier response. Furthermore, the generation of hot carriers with excess energy, as well as the formation of excitons and free charges in hybrid perovskites, have implications for the transport processes [16].

A versatile technique for studying the carriers' generation and transport in semiconductors during the excitation is the use of a scanning confocal photocurrent microscopy (SCPM) [17–23], which allows the current generated in different regions of the sample by optical excitation to be mapped. Additionally, fluorescence information can be simultaneously acquired, which makes it possible to analyze the radiative processes of the samples being studied. This technique has been widely employed to investigate the morphology and photocurrent generation in a wide range of semiconducting materials, including silicon [24], cadmium telluride [25], lead selenide [26], organic heterojunctions [27] and others [28]. In this context, we developed a confocal microscopy tool that allowed us to demonstrate the behavior of hybrid perovskites and octahedral molybdenum clusters as Fabry–Pérot cavities [29,30], as well as the behavior of perovskite nanowires as photonic waveguides [31], and to measure the enhanced photoresponse of perovskites containing subphthalocyanines [32]. In all the cases, the studies were realized on micrometer-size single crystals because they allow perturbing factors which are present in a poly-crystalline sample such as the grain boundaries to be disregarded. Parameters such as size, shape, crystal orientation and composition can have a significant influence, but generally, single crystals exhibit much better optoelectronic properties than their polycrystalline counterparts [33,34].

In this work, we take a step further by introducing a scanning excitation system and a punctual electrical probe that can be positioned freely on the sample area with micrometer precision in a top (and back) contact configuration (Figure 1). This approach is similar to the widely used technique of SCPM (all back contact) that allows estimating the photocarrier decay lengths under a certain bias. This decay length corresponds to the diffusion length (L_d) evaluating the sample region without the influence of the external electric field or under low bias [35,36]. This method offers advantages over other types of experiments where L_D is deduced from a combination of carrier diffusion (D) and lifetime measurements (τ), ($L_d = \sqrt{D\tau}$).

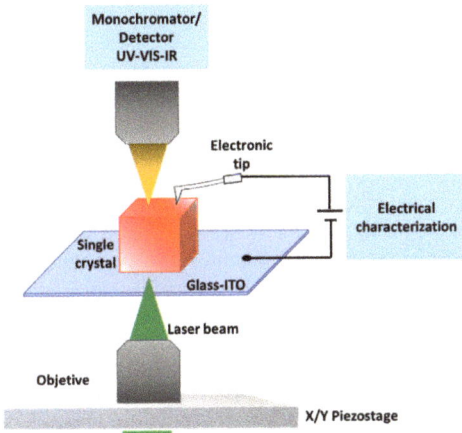

Figure 1. Schematic representation of the electric contact on the microcrystalline samples, showing the irradiation objective at the bottom part controlled with a piezoelectric stage.

There are some examples reported in the literature that study the diffusion length over single crystals using an SCPM system. For instance, Shreetu et al. demonstrated on millimeter-sized crystals that hybrid perovskites having iodine in the structure show a long in-plane charge carrier diffusion length of above 7–15 µm depending on the n value in the structure [10]. Regarding single crystals of perovskites containing bromide halide in their structure, Ganesh et al. reported diffusion length values of 13.3 µm and 13.8 µm for electrons and holes, respectively [37]. In addition, Zhang et al. studied the role of chlorine incorporation in the perovskite structure, resulting in diffusion lengths of several hundreds of micrometers [38]. In any case, the direct extraction of L_d requires several conditions to be fulfilled. For a 1D system, L_d must be much larger than the cross section and the distance between contacts is chosen to be larger than L_d [39] or the region to characterize is placed out of the electrode gap [35]. In this way, L_d from lithographically patterned strips, synthetized nanotubes or fibers could be extracted. In addition, other effects or parameters such as the influence of a bias voltage, the charge drift length, or the determination of the minority carriers have being studied [35]. In principle, this single exponential decay model of L_d could be directly extended to planar schemes by changing the electrode shape from an ideal single point to an infinite line [28]. This is a difficult condition to fulfill. However, assuming that the sample thickness is smaller than L_d, it has been proven that L_d could be extracted (employing a single exponential decay) regardless of the width of the planar sample [40]. On the other hand, this technique permits evaluating the inhomogeneity of different sections when a back-contact planar scheme is used [36]. In fact, it simplifies the study methodology because the sample is directly sensitized on the electrodes, formed by transparent conductive contacts with micrometer gaps, thus eliminating the need for a sample transfer process [35].

Herein, we present a comparative study of the photocurrent decay length in 3D and multidimensional 2D–3D perovskites planar micrometer-size single crystals that have bromide in their structure. Our aim is to compare various azimuth sections, focusing particularly on their possible dependency with the outer border distance. Our scheme does not fulfill the requirements of the single exponential L_d extraction, because the contact is a mixture of a point contact and a planar surface and our sample thickness (tens of µm) surpass L_d values (for the case of the multidimensional sample) in the literature; thus, it could not be assumed to be a 2D sample/measurement scheme. Nevertheless, our point probe scheme makes it possible to measure the photocurrent decay at various azimuthal angles. Since the L_d conditions are not fulfilled, a certain bias has been chosen (−1.25V) to acquire low-noise signals from samples with low photocurrent efficiency, such as multidimensional 2D–3D perovskites.

2. Experiment

2.1. Perovskite Synthesis

For the preparation of the single crystal hybrid perovskite, the antisolvent method was employed. Firstly, this method consists of dissolving the precursors in a small amount of DMF (1 mL). For 3D perovskites (MAPbBr$_3$), the composition consisted of 0.5 M methyl ammonium bromide (MABr) and 0.5 M lead bromide (PbBr$_2$), while for the 2D–3D perovskite ((PEA)$_2$(MA)$_{n-1}$Pb$_n$Br$_{3n+1}$, n = 10), it was 0.4 M MABr, 0.5 M PbBr$_2$ and 0.2 M phenylethyl ammonium bromide (PEABr). This mixture (20 µL) was placed on a conductive substrate (ITO on glass, supplied by Ossila (Sheffield, UK)) on a Teflon stage inside a sealed vial, and THF was used as the antisolvent (5 mL) (see Figure S6). The vessel was completely sealed and kept in the dark until small orange-pinkish crystals were observed. At least 24 h are required to obtain good-quality crystals larger than 20 µm. The crystals were manually selected using Gel-Pack (Hayward, CA, USA) and a micromanipulator Narishige (Tokyo, Japan) MMN-1 coupled to a Nikon (Tokyo, Japan) Eclipse LV100 microscope to perform the corresponding characterization. MABr and PbBr$_2$ were purchased from ABCR (Karlsruhe, Germany). THF and DMF were purchased from Sigma-Aldrich (St. Luis, MO, USA) and PEABr.

2.2. Experimental Set-Up for SCPC Measurements

With the purpose of studying the photoresponse of both 3D and multidimensional 2D–3D perovskites, we placed a selected crystal onto a glass substrate coated with indium-tin oxide (ITO), which served as the back-contact electrode (Figure 1). Then an electronic probe was positioned on the crystal. This established the top contact. The light of a laser with λ = 405 nm and P = 350 µW was chopped and focused onto the sample through the back-contact transparent electrode by means of a microscope objective that was mounted on a piezo system. The photocurrent signal from the sample was acquired with a lock-in amplifier through a transimpedance amplifier that provides the bias voltage as well. The photocurrent response was mapped by varying the objective position with the help of the piezo system. At the same time, the photoluminescence signal was acquired employing a photodiode connected to a second lock-in amplifier. This measurement makes it possible to check the crystal stability condition. With the aim of limiting the sample degradation and minimizing possible fluctuations in the response [41], the scanning time was minimized by setting the position control of the piezo system to the open-loop mode. In this case, the signals that drive the piezo actuator are not continuously corrected according to the sensed position (closed-loop scheme). As a result, the elapsed time for the characterization (scanning area of 80 µm × 80 µm) is reduced about three times, from 791.7 s (open loop) to 243.2 s (closed loop), acquiring only the forward scan direction of zig-zag paths. Despite the non-linear response of the piezo system, which may introduce some non-linear deviations in position, this technique is widely used due to its speed and the absence of (reliable) position sensors [42].

2.3. Characterization Techniques

XRD patterns of the powders were recorded on a Philips X'PERT diffractometer (Amsterdam, The Netherlands) that was equipped with a proportional detector and a secondary graphite monochromator. The data were collected stepwise over the range

2θ = 2–20°, at steps of 0.02°, an accumulation time of 20 s per step, using the Cu Kα radiation (λ = 1.54178 Å).

UV–Vis optical spectroscopy of the perovskite powders was carried out using a Cary 5G spectrophotometer (Santa Clara, CA, USA) and CaSO$_4$ as reference.

Field-emission scanning electron microscopy (FESEM) images were recorded with a Zeiss (Jena, Germany) Ultra 55 field FESEM apparatus.

Lifetime photoluminescence measurements on the single crystals were carried out using an inverted microscope, Nikon (Tokyo, Japan) Ti2-U, equipped with an XY motorized stage. The emission signal was transmitted through optical fibers to an Edinburgh Instruments (Livingston, UK) FLS1100 spectrofluorometer, which was coupled to a cooled photomultiplier (PMT-980). The measurements were performed at room temperature, utilizing a 405 nm excitation wavelength provided by a picosecond (ps) laser diode incorporated into the microscope. The lifetimes (τ) were calculated from the best fitting of the signal to a single-exponential decay (I(t) = I(0)exp(−t/τ)).

3. Results and Discussion

3.1. Crystal Characterization

The studied single crystals were characterized by XRD. In Figure 2A, the X-ray patterns correspond to the 3D structure perovskite, exhibiting characteristic peaks at 14.77, 29.95, and 45.74 degrees, which are assigned to the (100), (200), and (300) crystal planes, respectively [43–45]. This high-intensity XRD diffraction pattern confirms a high phase purity, indicating a preferred orientation in the (001) plane. For the 2D–3D perovskite, additional lower-angle diffraction peaks at 5.10, 10.51, 15.93, 21.17, 26.73, 32.26, 37.7, and 42.9 degrees are observed, corresponding to (001), (002), (003), (004), (005), (006), (007), and (008), respectively. These peaks are characteristic of the 2D phase within these multidimensional 2D–3D structures. Furthermore, they also indicate preferential growth in the (001) plane, suggesting that the 2D phase predominantly grows along the organic or inorganic layer planes [46–49].

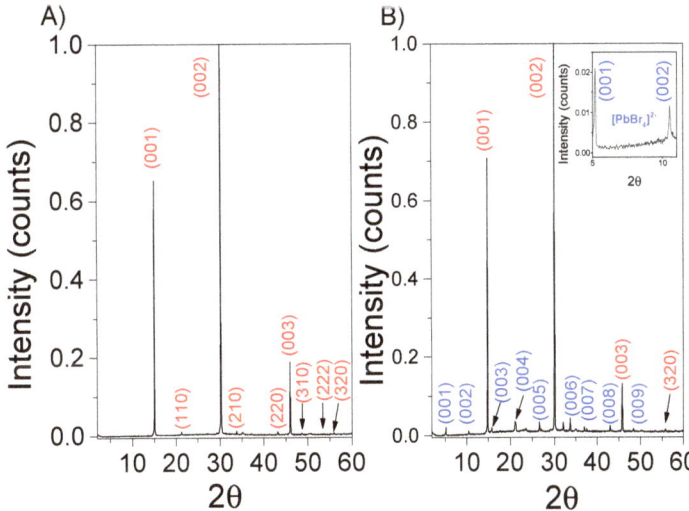

Figure 2. X-ray diffraction patterns for 3D (**A**) and 2D–3D perovskite (**B**). The inset shows a magnification of the low angles region that is characteristic of the layered phase. Red peaks correspond to the 3D phase and blue peaks to the 2D phase (PbBr$_4$)$^{2-}$.

The samples were also characterized by electron microscopy (see Figure 3). In both cases, cubic-shaped crystals with a size of approximately 50–100 μm can be observed. In the case of mixed 2D–3D structure (Figure 3B), the crystals exhibit a more random shape and size distribution, with the formation of lateral heterostructures attributed to the 2D phase. This effect is associated with the presence of the long-chain PEA cation during the crystal growth, which is consistent with previous studies [50].

Figure 3. FESEM image of crystalline 3D (**A**) and 2D–3D Perovskite (**B**).

The optical properties of the 3D and 2D–3D perovskites were first obtained by measuring the UV–Vis diffuse reflectance spectra and the photoluminescence in polycrystalline samples. The optical bandgap was calculated to be 2.18 and 2.22 eV for the 3D and 2D–3D, respectively, (Figure S1, ESI) through their corresponding Tauc Plots, which agrees with the previous reported values [49]. It should be noted that the bandgap of the 2D–3D perovskite presents a slight shift at a high energy value due dielectric quantum confinement effects corresponding to the 2D phase (see inset in Figure 4B) [51].

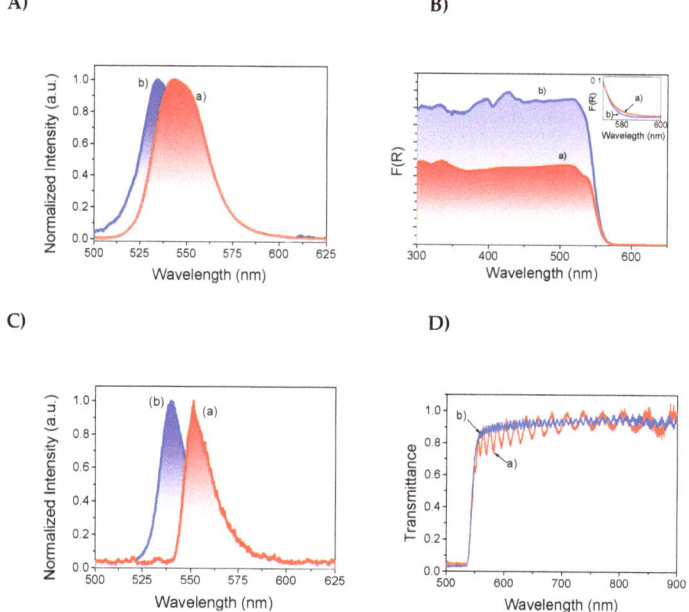

Figure 4. (**A**) Photoluminescence (PL) emission spectra of polycrystalline samples 3D (a) and 2D–3D (b). (**B**) Diffuse reflectance UV–Vis absorption spectra (plotted on the Kubelka–Munk function of the reflectance) F(R) of (a) 3D and (b) mixed 2D–3D perovskite; the inset shows a magnification of the absorption edge of both samples. (**C**) PL emission of single crystals measured with our SCPM and (**D**) transmittance spectrum of the single crystals of (a) 3D and (b) mixed 2D–3D perovskite. They have a thickness of 4 and 13 µm, respectively.

The steady-state luminescence on polycrystalline films shows the characteristic peak emission located at 545 and 535 nm for the 3D and 2D–3D perovskites, respectively (Figure 4A). It should be noted that these samples exhibit a broad emission band compared to the emission observed in selected single crystals, where the emission peaks are well defined (Figure 4C); this is one reason why single crystals are preferred for optoelectronic applications. The broadness of the emission band in polycrystalline films is generally attributed to the scattering of the light by the particles and the defects on the surface and interphases [32]. In addition, comparative temporal profiles of fluorescence decays for both 3D and 2D–3D perovskite single crystals (Figure S5, ESI) allow us to study changes in carrier lifetimes of these perovskites, resulting in values of 9.70 ns for 3D and 18.39 ns for 2D–3D perovskite, respectively. The longer fluorescence lifetime observed in 2D–3D perovskite, compared to 3D perovskite, can be attributed to the reduced nonradiative recombination due to the defect passivation effect of PEABr, which is consistent with previous reports. [52,53]. Nonetheless, and likely due to the limited carrier mobility in a screened layered structure, the prolonged lifetime of 2D–3D perovskite does not lead to greater effective decay length extraction when compared to 3D perovskites, as we will show in point 3.2.

The transmission spectra of the single crystals were also investigated (Figure 4D). The sharp transition from low to high transmittance values corresponds to the band gap and it is in good agreement with diffuse reflectance measurements (Figure 4B). The spectral oscillations in the transparency window are produced by Fabry–Pérot-type optical resonances occurring between opposite crystal faces in the measurement direction. This behavior, known as an optical cavity in hybrid perovskites, has already been reported [54]. We took a step forward by developing a model capable of determining optical properties, such as the refractive index and dielectric constant, as a function of the crystal size, PL and transmission [29,30,55]. According to this model, one of the main factors influencing the amplitude and periodicity of the oscillations associated with optical resonances is the crystal thickness. This way, the 3D crystal, with a thickness (measured by optical profilometry means) of 4 µm gives rise to oscillations with much longer periodicity than those of the 2D crystal, which has a thickness of 13 µm. Deviations in the refractive index between both crystals could additionally influence this phenomenon, but we considered them negligible in comparison with the thickness effect. In any case, the appearance of oscillations in the spectra is a sign of the good quality of the crystals, since they are very sensitive to defects.

3.2. Scanning Photocurrent Microscopy

Single perovskite crystals were measured and analyzed with an SPCM technique. Figure 5 shows the photoluminescence (PL) and photocurrent (I_{SC}) maps for a multidimensional 2D–3D perovskite crystal (panels (a) and (b)) and for a 3D perovskite crystal (panels (c) and (d)), respectively. The sample limits (panels (e) and (f), corresponding to 2D–3D and 3D samples, respectively) were extracted directly from the image scans. For that purpose, a quadrilateral geometry was fitted to a set of points, extracted by evaluating the derivative of the map signals. Then, we used those maps with best defined sample edges (panels (a) and (d) for the multidimensional 2D–3D and 3D crystals, respectively). Panels (g) and (h) show the photocurrent profiles extracted from the I_{SC} maps through different directions indicated by the colored lines in panels (e) and (f), respectively. The data cover from short (SPED) to long probe-to-edge distances (LPED) and reveal notable differences among the profile curves. Considering that the exponential decay value represents the distance from the collection electrode at which the intensity decays a factor of $e^{-1} \approx 0.368$ of the maximum value, we evaluated the intersection of e^{-1} with all the extracted profiles [panels (g) and h] (see dashed horizontal lines). We employed this method for extracting an effective-like photocurrent decay length (ELPD) among complex decay data that could be influenced by several factors such as different material responses, proximity of sample edges and the presence of defects and inhomogeneities. In fact, the ELPD values in panel (i) are represented against the distance from the collecting punctual probe to the sample border in the direction of the corresponding decay profile. This makes it possible to establish an interesting comparative evaluation. In general, the ELPD tendency is to rise with increasing distance to the border, where recombination losses are more prominent. This tendency seems to be modulated by the rest of the surrounding rims, which define the geometry of the sample itself, because the growing tendency of ELPD is flattened at short distances for both the 2D–3D and 3D crystals (pink to light blue dots) and at an intermediate distance for the 3D crystal (yellow dots), and it is boosted at intermediate and long distances for the 2D–3D and 3D crystal, respectively. This flatness/boost tendency of ELPD could be explained, firstly, by the proximity/remoteness of the path to all the external borders, and secondly, by the angle defined between the cross section to the path and the corresponding intersected outer rim, the tangent/normal case being the worst/best scenario for a long ELPD. Interestingly, in the 3D sample, under LPED conditions, a protuberance emerges, creating an almost flat region, resulting in a giant boost. This effect shows a huge difference between samples across all panels of the Figure 5, where the ELPD of the 3D is greater than that of multidimensional 2D–3D, as expected. This may confirm the assumption of the lower photon harvesting efficiency of the 2D–3D in contrast to the 3D counterparts, attributed to the presence of interlayer screening and higher exciton binding energy.

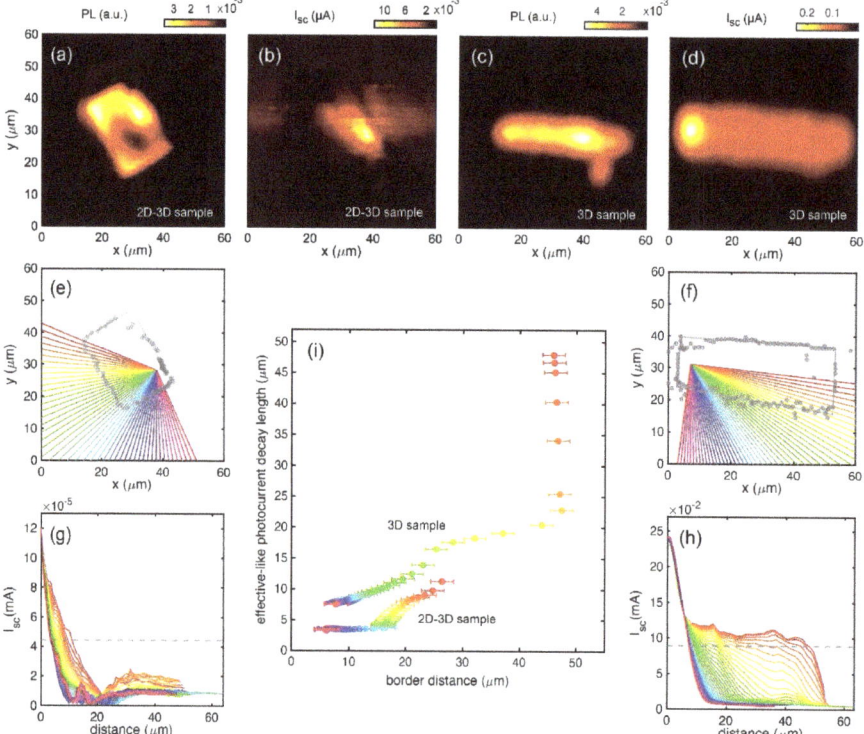

Figure 5. Scanning photocurrent microscopy images and profile analysis. (**a**) PL and (**b**) I_{sc} images of a multidimensional 2D–3D sample. (**c**) PL and (**d**) I_{sc} images of a 3D sample. (**e**,**f**) Edge extraction and (**g**,**h**) photocurrent sections of the 2D–3D and 3D samples, respectively. (**i**) Effective-like photocurrent length extracted from the dashed lines (e^{-1}) of (**g**,**h**). The different colors correspond to the data (**g**–**i**) extracted from different angular slices (**e**,**f**).

In order to extract more precise data from the photocurrent measurements and support the previous analysis, two I_{SC} profiles corresponding to SPED and LPED were fitted to phenomenological models for both crystal types. They correspond to pink and red dots in Figure 5i, with edge distances of 6.02 and 26.42 µm for the 2D–3D case (red curves in Figure 6a,b, respectively), and 7.78 and 46.02 µm for the 3D case (red curves in Figure 6c,d, respectively). The fitting model of the 2D–3D sample consists of a single exponential function in the form of $J \propto e^{\frac{-x}{D}}$, where D is a fitting decay parameter. The fitted curves (black lines) agree well with the experiment except for very low I_{SC} values that were disregarded (dashed lines), where the signal slightly oscillates due to some unmeaningful factors (e.g., the presence of scattered light). The results show that L_D is clearly longer (in fact, it doubles its magnitude) for the LPED (10.71 ± 0.08 µm) than for the SPED (4.08 ± 0.042 µm) profile, which could be attributed to the large difference in the edge-to-tip distance.

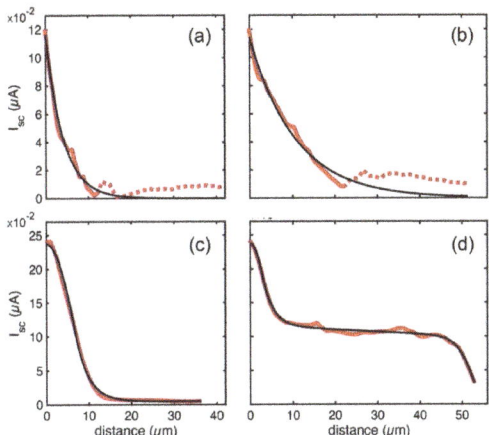

Figure 6. Model fit to experimental photocurrent profiles. (**a,b**) Experimental (red curves) and fitted (black curves) profiles for the 2D–3D sample corresponding to a short (STED) and a long (LTED) distance to the collection tip electrode, respectively. The dashed sections indicate experimental points that were disregarded for the fit process. (**c,d**) The same as (**a,b**) but for the 3D sample.

On the other hand, more complex models with additional parameters were required in the 3D case (Figure 6c,d) because the experimental curves (red lines) display richer profiles. Firstly, both SPED and LPED present a smooth peak at zero distance for which a simple Gaussian function that extends from the origin (0 µm) to its half width was used. Secondly, an exponential function accounting for the fast decay of I_{SC} ($J \propto e^{\frac{-x}{D_1}}$) was added. Thirdly, a small offset was summed to the SPED profile (Figure 6c), while a second, long exponential decay ($J \propto e^{\frac{-x}{D_2}}$) was considered for the LPED profile (Figure 6d). Strictly speaking, such a long decay should be present in all the profiles of the 3D crystal; however, we think it is hidden when the sample borders are close to the collection probe because the borders may deactivate the charge carriers. Finally, another half-Gaussian curve that accounts for the sample edge was added for the LPED profile.

We hypothesize that the initial Gaussian peak at the origin is associated with the electric contact of the probe with the crystal surface. At the interface, the higher metal/perovskite built-in potential allows for easier dissociation of excitons. The fits yielded values for the gaussian half-width of 6.55 ± 0.41 µm and 2.92 ± 0.53 µm for the SPED and LPED profiles, respectively. We attribute the presence/absence or differences in the symmetry of this peak to experimental conditions related to the contact of the electronic probe. Indeed, the peak at the origin is not present in the 2D–3D crystal, likely due to its fragility compared to the 3D perovskite [56,57].

The short exponential decays yielded quite similar D_1 values for the SPED and LPED profiles (3.06 ± 0.03 µm and 2.35 ± 0.05 µm, respectively), while the large exponential decay for LPED gave a D_2 value of 328.88 ± 13.73 µm, and the center position of the second Gaussian was fitted at 48.89 µm ± 0.53, which is quite similar to the assumed border distance: 46.02 µm. With regard to the D_1, in the proximity of both the source and drain contacts, the current could be increased because the photogenerated electrons and holes show the same probability to reach the contacts [35]. On the other hand, the large D_2 value could be the fingerprint of the higher-efficiency photocurrent harvesting of the 3D perovskite with respect to the multidimensional one. Although the extracted values of decay lengths could not be directly comparable to the diffusion lengths reported in the literature because of the sample and experimental conditions in our scheme, the relationship between the values of D_1 and D_2 exhibits significant differences, with the 3D perovskite case clearly showing a larger value, as expected.

4. Conclusions

In summary, this study demonstrates the role of the composition of hybrid halide perovskites in their optoelectronic properties, particularly the influence on the spatial dynamics of the photogenerated transport charges along the microcrystals. Two types of single crystals of perovskites were investigated based on the incorporated organic cation: conventional 3D structures and multidimensional 2D–3D structures. A scanning photocurrent microscopy technique based on a probe tip shows different charge distributions over the perovskite single crystals; the 3D perovskites exhibited two distinct carrier transport regimes. In contrast, the transport of carriers in multidimensional 2D–3D structures was primarily dominated by a unique regime. All of them are influenced by the presence of the border as a recombination source. Overall, this study highlights the significance of the dimensionality of perovskite materials in their optoelectronic and transport properties. Furthermore, studying single crystals provides a valuable platform for exploring these fundamental characteristics.

Supplementary Materials: The following supporting information can be downloaded at: https://www.mdpi.com/article/10.3390/nano13182570/s1, Figure S1: Tauc plots of 3D and 2D–3D perovskites; Figure S2: Schematic representation of the homemade SPCM used; Figure S3: Optical microscope images of single crystals of 3D and 2D–3D perovskites; Figure S4: Optical microscope image showing the electronic contact on a single crystal; Figure S5: PL decays lifetime of single crystals of 3D and 2D–3D multidimensional perovskites; Figure S6: Photography of the antisolvent system used for the growth of the single crystals.

Author Contributions: P.A. conceived the idea of the comparative between 3D and 2D–3D perovskites. E.S-S., R.G.-A. and P.A. planned the experiments. R.G.-A. carried out the synthesis. E.S.-S. carried out the optical and electrical measurements. F.R.-M. and R.F. built the SPCM setup. E.S.-S. and F.R.-M. analyzed the data. F.R.-M., R.F. and P.A. supervised the work. All the authors wrote the manuscript. All authors have read and agreed to the published version of the manuscript.

Funding: This research was supported by the project MCIN/AEI/PID2021-123163OB-I00, and ERDF "A way of making Europe" by the European Union, as well as the Severo Ochoa Centre of Excellence program (CEX2021-001230-S), Ministry for Science and Innovation (MCIN) of Spain.

Data Availability Statement: The data supporting the findings of this study are available from the corresponding authors upon reasonable request.

Acknowledgments: The authors thank the fundings entities. The authors also thank the Electron Microscopy Service of the Universitat Politècnica de Valencia for their support in FESEM image acquisition.

Conflicts of Interest: The authors declare no conflict of interest.

References

1. Bati, A.S.R.; Zhong, Y.L.; Burn, P.L.; Nazeeruddin, M.K.; Shaw, P.E.; Batmunkh, M. Next-generation applications for integrated perovskite solar cells. *Commun. Mater.* **2023**, *4*, 2. [CrossRef]
2. Kieslich, G.; Sun, S.; Cheetham, A.K. An extended Tolerance Factor approach for organic–inorganic perovskites. *Chem. Sci.* **2015**, *6*, 3430–3433. [CrossRef] [PubMed]

3. Liu, Y.; Guo, W.; Hua, L.; Zeng, X.; Yang, T.; Fan, Q.; Ma, Y.; Gao, C.; Sun, Z.; Luo, J. Giant Polarization Sensitivity via the Anomalous Photovoltaic Effect in a Two-Dimensional Perovskite Ferroelectric. *J. Am. Chem. Soc.* **2023**, *145*, 16193–16199. [CrossRef] [PubMed]
4. Han, S.; Li, L.; Ji, C.; Liu, X.; Wang, G.-E.; Xu, G.; Sun, Z.; Luo, J. Visible-Photoactive Perovskite Ferroelectric-Driven Self-Powered Gas Detection. *J. Am. Chem. Soc.* **2023**, *145*, 12853–12860. [CrossRef] [PubMed]
5. Pham, P.V.; Bodepudi, S.C.; Shehzad, K.; Liu, Y.; Xu, Y.; Yu, B.; Duan, X. 2D Heterostructures for Ubiquitous Electronics and Optoelectronics: Principles, Opportunities, and Challenges. *Chem. Rev.* **2022**, *122*, 6514–6613. [CrossRef] [PubMed]
6. Mao, L.; Stoumpos, C.C.; Kanatzidis, M.G. Two-Dimensional Hybrid Halide Perovskites: Principles and Promises. *J. Am. Chem. Soc.* **2019**, *141*, 1171–1190. [CrossRef]
7. Mitzi, D.B.; Chondroudis, K.; Kagan, C.R. Organic-inorganic electronics. *IBM J. Res. Dev.* **2001**, *45*, 29–45. [CrossRef]
8. Frohna, K.; Stranks, S.D. 7–Hybrid Perovskites for Device Applications. In *Handbook of Organic Materials for Electronic and Photonic Devices*, 2nd ed.; Ostroverkhova, O., Ed.; Woodhead Publishing: Cambridge, UK, 2019; pp. 211–256. [CrossRef]
9. Sherkar, T.S.; Momblona, C.; Gil-Escrig, L.; Ávila, J.; Sessolo, M.; Bolink, H.J.; Koster, L.J.A. Recombination in Perovskite Solar Cells: Significance of Grain Boundaries, Interface Traps, and Defect Ions. *ACS Energy Lett.* **2017**, *2*, 1214–1222. [CrossRef]
10. Shrestha, S.; Li, X.; Tsai, H.; Hou, C.-H.; Huang, H.-H.; Ghosh, D.; Shyue, J.-J.; Wang, L.; Tretiak, S.; Ma, X.; et al. Long carrier diffusion length in two-dimensional lead halide perovskite single crystals. *Chem* **2022**, *8*, 1107–1120. [CrossRef]
11. Turedi, B.; Lintangpradipto, M.N.; Sandberg, O.J.; Yazmaciyan, A.; Matt, G.J.; Alsalloum, A.Y.; Almasabi, K.; Sakhatskyi, K.; Yakunin, S.; Zheng, X.; et al. Single-Crystal Perovskite Solar Cells Exhibit Close to Half A Millimeter Electron-Diffusion Length. *Adv. Mater.* **2022**, *34*, e2202390. [CrossRef]
12. Tian, W.; Zhao, C.; Leng, J.; Cui, R.; Jin, S. Visualizing Carrier Diffusion in Individual Single-Crystal Organolead Halide Perovskite Nanowires and Nanoplates. *J. Am. Chem. Soc.* **2015**, *137*, 12458–12461. [CrossRef] [PubMed]
13. Misra, R.K.; Ciammaruchi, L.; Aharon, S.; Mogilyansky, D.; Etgar, L.; Visoly-Fisher, I.; Katz, E.A. Effect of Halide Composition on the Photochemical Stability of Perovskite Photovoltaic Materials. *ChemSusChem* **2016**, *9*, 2572–2577. [CrossRef] [PubMed]
14. Guo, Z.; Wan, Y.; Yang, M.; Snaider, J.; Zhu, K.; Huang, L. Long-range hot-carrier transport in hybrid perovskites visualized by ultrafast microscopy. *Science* **2017**, *356*, 59–62. [CrossRef] [PubMed]
15. Yang, Y.; Ostrowski, D.P.; France, R.M.; Zhu, K.; van de Lagemaat, J.; Luther, J.M.; Beard, M.C. Observation of a hot-phonon bottleneck in lead-iodide perovskites. *Nat. Photonics* **2016**, *10*, 53–59. [CrossRef]
16. D'innocenzo, V.; Grancini, G.; Alcocer, M.J.P.; Kandada, A.R.S.; Stranks, S.D.; Lee, M.M.; Lanzani, G.; Snaith, H.J.; Petrozza, A. Excitons versus free charges in organo-lead tri-halide perovskites. *Nat. Commun.* **2014**, *5*, 3586. [CrossRef]
17. Lang, D.V.; Henry, C.H. Scanning photocurrent microscopy: A new technique to study inhomogeneously distributed recombination centers in semiconductors. *Solid-State Electron.* **1978**, *21*, 1519–1524. [CrossRef]
18. O'dea, J.R.; Brown, L.M.; Hoepker, N.; Marohn, J.A.; Sadewasser, S. Scanning probe microscopy of solar cells: From inorganic thin films to organic photovoltaics. *MRS Bull.* **2012**, *37*, 642–650. [CrossRef]
19. Hao, Y.; He, C.; Xu, J.; Bao, Y.; Wang, H.; Li, J.; Luo, H.; An, M.; Zhang, M.; Zhang, Q.; et al. High-Performance van der Waals Photodetectors Based on 2D Ruddlesden–Popper Perovskite/MoS$_2$ Heterojunctions. *J. Phys. Chem. C* **2022**, *126*, 16349–16356. [CrossRef]
20. Ha, D.; Yoon, Y.; Park, I.J.; Cantu, L.T.; Martinez, A.; Zhitenev, N. Nanoscale Characterization of Photocurrent and Photovoltage in Polycrystalline Solar Cells. *J. Phys. Chem. C* **2023**, *127*, 11429–11437. [CrossRef]
21. Xu, J.; Li, J.; Wang, H.; He, C.; Li, J.; Bao, Y.; Tang, H.; Luo, H.; Liu, X.; Yang, Y. A Vertical PN Diode Constructed of MoS$_2$/CsPbBr$_3$ Heterostructure for High-Performance Optoelectronics. *Adv. Mater. Interfaces* **2022**, *9*, 2101487. [CrossRef]
22. Fang, F.; Wan, Y.; Li, H.; Fang, S.; Huang, F.; Zhou, B.; Jiang, K.; Tung, V.; Li, L.-J.; Shi, Y. Two-Dimensional Cs$_2$AgBiBr$_6$/WS$_2$ Heterostructure-Based Photodetector with Boosted Detectivity via Interfacial Engineering. *ACS Nano* **2022**, *16*, 3985–3993. [CrossRef] [PubMed]
23. Wen, X.; Jia, B. New insight into carrier transport in 2D layered perovskites. *Chem* **2022**, *8*, 904–906. [CrossRef]
24. Ahn, Y.; Dunning, J.; Park, J. Scanning Photocurrent Imaging and Electronic Band Studies in Silicon Nanowire Field Effect Transistors. *Nano Lett.* **2005**, *5*, 1367–1370. [CrossRef]
25. Leite, M.S.; Abashin, M.; Lezec, H.J.; Gianfrancesco, A.; Talin, A.A.; Zhitenev, N.B. Nanoscale Imaging of Photocurrent and Efficiency in CdTe Solar Cells. *ACS Nano* **2014**, *8*, 11883–11890. [CrossRef] [PubMed]
26. Otto, T.; Miller, C.; Tolentino, J.; Liu, Y.; Law, M.; Yu, D. Gate-Dependent Carrier Diffusion Length in Lead Selenide Quantum Dot Field-Effect Transistors. *Nano Lett.* **2013**, *13*, 3463–3469. [CrossRef] [PubMed]
27. Lombardo, C.J.; Glaz, M.S.; Ooi, Z.-E.; Vanden Bout, D.A.; Dodabalapur, A. Scanning photocurrent microscopy of lateral organic bulk heterojunctions. *Phys. Chem. Chem. Phys.* **2012**, *14*, 13199–13203. [CrossRef]
28. Graham, R.; Yu, D. Scanning photocurrent microscopy in semiconductor nanostructures. *Mod. Phys. Lett. B* **2013**, *27*, 1330018. [CrossRef]
29. Ramiro-Manzano, F.; García-Aboal, R.; Fenollosa, R.; Biasi, S.; Rodriguez, I.; Atienzar, P.; Meseguer, F. Optical properties of organic/inorganic perovskite microcrystals through the characterization of Fabry–Pérot resonances. *Dalton Trans.* **2020**, *49*, 12798–12804. [CrossRef]
30. Segura-Sanchis, E.; Fenollosa, R.; Rodriguez, I.; Molard, Y.; Cordier, S.; Feliz, M.; Atienzar, P. Octahedral Molybdenum Cluster-Based Single Crystals as Fabry–Pérot Microresonators. *Cryst. Growth Des.* **2022**, *22*, 60–65. [CrossRef]

31. Rodriguez, I.; Fenollosa, R.; Ramiro-Manzano, F.; García-Aboal, R.; Atienzar, P.; Meseguer, F.J. Groove-assisted solution growth of lead bromide perovskite aligned nanowires: A simple method towards photoluminescent materials with guiding light properties. *Mater. Chem. Front.* **2019**, *3*, 1754–1760. [CrossRef]
32. García-Aboal, R.; García, H.; Remiro-Buenamañana, S.; Atienzar, P. Expanding the photoresponse of multidimensional hybrid lead bromide perovskites into the visible region by incorporation of subphthalocyanine. *Dalton Trans.* **2021**, *50*, 6100–6108. [CrossRef]
33. Lou, Y.; Zhang, S.; Gu, Z.; Wang, N.; Wang, S.; Zhang, Y.; Song, Y. Perovskite single crystals: Dimensional control, optoelectronic properties, and applications. *Mater. Today* **2023**, *62*, 225–250. [CrossRef]
34. Rong, S.-S.; Faheem, M.B.; Li, Y.-B. Perovskite single crystals: Synthesis, properties, and applications. *J. Electron. Sci. Technol.* **2021**, *19*, 100081. [CrossRef]
35. Xiao, R.; Hou, Y.; Fu, Y.; Peng, X.; Wang, Q.; Gonzalez, E.; Jin, S.; Yu, D. Photocurrent Mapping in Single-Crystal Methylammonium Lead Iodide Perovskite Nanostructures. *Nano Lett.* **2016**, *16*, 7710–7717. [CrossRef] [PubMed]
36. Yang, B.; Chen, J.; Shi, Q.; Wang, Z.; Gerhard, M.; Dobrovolsky, A.; Scheblykin, I.G.; Karki, K.J.; Han, K.; Pullerits, T. High Resolution Mapping of Two-Photon Excited Photocurrent in Perovskite Microplate Photodetector. *J. Phys. Chem. Lett.* **2018**, *9*, 5017–5022. [CrossRef] [PubMed]
37. Ganesh, N.; Ghorai, A.; Krishnamurthy, S.; Banerjee, P.; Narasimhan, K.L.; Ogale, S.B.; Narayan, K.S. Impact of trap filling on carrier diffusion in MAPbBr$_3$ single crystals. *Phys. Rev. Mater.* **2020**, *4*, 084602. [CrossRef]
38. Zhang, F.; Yang, B.; Li, Y.; Deng, W.; He, R. Extra long electron–hole diffusion lengths in CH$_3$NH$_3$PbI$_{3-x}$Cl$_x$ perovskite single crystals. *J. Mater. Chem. C* **2017**, *5*, 8431–8435. [CrossRef]
39. Fu, D.; Zou, J.; Wang, K.; Zhang, R.; Yu, D.; Wu, J. Electrothermal Dynamics of Semiconductor Nanowires under Local Carrier Modulation. *Nano Lett.* **2011**, *11*, 3809–3815. [CrossRef]
40. Wei, Y.-C.; Chu, C.-H.; Mao, M.-H. Minority carrier decay length extraction from scanning photocurrent profiles in two-dimensional carrier transport structures. *Sci. Rep.* **2021**, *11*, 21863. [CrossRef]
41. Teng, P.; Reichert, S.; Xu, W.; Yang, S.-C.; Fu, F.; Zou, Y.; Yin, C.; Bao, C.; Karlsson, M.; Liu, X.; et al. Degradation and self-repairing in perovskite light-emitting diodes. *Matter* **2021**, *4*, 3710–3724. [CrossRef]
42. Huang, B.; Clark, G.; Navarro-Moratalla, E.; Klein, D.R.; Cheng, R.; Seyler, K.L.; Zhong, D.; Schmidgall, E.; McGuire, M.A.; Cobden, D.H.; et al. Layer-dependent ferromagnetism in a van der Waals crystal down to the monolayer limit. *Nature* **2017**, *546*, 270–273. [CrossRef] [PubMed]
43. Tisdale, J.T.; Smith, T.; Salasin, J.R.; Ahmadi, M.; Johnson, N.; Ievlev, A.V.; Koehler, M.; Rawn, C.J.; Lukosi, E.; Hu, B. Precursor purity effects on solution-based growth of MAPbBr$_3$ single crystals towards efficient radiation sensing. *CrystEngComm* **2018**, *20*, 7818–7825. [CrossRef]
44. Peng, W.; Wang, L.; Murali, B.; Ho, K.-T.; Bera, A.; Cho, N.; Kang, C.-F.; Burlakov, V.M.; Pan, J.; Sinatra, L.; et al. Solution-Grown Monocrystalline Hybrid Perovskite Films for Hole-Transporter-Free Solar Cells. *Adv. Mater.* **2016**, *28*, 3383–3390. [CrossRef] [PubMed]
45. Shen, H.; Nan, R.; Jian, Z.; Li, X. Defect step controlled growth of perovskite MAPbBr$_3$ single crystal. *J. Mater. Sci.* **2019**, *54*, 11596–11603. [CrossRef]
46. Di, J.; Li, H.; Chen, L.; Zhang, S.; Hu, Y.; Sun, K.; Peng, B.; Su, J.; Zhao, X.; Fan, Y.; et al. Low Trap Density Para-F Substituted 2D PEA$_2$PbX$_4$ (X = Cl, Br, I) Single Crystals with Tunable Optoelectrical Properties and High Sensitive X-ray Detector Performance. *Research* **2022**, *2022*, 9768019. [CrossRef] [PubMed]
47. Dhanabalan, B.; Leng, Y.-C.; Biffi, G.; Lin, M.-L.; Tan, P.-H.; Infante, I.; Manna, L.; Arciniegas, M.P.; Krahne, R. Directional Anisotropy of the Vibrational Modes in 2D-Layered Perovskites. *ACS Nano* **2020**, *14*, 4689–4697. [CrossRef]
48. Tien, C.-H.; Lee, K.-L.; Tao, C.-C.; Lin, Z.-Q.; Lin, Z.-H.; Chen, L.-C. Two-Dimensional (PEA)$_2$PbBr$_4$ Perovskites Sensors for Highly Sensitive Ethanol Vapor Detection. *Sensors* **2022**, *22*, 8155. [CrossRef]
49. Cohen, B.-E.; Wierzbowska, M.; Etgar, L. High Efficiency and High Open Circuit Voltage in Quasi 2D Perovskite Based Solar Cells. *Adv. Funct. Mater.* **2017**, *27*, 1604733. [CrossRef]
50. Yang, T.; Li, F.; Lin, C.-H.; Guan, X.; Yao, Y.; Yang, X.; Wu, T.; Zheng, R. One-pot solution synthesis of 2D-3D mixed-dimensional perovskite crystalline lateral heterostructures. *Cell Rep. Phys. Sci.* **2023**, *4*, 101447. [CrossRef]
51. Jiang, Y.; Yuan, J.; Ni, Y.; Yang, J.; Wang, Y.; Jiu, T.; Yuan, M.; Chen, J. Reduced-Dimensional α-CsPbX$_3$ Perovskites for Efficient and Stable Photovoltaics. *Joule* **2018**, *2*, 1356–1368. [CrossRef]
52. Li, S.; Hu, L.; Zhang, C.; Wu, Y.; Liu, Y.; Sun, Q.; Cui, Y.; Hao, Y.; Wu, Y. In situ growth of a 2D/3D mixed perovskite interface layer by seed-mediated and solvent-assisted Ostwald ripening for stable and efficient photovoltaics. *J. Mater. Chem. C* **2020**, *8*, 2425–2435. [CrossRef]
53. Yoo, H.-S.; Park, N.-G. Post-treatment of perovskite film with phenylalkylammonium iodide for hysteresis-less perovskite solar cells. *Sol. Energy Mater. Sol. Cells* **2018**, *179*, 57–65. [CrossRef]
54. Zhang, Y.; Lim, C.-K.; Dai, Z.; Yu, G.; Haus, J.W.; Zhang, H.; Prasad, P.N. Photonics and optoelectronics using nano-structured hybrid perovskite media and their optical cavities. *Phys. Rep.* **2019**, *795*, 1–51. [CrossRef]
55. García-Aboal, R.; Fenollosa, R.; Ramiro-Manzano, F.; Rodríguez, I.; Meseguer, F.; Atienzar, P. Single Crystal Growth of Hybrid Lead Bromide Perovskites Using a Spin-Coating Method. *ACS Omega* **2018**, *3*, 5229–5236. [CrossRef] [PubMed]

56. Kim, D.; Vasileiadou, E.S.; Spanopoulos, I.; Kanatzidis, M.G.; Tu, Q. In-Plane Mechanical Properties of Two-Dimensional Hybrid Organic–Inorganic Perovskite Nanosheets: Structure–Property Relationships. *ACS Appl. Mater. Interfaces* **2021**, *13*, 31642–31649. [CrossRef]
57. Rathore, S.; Leong, W.L.; Singh, A. Mechanical properties estimation of 2D–3D mixed organic-inorganic perovskites based on methylammonium and phenylethyl-ammonium system using a combined experimental and first-principles approach. *J. Alloys Compd.* **2023**, *936*, 168328. [CrossRef]

Disclaimer/Publisher's Note: The statements, opinions and data contained in all publications are solely those of the individual author(s) and contributor(s) and not of MDPI and/or the editor(s). MDPI and/or the editor(s) disclaim responsibility for any injury to people or property resulting from any ideas, methods, instructions or products referred to in the content.

Article

Surface-Enhanced Raman Spectroscopy (SERS) Investigation of a 3D Plasmonic Architecture Utilizing Ag Nanoparticles-Embedded Functionalized Carbon Nanowall

Chulsoo Kim [1], Byungyou Hong [2] and Wonseok Choi [1,*]

[1] Department of Electrical Engineering, Hanbat National University, Daejeon 34158, Republic of Korea; msdkcs1@gmail.com
[2] School of Electronic and Electrical Engineering, Sungkyunkwan University, Suwon 16419, Republic of Korea; byhong@skku.edu
* Correspondence: wschoi@hanbat.ac.kr

Abstract: Surface-enhanced Raman scattering (SERS) is a highly sensitive technique for detecting DNA, proteins, and single molecules. The design of SERS substrates plays a crucial role, with the density of hotspots being a key factor in enhancing Raman spectra. In this study, we employed carbon nanowall (CNW) as the nanostructure and embedded plasmonic nanoparticles (PNPs) to increase hotspot density, resulting in robust Raman signals. To enhance the CNW's performance, we functionalized it via oxygen plasma and embedded silver nanoparticles (Ag NPs). The authors evaluated the substrate using rhodamine 6G (R6G) as a model target molecule, ranging in concentration from 10^{-6} M to 10^{-10} M for a 4 min exposure. Our analysis confirmed a proportional increase in Raman signal intensity with an increase in concentration. The CNW's large specific surface area and graphene domains provide dense hotspots and high charge mobility, respectively, contributing to both the electromagnetic mechanism (EM) and the chemical mechanism (CM) of SERS.

Keywords: surface-enhanced Raman scattering (SERS); plasmonic nanoparticles (PNPs); carbon nanowall (CNW); silver nanoparticles (Ag NPs); rhodamine 6G (R6G)

Citation: Kim, C.; Hong, B.; Choi, W. Surface-Enhanced Raman Spectroscopy (SERS) Investigation of a 3D Plasmonic Architecture Utilizing Ag Nanoparticles-Embedded Functionalized Carbon Nanowall. *Nanomaterials* **2023**, *13*, 2617. https://doi.org/10.3390/nano13192617

Academic Editor: Thomas Dippong

Received: 31 August 2023
Revised: 20 September 2023
Accepted: 21 September 2023
Published: 22 September 2023

Copyright: © 2023 by the authors. Licensee MDPI, Basel, Switzerland. This article is an open access article distributed under the terms and conditions of the Creative Commons Attribution (CC BY) license (https://creativecommons.org/licenses/by/4.0/).

1. Introduction

The phenomenon of surface-enhanced Raman scattering (SERS), initially by authors including Fleischmann in 1974, has undergone extensive study over the past 5 decades [1]. It has found applications in various fields, including bio-analysis, chemical analysis, and biomedical sciences. Traditional methods like enzyme-linked immunosorbent assay (ELISA), high-performance liquid chromatography (HPLC), and isotope dilution mass spectrometry (IDMS), while reliable, are often expensive and time consuming. However, the current era demands high-sensitivity, cost-effective, and rapid analytical methods. SERS, a label-free detection method known for its high sensitivity, resolution, affordability, and user-friendly interface, has emerged as a promising solution for bioanalytical applications. This technique enables chemical fingerprinting through a specific spectral signature, making it suitable for analytical methods requiring ultra-trace detection limits in parts per million (ppm) and parts per billion (ppb) units [2]. SERS operates through two mechanisms: the electromagnetic mechanism (EM) and the chemical mechanism (CM) [3]. EM is attributed to the phenomenon of local surface plasmon resonance, which occurs when specific metal nanostructures like plasmonic nanoparticles, nanospheres, and nanoshells are exposed to light of a specific frequency [4,5]. Incident photons induce vibrations of free electrons on the surface of these nanostructures, generating a local electromagnetic field known as a "hot spot". In contrast, CM is primarily related to the electron arrangement at the interface between the detection material and plasmonic nanoparticles, involving electron transitions between the highest occupied molecular orbital (HOMO) and the lowest unoccupied molecular orbital (LUMO). However, the influence of CM in SERS is relatively limited [6], with

EM being the dominant factor. Given the critical role of providing high-density hotspots in SERS, researchers have explored three-dimensional nano-architectures (3D NAs) like nanopyramids and metal–organic frameworks. However, the fabrication of 3D NAs often involves complex processes such as imprinting and lithography, and it may rely on specific materials with limitations in terms of mechanical properties and reproducibility. In this study, we propose the use of functionalized carbon nanowall (FCNW) as a scaffold for silver nanoparticles (Ag NPs) in a 3D NA structure. FCNWs can be synthesized in a one-step process using plasma-enhanced chemical vapor deposition (PECVD) [7]. The pristine material, carbon nanowall (CNW), consists of vertically aligned graphene domains and has demonstrated excellent performance in various applications, including gas sensors and field-effect transistors, due to its high electron mobility, mechanical strength, and chemical stability [8–10]. In a previous study, the synthesis of CNW and Ag NPs in a 3D lamellar structure showed promise as a SERS substrate. Although the plasmons of graphene domains in CNWs occur in the terahertz frequency range, they do not directly contribute to localized surface plasmon resonance (LSPR) enhancement. However, they provide high-density hot spots due to the large specific surface area of CNW [11]. Furthermore, as reported in the literature, graphene and graphene oxide can enhance the CM, and the high electron mobility of CNW contributes to a low CM ratio [12]. The combined enhancement of EM and CM has a synergistic effect on SERS improvement. Compared to CNW, FCNW offers significant enhancements in CNW-metal/molecule bonding, and the introduction of oxygen groups through plasma functionalization results in more stable SERS substrates. In this study, we successfully collected a sensitive SERS signal for rhodamine 6G (R6G) using the proposed Ag NP-embedded FCNW substrate. This study demonstrates the potential of Ag NP-embedded FCNWs as a SERS substrate for sensitive in situ detection of R6G.

2. Materials and Methods

2.1. Reagents and Materials

Trichloroethylene (EP grade) for wafer cleaning was purchased from Junsei Chemical Co., Ltd. Methanol (Tokyo, Japan) (ACS grade) and acetone (ACS grade) was purchased from Honeywell International, Inc. (Mecklenburg County, NC, USA) and used without further purification. R6G, used as a model target molecule, was obtained from Sigma-Aldrich (St. Louis, MO, USA).

2.2. Fabrication of Ag Nanoparticles–Embedded Functionalized Carbon Nanowall

To use CNW as a SERS substrate, first, CNW was grown on a Si substrate by microwave plasma-enhanced chemical vapor deposition (ASTeX-type, MPECVD, Woosin CryoVac (Uiwang-si, Republic of Korea), 2.45 GHz microwave). Hydrogen gas (purity 99.9999%) was first injected into the chamber to form a hydrogen atmosphere, and then methane gas (purity 99.999%), a precursor of CNW, was injected. The gas ratio in the chamber was $H_2:CH_4$ = 40:40 sccm until 90 s (step of the carbon particulate formation and nanographene sheets generation) into the process, and then $H_2:CH_4$ = 40:20 sccm. Manipulation of these gas ratios is an important factor in determining the CNW growth rate and graphitization properties. After CNW growth, it was cooled to room temperature at a high vacuum (3×10^{-2} Torr). Secondly, functionalization was performed while maintaining a vacuum in the chamber. This is an in situ process for the synthesis and functionalization of CNW. A plasma ball was formed at 300 °C with 500 W of microwave power while injecting 20 sccm of oxygen, which lasted for 20 s. In this case, oxygen radicals can cause destruction or etching of CNW, and the process time is important because a short time reduces the CNW branch removal rate. In the functionalization step, oxygen radicals remove the carbon branches present in the CNW and simultaneously impart functional groups to the graphene domains, thereby increasing the deposition rate of Ag NPs. The embedding of Ag NPs into the FCNW was performed via a radio frequency (RF) magnetron sputtering system (ITS, PG600A600W, Daejeon, Republic of Korea). At this time, 80 sccm of argon gas was injected, and a Ag target (purity of 99.999%) was used. To help understand the process, a schematic diagram of the process is shown in Scheme 1. R6G was adopted as a model target molecule for SERS signal detection to evaluate the performance of the Ag NPs-embedded FCNW substrate. SERS substrates are prepared by soaking in a solution of R6G molecules (10^{-6}–10^{-10} M) for 1–5 min, followed by final drying with a nitrogen injection.

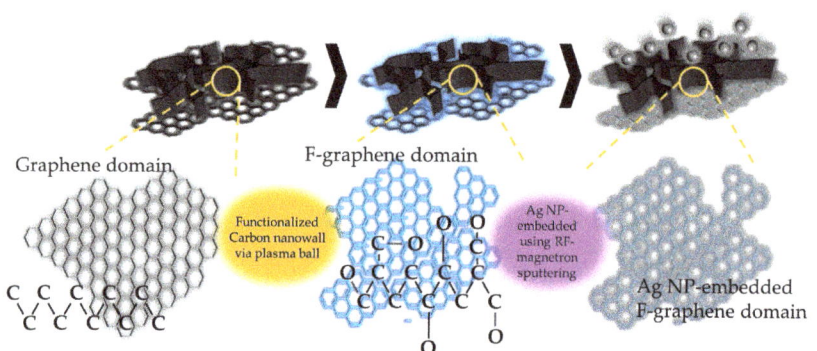

Scheme 1. Schematic diagram of Ag NP-embedded FCNW fabrication.

2.3. Characterization and Analysis of Materials

All Raman spectra, including the SERS analysis, were analyzed using Raman spectroscopy (HORIBA, Osaka, Japan, LabRAM HR-800). The laser power was 3 mW, the excitation wavelength was 532 nm, and a × 50 objective with NA = 0.5 was used. The morphological characteristics were analyzed by using a field-emission scanning electron microscopy (FE-SEM, HITACHI, Tokyo, Japan, S-4800) at 15 kV. Surface activation energy analyses were performed by using a water contact angle (WCA) analyzer (SEO, Phoenix MT, Seoul, Republic of Korea). Characterization of the pore distribution of the samples was performed using mercury intrusion porosimetry (Micromeritics, Norcross, GA, USA, AutoPore 9520).

3. Results and Discussion

3.1. Raman Shift of CNW and FCNW

The carbon molecules that constituted the graphene domains show the Raman active bands in the spectrum. Representatively, there are D-peak, G-peak, and 2D-peak near 1350, 1570, and 2700 cm^{-1}, respectively [13]. The D-peak is due to out-of-plane vibrations at defects in carbon atoms, whereas the G-peak is due to in-plane vibrations attributed to sp^2 bonding in carbon [14]. Figure 1a,c show the Raman spectrum of a pristine CNW and the I_D/I_G ratio, respectively. CNW has many edges, a high D-peak is observed, and a sharp G-peak is observed because it is a material composed of sp^2 combined carbon atoms [13–15]. The 2D-peak can also be attributed to the double resonance of carbon atoms and is also observed in CNW, confirming that CNW has a graphene-based 3D architecture [16]. In this paper, the growth time varied from 300 to 900 s. Among them, the sharpest 2D-peak was observed at 600 s, suggesting that the graphene sheets with the fewest number of layers were vertically oriented [17,18]. In the Raman spectrum, FCNW increased the ratio of D-peak, which may be due to the modification of the hexagonal carbon ring by oxygen radicals or the multiple of functional groups. However, the highest 2D-peak was also observed for the 600 s sample. Figure 2 shows that the scenarios in (b) and (d) can also be known through the I_{2D}/I_G ratio [19]. Disorder and defects in carbon materials can be evaluated easily through the I_D/I_G ratio, but they are not completely reliable considering the ambiguity of the analysis due to the overlapping of the D-peak and G-peak.

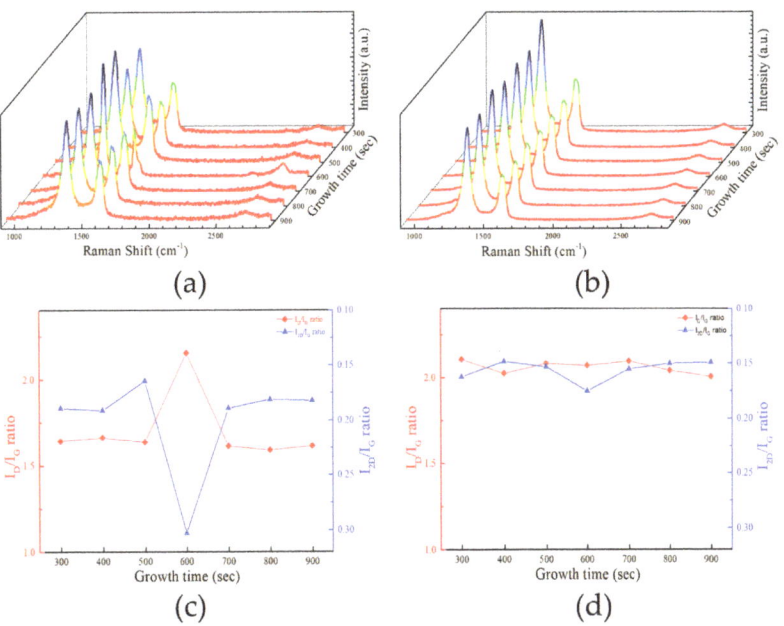

Figure 1. Raman spectra of (**a**) CNW and (**b**) FCNW, growth time of 300–900 s, (**c**) and (**d**) I_D/I_G and I_{2D}/I_G ratios corresponding to (**a**) and (**b**), respectively.

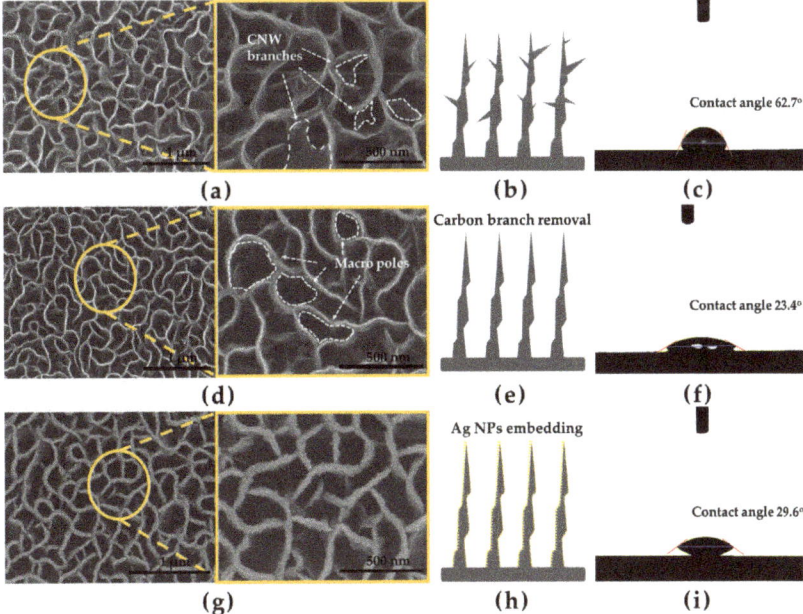

Figure 2. The top-view FE-SEM image, schematic images, and WCA images in order as CNW: (**a**–**c**); FCNW: (**d**–**f**); Ag NPs-embedded FCNW: (**g**–**i**), respectively.

3.2. Morphological Characterization for Plasmonic Nanoparticles

CNWs have a bottom layer of approximately 20 or more layers of graphene, which grows vertically at the interface of localized graphene islands [20]. In our study, if nanographene is formed about 90 s after the start of the process and the methane gas ratio is maintained at 40 sccm, it may be advantageous in terms of the growth rate, but the yield of amorphous carbon branches also increases. To reduce the yield of carbon branches, the methane gas flow rate was reduced by half after 90 s of the process, but it was not possible to completely control the growth of carbon branches. Figure 2a shows FE-SEM images of the pristine CNW surface. The amorphous carbon branches, consisting of sp^3 bonds, significantly contribute to the increase in the D-peak in the Raman spectrum, and, in this study, they were removed in the functionalization step using oxygen plasma [21]. It should be noted that during the functionalization step, destruction of the nanowall occurred due to the saturation of the oxygen functional groups. Figure 2b shows that physical destruction of the nanowall did not occur, round edges were maintained, and FCNWs with carbon branches removed were observed. The WCA of the CNW was 62.7 degrees, whereas the FCNW decreased to 23.4 degrees (Figure 2c,f). The formation of functional groups by oxygen plasma generates a Laplace pressure in the same direction as gravity due to an increase in surface energy, thereby lowering the surface WCA [22–24]. Functional groups present on the FCNW basal plane and edge create sites for potential interactions with Ag NPs. The charged functional groups of the FCNW can cause electrostatic interactions or coordination bonds with Ag NPs, which can consequently increase the rate of Ag NPs-embedding. The functional group is not directly involved in SERS. Ag NPs were embedded into the FCNW by RF-magnetron sputtering, and the corresponding images are shown in Figure 2g. The increase in FCNW edge thickness is due to the insertion of Ag NPs, which can induce changes in surface energy. The corresponding schematic illustration and WCA analysis results are included in Figure 2h,i, respectively.

The surface morphological characteristics of CNW and FCNW are differentiated as shown in Figure 3a, due to the removal of amorphous carbon branches by the oxygen plasma ball in the functionalization step. Based on the interface, the left side is the FCNW area and the right side is the CNW area. CNW has a variety of edge shapes. In previous studies, shapes and patterns such as zigzags and rounds were found [25,26]. Although not reported in the literature, the round shape contains the largest pores and is suitable for nanoparticle embedding. The porosity distribution can be seen in Figure 3(b-1–b-3). Pores exist on the surface of the CNW and the FCNW, and the porosity distribution is an essential factor when embedding nanoparticles. The pore distribution of 100 to 200 nm is dominant, but sometimes the pore diameter is smaller than 100 nm or larger than 200 nm. Pores with a diameter of less than 100 nm are predominantly deposited only on CNW edges during Ag NPs-embedding using sputtering. In addition, Ag NPs-embedding in pores with a diameter exceeding 200 nm has a high probability of being inserted between pores to form a thin film. In this paper, the authors have determined that a pore diameter within the range of 100 to 200 nm is optimal for sputtering. Looking at the pore diameter distribution in Figure 3c, it was confirmed that the overall diameter increased, which was due to the removal of carbon branches. The decrease in pore diameter in Figure 3(b-3) is due to the insertion of Ag NPs into the FCNW edge according to the distribution of pore diameters. Nevertheless, it can be seen from the Gaussian fitting curves in Figure 3(b-1–b-3) that pores with an average diameter of 100–200 nm dominate, as the authors intended.

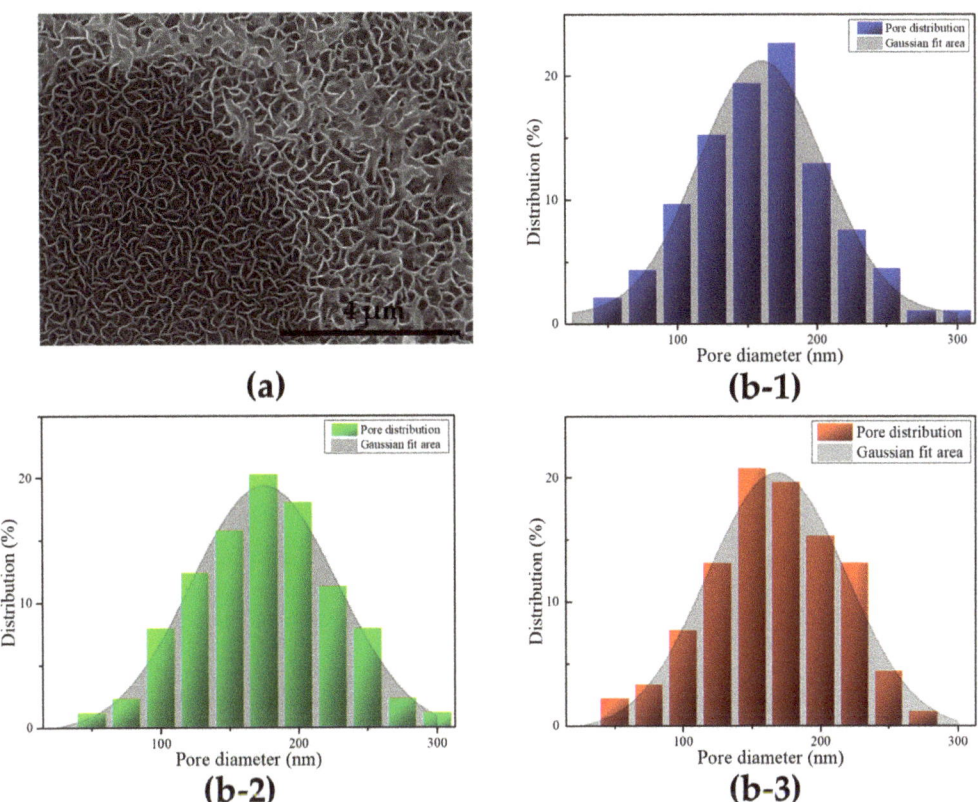

Figure 3. (**a**) The top-view FE-SEM images of the interface of CNW and FCNW; porosity distribution histogram of (**b-1**) CNW, (**b-2**) FCNW, and (**b-3**) Ag NPs-embedded FCNW.

3.3. SERS Activation on the Ag NPs-Embedded FCNW

Figure 4a shows the molecular formula of R6G and the Raman spectrum of pristine CNW exposed to high concentrations of R6G. SERS substrates using a pristine CNW are capable of identifying high concentrations of R6G, but they exhibit low intensity. This can be inferred to be due to the slight contribution of CM due to the high electron affinity [27], which is a physical property of CNW, and molecular detection at low concentrations is limited. The Ag NPs-embedded FCNW substrate shows an exposure time response of 1–5 min with R6G used as a model target molecule at a certain concentration (10^{-7} M), as shown in Figure 4b. The peak around 616 cm^{-1} is due to C-C-C ring in-plane vibrations, and the peak at 778 cm^{-1} is due to C-H out-of-plane vibrations [28,29]. In addition, the peaks observed at near 1365, 1514, 1577, and 1654 cm^{-1} are aromatic stretching modes present in the R6G molecule, and they are considered representative Raman spectra of R6G [30]. The intensity of the peaks decreased according to the increased exposure time described in Figure 4b, because the dynamic behavior of R6G molecules adsorbed on the surface of Ag NPs was dependent on the exposure time [31].

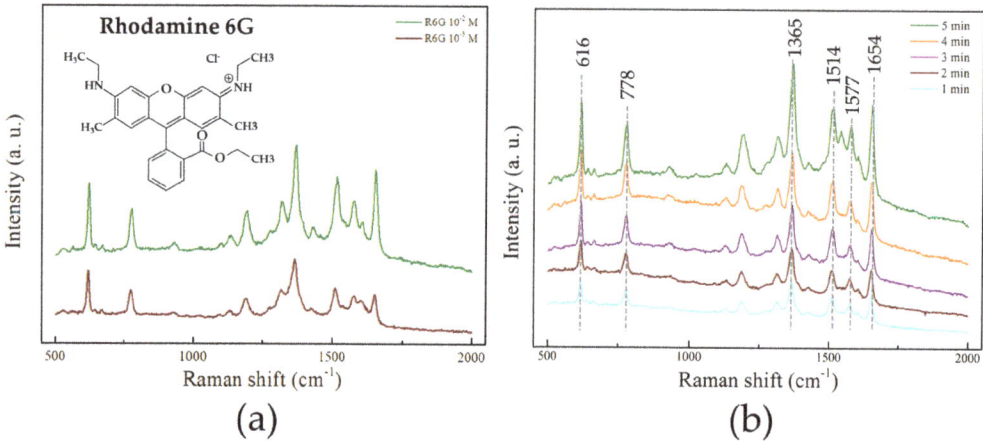

Figure 4. Raman spectra: (**a**) pristine CNW was exposed to a high-concentration R6G solution (the R6G molecular formula is included); (**b**) Ag NPs-embedded FCNW was exposed to a 10^{-7} M concentration R6G aqueous solution with exposure time increased by 1 min from 1 to 5 min.

Raman mapping is a robust tool for visualizing the heterogeneity in intensity differences across substrates. Figure 5c illustrates the Raman spectral mapping images corresponding to various substrates. The exposure time for all experiments was fixed at 4 min with a concentration of a 10^{-7} M R6G aqueous solution. Raman mapping images obtained from pristine CNW exhibit high localization and limited intensity, potentially explaining the remarkably low R6G detection limit. This observation is further supported by the Raman spectrum in Figure 5a. And Table 1 shows samples corresponding to Figure 5a. Although pristine CNWs do not directly affect the detection of model target molecules, they can be used as architectures with a large specific surface area for providing a high density of SERS active sites. Conversely, CNW embedded with Ag NPs, despite having a low density, exhibits Raman mapping intensity capable of identifying R6G. This phenomenon can be attributed to the high plasmonic activity of Ag NPs embedded within the extensive specific surface area of CNW. Raman mapping images of Ag NPs-embedded FCNW substrates display exceptionally high intensity. The presence of oxygen functional groups in FCNW contributes significantly to increased Ag NPs embedding yield through enhanced van der Waals bonds, hydrogen bonds, or other interactions with molecules [32–35]. The resulting high-density plasmons, stemming from the increased Ag NPs embedding, emerge as the dominant factor responsible for SERS enhancement. A schematic illustration to aid reader comprehension is included in Figure 5b. Ag NPs-embedded FCNW (samples 1–5) showed very strong SERS signals. Non-functionalized CNW (samples 6–10) with embedded Ag NPs showed lower peaks than Ag NPs-embedded FCNW. It has been confirmed that carbon branches are considered to be a very limiting factor for the embedding of Ag NPs. Pristine CNW (samples 11–15) display minimal peaks, with the target molecule remaining unidentified. This further suggests that the contribution of CNW's CM is minimal, aligning with the findings from Raman mapping. The SERS enhancement factor (EF) was calculated using Equation (1) [36]:

$$EF = \frac{I_{SERS}/N_{SERS}}{I_{REF}/N_{REF}} \quad (1)$$

Here, I_{SERS} and I_{REF} represent the intensity in the Raman spectrum at 1365 cm^{-1} for R6G adsorbed on SERS and reference substrate, respectively. N_{SERS} and N_{REF} denote the number of R6G molecules absorbed on SERS and reference substrates, respectively. The SERS EF of Ag NPs-embedded FCNW was determined to be 6.817×10^7. These findings highlight that, when compared to CNW and Ag NPs CNW, Ag NPs-embedded FCNW forms high-density plasmons, significantly influencing electromagnetic (EM) enhancement and contributing to the SERS signal enhancement.

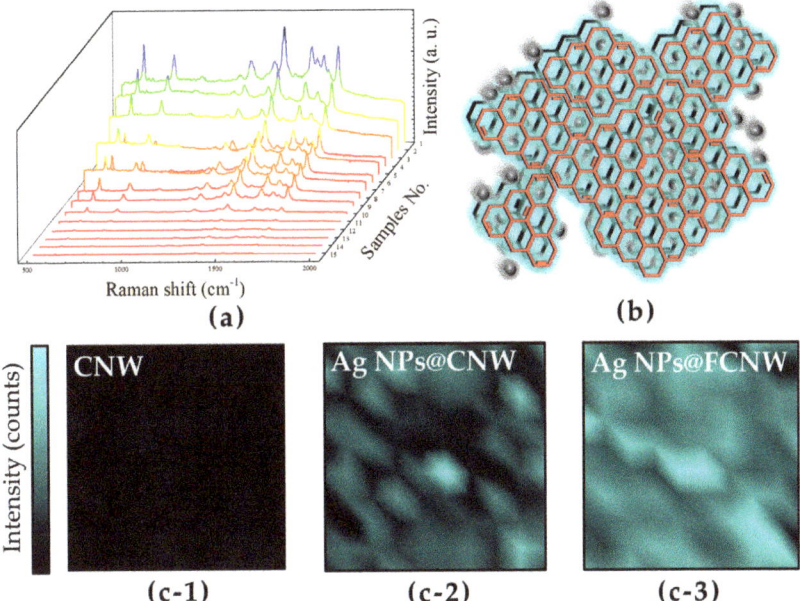

Figure 5. (a) Raman spectra for CNW, Ag NPs-embedded CNW and Ag NPs-embedded FCNW; (b) schematic diagram of SERS active mode of Ag NPs-embedded FCNW; Raman mapping images in a 50 µm square area of samples (c-1) CNW, (c-2) Ag NPs-embedded CNW, and (c-3) Ag NPs-embedded FCNW.

Table 1. Samples numbering table corresponding to Figure 5a.

Samples	10^{-6} M	10^{-7} M	10^{-8} M	10^{-9} M	10^{-10} M
Ag@FCNW	1	2	3	4	5
Ag@CNW	6	7	8	9	10
CNW	11	12	13	14	15

4. Conclusions

Incorporating 3D nanoarchitectures (3D NA) with plasmonic nanoparticles (NPs) has been shown to yield high-level surface-enhanced Raman scattering (SERS) signals, making it a promising SERS substrate. The integration of nanoparticles into complex NA structures, such as carbon nanowall (CNW), demands advanced processing techniques. Nonetheless, our successful fabrication of Ag NPs-embedded FCNW using oxygen plasma underscores the viability of this approach. Our study demonstrates that the combination of FCNW and Ag NPs exhibits two synergistic effects as SERS substrates. Firstly, plasmonic NPs can be densely deposited on the ample specific surface area of FCNW, resulting in the distribution of dense hotspots. This, in turn, significantly contributes to electromagnetic (EM) enhancement by promoting more plasmon formation. Secondly, the binding of Ag NPs to R6G, a model target molecule, facilitates charge transfer with high probability, making a slight but notable contribution to chemical (CM) enhancement. In conclusion, the Raman peak of R6G exhibited a robust SERS activity signal on the Ag NPs-embedded FCNW substrate. This innovative approach holds great potential for applications in the fields of bioanalysis and chemical analysis, paving the way for further SERS proof-of-concept research.

Author Contributions: Conceptualization, C.K. and W.C.; methodology, C.K.; software, C.K.; validation, C.K.; formal analysis, C.K.; investigation, B.H.; resources, C.K.; data curation, C.K.; writing—original draft preparation, C.K.; writing—review and editing, C.K.; visualization, B.H.; supervision, W.C.; project administration, C.K. and W.C. All authors have read and agreed to the published version of the manuscript.

Funding: This research received no external funding.

Data Availability Statement: The data that support the findings of this study are available from the corresponding author upon reasonable request.

Acknowledgments: This work was supported by the research fund of Hanbat National University in 2022.

Conflicts of Interest: The authors declare no conflict of interest.

References

1. Fleischmann, M.; Hendra, P.J.; McQuillan, A.J. Raman spectra of pyridine adsorbed at a silver electrode. *Chem. Phys. Lett.* **1974**, *26*, 163–166. [CrossRef]
2. Kneipp, K.; Wang, Y.; Kneipp, H.; Perelman, L.T.; Itzkn, I.; Ramachandra, R.; Feld, M.S. Single Molecule Detection Using Surface-Enhanced Raman Scattering (SERS). *Phys. Rev. Lett.* **1997**, *78*, 1667–1670. [CrossRef]
3. Garcia-Vidal, F.J.; Pendry, J.B. Collective theory for surface enhanced Raman scattering. *Phys. Rev. Lett.* **1996**, *77*, 1163. [CrossRef] [PubMed]
4. Yuen, C.; Zheng, W. Huang Optimization of extinction efficiency of gold-coated polystyrene bead substrates improves surface-enhanced Raman scattering effects by post-growth microwave heating treatment. *J. Raman Spectrosc.* **2010**, *41*, 361–478.
5. Zhang, X.; Wei, Z.Y.; Sui, H.; Cong, Q.; Wang, X.; Zhao, B. Charge-Transfer Effect on Surface-Enhanced Raman Scattering (SERS) in an Ordered Ag NPs/4-Mercaptobenzoic Acid/TiO_2 System. *J. Phys. Chem.* **2015**, *119*, 22439–22444. [CrossRef]
6. Chen, H.; Shao, L.; Li, Q.; Wang, J. Gold Nanorods and Their Plasmonic Properties. *Chem. Soc. Rev.* **2013**, *42*, 2679–2724. [CrossRef]
7. Khan, A.; Kumar, R.R.; Cong, J.; Imran, M.; Yang, D.; Yu, X. CVD Graphene on Textured Silicon: An Emerging Technologically Versatile Heterostructure for Energy and Detection Application. *Adv. Mater. Interfaces* **2022**, *9*, 2100977. [CrossRef]
8. Lee, S.; Kwon, S.; Kim, K.; Kang, H.; Ko, J.M.; Choi, W. Preparation of Carbon nanowall and Carbon Nanotube for Anode Material of Lithium-Ion Battery. *Molecules* **2021**, *26*, 6950. [CrossRef]
9. Mori, S.; Ueno, T.; Suzuki, M. Synthesis of carbon nanowalls by plasma-enhanced chemical vapor deposition in a CO/H_2 microwave discharge system. *Diamond Relat. Mater.* **2011**, *20*, 1129–1132. [CrossRef]
10. Zheng, W.; Zhao, X.; Fu, W. Review of Vertical Graphene and its Applications. *ACS Appl. Mater. Interfaces* **2021**, *13*, 9561–9579. [CrossRef]
11. Kim, C.; Kim, K.; Kwon, S.; Kang, H.; Hong, B.; Choi, W. Innovative Variation in the Morphological Characteristics of Carbon Nanowalls Grown on a Molybdenum Disulfide Interlayer. *Nanomaterials* **2022**, *12*, 4334. [CrossRef] [PubMed]
12. Zhang, Z.; Lee, T.; Haque, M.F.; Leem, J.; Hsieh, E.Y.; Nam, S.W. Plasmonic sensors based on graphene and graphene hybrid materials. *Nano Converg.* **2022**, *9*, 28. [CrossRef] [PubMed]
13. Thodkar, K.; Gramm, F. Enhanced Mobility in Suspended Chemical Vapor-Deposited Graphene Field-Effect Devices in Ambient Conditions. *ACS Appl. Mater. Interfaces* **2023**, *15*, 31. [CrossRef] [PubMed]
14. Khan, A.; Cong, J.; Kumar, R.R.; Ahmed, S.; Yang, D.; Yu, X. Chemical Vapor Deposition of Graphene on self-Limited SiC Interfacial Layers Formed on Silicon Substrates for Heterojunction Devices. *ACS Appl. Nano Mater.* **2022**, *5*, 17544–17555. [CrossRef]
15. Ghodke, S.; Murashima, M.; Christy, D.; Nong, N.V.; Ishikawa, K.; Oda, O.; Umehara, N.; Hori, M. Mechanical properties of maze-like carbon nanowalls synthesized by the radial injection plasma enhanced chemical vapor deposition method. *Mater. Sci. Eng. A* **2023**, *862*, 144428. [CrossRef]
16. Zhou, T.; Xu, C.; Ren, W. Grain-Boundary-Induced Ultrasensitive Molecular Detection of Graphene Film. *Nano Lett.* **2022**, *23*, 9380–9388. [CrossRef]
17. Bohlooli, F.; Anagri, A.; Mori, S. Development of carbon-based metal free electrochemical sensor for hydrogen peroxide by surface modification of carbon nanowalls. *Carbon* **2022**, *196*, 327–336. [CrossRef]
18. Ferrari, A.C.; Meyer, J.C.; Scardaci, V.; Casiraghi, C.; Lazzeri, M.; Mauri, F.; Piscanec, S.; Jiang, D.; Novoselov, K.S.; Roth, S.; et al. Raman Spectrum of Graphene and Graphene Layers. *Phys. Rev. Lett.* **2006**, *97*, 187401. [CrossRef]
19. Cong, J.; Khan, A.; Li, J.; Wang, Y.; Xu, M.; Yang, D.; Yu, X. Direct Growth of Graphene Nanowalls on Silicon Using Plasma-Enhanced Atomic Layer Deposition for High-Performance Si-Based Infrared Photodetectors. *ACS Appl. Electron. Mater.* **2021**, *3*, 5048–5058. [CrossRef]

20. Sun, J.; Rattanasawatesun, T.; Tang, P.; Bi, Z.; Pandit, S.; Lam, L.; Wasen, C.; Erlandsson, M.; Bokarewa, M.; Dong, J.; et al. Insights into the Mechanism for Vertical Graphene Growth by Plasma-Enhanced Chemical Vapor Deposition. *ACS Appl. Mater. Interfaces* **2022**, *14*, 7152–7160. [CrossRef]
21. Li, J.; Li, M.; Zhou, L.L.; Lang, S.Y.; Lu, H.Y.; Wang, D.; Chen, C.F.; Wan, L.J. Click and Patterned Functionalization of Graphene by Diels-Alder Reaction. *J. Am. Chem. Soc.* **2016**, *138*, 7448–7451. [CrossRef] [PubMed]
22. Choi, H.; Kwon, S.; Lee, S.; Kim, Y.; Kang, H.; Kim, J.H.; Choi, W. Innovative Method Using Adhesive Force for Surface Micromachining of Carbon Nanowall. *Nanomaterials* **2020**, *10*, 1978. [CrossRef] [PubMed]
23. Li, H.; Yang, B.; Yu, B.; Huang, N.; Liu, L.; Lu, J.; Jiang, X. Graphene-coated Si nanowires as substrates for surface-enhanced Raman scattering. *Appl. Surf. Sci.* **2021**, *541*, 148486. [CrossRef]
24. Choi, H.; Kwon, S.; Kim, J.H.; Choi, W. Analysis of plasma-grown carbon oxide and reduced-carbon-oxide nanowalls. *RSC Adv.* **2020**, *10*, 9761–9767. [CrossRef]
25. Kwon, S.; Choi, H.; Lee, S.; Kim, Y.; Choi, W.; Kang, H. Solubility of modified catalyst-free carbon nanowall with organic solvents. *Appl. Surf. Sci.* **2020**, *529*, 147161. [CrossRef]
26. Kwon, S.; Kim, C.; Kim, K.; Jung, H.; Kang, H. Effect of Ag NPs-decorated carbon nanowalls with integrated Ni-Cr alloy microheater for sensing ammonia and nitrogen dioxide gas. *J. Alloys Compd.* **2023**, *932*, 167551. [CrossRef]
27. Prasad, A.; Chaichi, A.; Mahigir, A.; Sahu, S.P.; Ganta, D.; Veronis, G.; Gartia, M.R. Ripple mediated surface enhanced Raman spectroscopy on graphene. *Carbon* **2020**, *157*, 525–536. [CrossRef]
28. He, X.N.; Gao, Y.; Mahjouri-Samani, M.; Black, P.N.; Allen, J.; Mitchell, M.; Zhou, W.X.Y.S.; Jiang, L.; Lu, Y.F. Surface-enhanced Raman spectroscopy using gold-coated horizontally aligned carbon nanotubes. *Nanotechnology* **2012**, *23*, 205702. [CrossRef]
29. Zhao, Y.; Xie, Y.; Bao, Z.; Tsang, Y.H.; Xie, L.; Chai, Y. Enhanced SERS Stability of R6G Molecules with Monolayer Graphene. *J. Phys. Chem. C* **2014**, *118*, 11827–11832. [CrossRef]
30. Barros, A.; Shimizu, F.M.; Oliveira, C.S.; Sigoil, F.A.; Santos, D.P.; Mazali, I.O. Dynamic Behavior of Surface-Enhanced Raman Spectra for Rhodamine 6G Interacting with Gold Nanorods: Implication for Analyses under Wet versus Dry Conditions. *ACS Apple. Mater. Interfaces* **2020**, *3*, 8138–8147. [CrossRef]
31. Fedoseeva, M.; Letrun, R.; Vauthey, E. Excited-State Dynamics of Rhodamine 6G in Aqueous Solution and at the Dodecane/Water Interface. *J. Phys. Chem. B* **2014**, *118*, 5184–5193. [CrossRef] [PubMed]
32. Medhekar, N.V.; Ramasubramaniam, A.; Ruoff, R.S.; Shenoy, V.B. Hydrogen Bond Networks in Graphene Oxide Composite Paper: Structure and Mechanism Properties. *ACS Nano* **2010**, *4*, 2300–2306. [CrossRef]
33. Yildiz, G.; Warberg, M.B.; Awaja, F. Graphene and graphene oxide for bio-sensing: General properties and the effects of graphene ripples. *Acta Biomater.* **2021**, *131*, 62–79. [CrossRef] [PubMed]
34. Lu, C.; Liu, Y.; Ying, Y.; Liu, J. Comparison of MoS_2, WS_2, and Graphene Oxide for DNA Adsorption and Sensing. *Langmuir* **2017**, *33*, 630–637. [CrossRef] [PubMed]
35. Choi, S.; Kim, C.; Suh, J.M.; Jang, H.W. Reduced graphene oxide-based materials for electrochemical energy conversion reactions. *Carbon Energy* **2019**, *1*, 85–108. [CrossRef]
36. Le Ru, E.C.; Blackie, E.; Meyer, M.; Etchegoin, P.G. Etchegoin Surface Enhancedment Raman Scattering Enhacement Factors: A Comprehensive Study. *J. Phys. Chem. C* **2007**, *111*, 13794–13803. [CrossRef]

Disclaimer/Publisher's Note: The statements, opinions and data contained in all publications are solely those of the individual author(s) and contributor(s) and not of MDPI and/or the editor(s). MDPI and/or the editor(s) disclaim responsibility for any injury to people or property resulting from any ideas, methods, instructions or products referred to in the content.

Review

Recent Breakthroughs in Using Quantum Dots for Cancer Imaging and Drug Delivery Purposes

Aisha Hamidu [1], William G. Pitt [2] and Ghaleb A. Husseini [3,4,*]

1. Biomedical Engineering Program, College of Engineering, American University of Sharjah, Sharjah P.O. Box 26666, United Arab Emirates; g00087960@alumni.aus.edu
2. Department of Chemical Engineering, Brigham Young University, Provo, UT 84602, USA; pitt@byu.edu
3. Materials Science and Engineering Program, College of Arts and Sciences, American University of Sharjah, Sharjah P.O. Box 26666, United Arab Emirates
4. Department of Chemical and Biological Engineering, College of Engineering, American University of Sharjah, Sharjah P.O. Box 26666, United Arab Emirates
* Correspondence: ghusseini@aus.edu

Abstract: Cancer is one of the leading causes of death worldwide. Because each person's cancer may be unique, diagnosing and treating cancer is challenging. Advances in nanomedicine have made it possible to detect tumors and quickly investigate tumor cells at a cellular level in contrast to prior diagnostic techniques. Quantum dots (QDs) are functional nanoparticles reported to be useful for diagnosis. QDs are semiconducting tiny nanocrystals, 2–10 nm in diameter, with exceptional and useful optoelectronic properties that can be tailored to sensitively report on their environment. This review highlights these exceptional semiconducting QDs and their properties and synthesis methods when used in cancer diagnostics. The conjugation of reporting or binding molecules to the QD surface is discussed. This review summarizes the most recent advances in using QDs for in vitro imaging, in vivo imaging, and targeted drug delivery platforms in cancer applications.

Keywords: quantum dots; functionalization; in vitro imaging; in vivo imaging; drug delivery

Citation: Hamidu, A.; Pitt, W.G.; Husseini, G.A. Recent Breakthroughs in Using Quantum Dots for Cancer Imaging and Drug Delivery Purposes. *Nanomaterials* **2023**, *13*, 2566. https://doi.org/10.3390/nano13182566

Academic Editor: Thomas Dippong

Received: 24 August 2023
Revised: 11 September 2023
Accepted: 12 September 2023
Published: 15 September 2023

Copyright: © 2023 by the authors. Licensee MDPI, Basel, Switzerland. This article is an open access article distributed under the terms and conditions of the Creative Commons Attribution (CC BY) license (https://creativecommons.org/licenses/by/4.0/).

1. Introduction

Cancer is a group of diseases characterized by the rapid growth of abnormal cells within the body. In most cancer cases, the mutations or changes in the expression of proto-oncogenes, tumor suppressor genes, and DNA repair genes are responsible for cancer development [1]. The majority of cancers are attributed to genetic (mutations, hormones, immune conditions) or environmental (radiation, chemicals, pollutants) factors, in addition to indicators of an unhealthy lifestyle (poor diet, tobacco smoking) [2,3]. Furthermore, the risk of cancer increases significantly with increasing age.

Cancer is one of the leading causes of death worldwide. According to the World Health Organisation (WHO), the number of cancer deaths was nearly 10 million in 2020 [4–6]. The number of new cases is estimated to be 28.4 million by 2040 [7]. The fight against cancer remains one of the most significant issues facing the world. Current conventional means to battle cancer have significant drawbacks, including but not limited to toxicity and non-specificity of conventional chemotherapeutics [8]. Early detection and intervention have a significant positive impact on patient outcomes.

In recent decades, research into and applications of nanomedicine have grown significantly, especially in cancer diseases [9–14]. Such research has shown great potential to overcome previous challenges relating to early tumor detection, accurate diagnoses, and individualized treatment [15–17]. The primary benefit of nanomedicine in cancer therapy is the tiny size of nanoparticles, which allows them to function at the molecular level, thereby enhancing diagnosis and improving the chances of achieving innovative targeting strategies at the molecular level [18–22]. For example, some nanoparticles work by binding to cancer biomarkers such as circulating tumor cells, circulating tumor DNA, exosomes, and specific cancer-associated proteins [23–25].

Nanometer-scale materials (1–100 nm) display intriguing properties due to their small size [26]. These novel properties are often due to the quantum confinement and surface effects affected by their small size [27]. The quantum confinement effect confines moving electrons within a small volume, producing unique optical and electronic effects. As for surface effects, the chemical reactivity of the surface usually increases as the size decreases, while the melting point usually decreases [28,29]. These novel optical and thermal properties of nanomaterials can be useful for both in vivo and in vitro applications via the active interaction with molecular components at the cellular level.

Several nanoparticles [30–33] have been investigated for cancer diagnosis and therapy. Nowadays, quantum dots (QDs), often referred to as "artificial atoms", are a hot topic in cancer nanomedicine. They were first described in 1981 by Alexey Ekimov [34]. QDs are made of a relatively small number of atoms (from 100–10,000 atoms) of semiconductor materials of groups II–VI, III–IV, and IV–VI elements in the periodic table [35]. Their tiny dimension leads to their characterization as "dots", while "quantum" is due to their properties and behavior being described extensively by quantum mechanics [36].

Quantum dots (QDs) are nanoscale nanomaterials that are said to be zero-dimensional because charge carriers are confined so tightly in three directions [37,38]. Many of their unique properties arise because semiconducting nanocrystals from 2–10 nm diameter are smaller than or equal to their exciton Bohr radius [39–43].

The unique electronic properties of QDs result from the particle size and shape, which can be manipulated for diagnostic purposes. When a QD is excited by an energy photon hv (the absorption of light), electrons from the valence band (lower energy level) jump to the conduction band (a higher energy level), resulting in an electron–hole pair called an "exciton". As they return to the lowest energy state (ground state), electrons and holes recombine and release energy or light in the form of single photons [44,45]. The crystal's size, composition, and shape determine the wavelength (color) of light that will be released [46]. The larger size QDs emit orange or red wavelengths, while smaller QDs emit shorter blue or green wavelengths. Consequently, the specific tuning of these optical properties (how the QD absorbs and emits energy) can be manipulated to produce distinctive colors by changing the size and shape of the dot [47].

QD semiconducting nanocrystals have an intrinsic band gap, and when light is absorbed, electrons are bridged by excitation. They differ from bulk semiconducting materials due to their inability to create continuous valence and conduction bands, due to the finite number of atoms in a small cluster. Instead, an electronic structure is produced by QDs that is analogous to the discrete electronic states seen in single atoms. Hence, they are also called 'artificial atoms' because of their discrete electronic states. As a QD becomes smaller, the band gap becomes larger. That is, there is an increase in the energy level between the higher valence band and the lower conduction band. More energy is further required to excite the dot, and correspondingly, more energy is released when it returns to the ground state [48,49].

QDs are currently studied by many researchers looking to take advantage of their unique optical properties, such as high fluorescence, excellent resistance to photobleaching, small size, and biocompatibility. These properties make them preferable fluorophores compared to conventional organic dyes with broad emission bands that can fade over time [50]. They have generated considerable interest in bioimaging and fluorescence labeling (in vitro and in vivo). Moreover, by adjusting their size and composition, their emission wavelength can be tuned from visible to infrared wavelengths [51,52], which could be useful for in vivo imaging, such as in sentinel lymph node mapping for image-guided surgery.

The surface modification of QDs gives them a potential tool in cancer imaging. The attachment of certain biomolecules (e.g., peptides, antibodies, or small molecules) to QDs can be used in cancer detection and bioimaging [51]. For example, Brunetti et al. created near-infrared (NIR) QDs functionalized with NT4 cancer-selective tetra-branched peptides that were used to produce their specific uptake and selective accumulation at the site of colon cancer [53]. Elsewhere, QDs were reported to aid in revealing in vivo drug release and drug targeting [54,55]. The potential that QDs offer in the fight against cancer is promising.

Inspired by the exceptional features of QDs and the extensive research on their potential and advancement in the field, this review presents basic insights into the properties of QDs and summarizes the different synthesis methods for their production. Then, we discuss the functionalization of QDs, their applications in cancer management, and their cytotoxicity issues, emphasizing the recent research progress mainly in the last 6 years. We guide the reader through the advancements of QDs as a potential cancer imaging and therapy tool with the hope of bridging the gap and leading to novel discoveries in QDs potential in the field of cancer.

2. Structural and Optical Properties of QDs

QDs have a structure comprising a core, shell, and sometimes a surface coating, which provides high stability in photo and chemical behaviors, surface activation, and photoluminescence quantum yield. The core is comprised of semiconductor material (e.g., CdSe, CdTe) in a crystal configuration upon which the excitation wavelengths and fluorescence emission are dependent. That core is stabilized by the shell structure that surrounds it. The shell affects the decay kinetics, photostability, and fluorescence quantum yield. A surface layer that can include organic molecules regulates the stability, dispersibility, and potential biological interactions. Initially, when prepared, QDs are generally hydrophobic because they lack surface moieties that form hydrogen bonds; however, hydrophilic molecules or polymers can be attached to confer dispersibility in water. For example, the stability of QDs in water can be increased by the attachment or adsorption of amphiphilic polymers with ionizable functional groups. Figure 1 shows a stylized illustration of a QD.

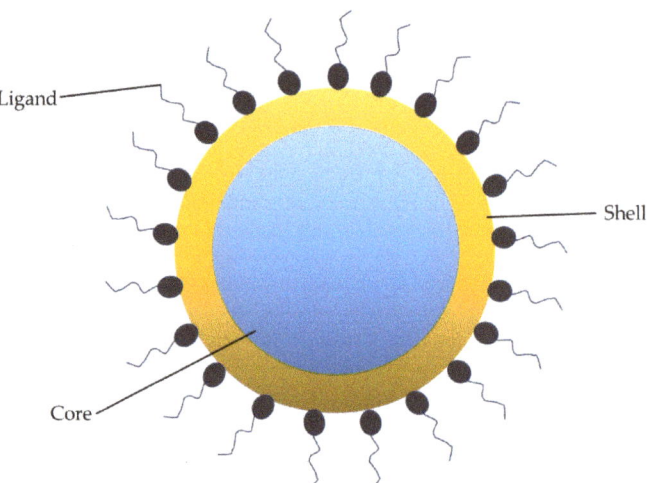

Figure 1. Structure of a QD showing the core/shell/ligand.

As mentioned, the structures of typical QDs are core or core/shell structures. Examples of core QD structures include cadmium telluride (CdTe), while core/shell QD structures include CdSe/ZnS or CdTe/CdS, whose properties can be further enhanced via different surface coatings. The electroluminescence and optical properties of the QD core can be manipulated by altering the sizes of the QD core and shell. Furthermore, core/shell QDs having a shell band gap larger than the core band gap give rise to the electroluminescence properties related to exciton decay by radiative processes [56,57].

Quantum Dots exhibit valuable optoelectronic properties due to the quantum confinement effect. These properties include broad absorption spectra, high fluorescence, strong photostability, and size-tunable emission. Larger QDs with large densities of states and band-overlapping structures possess broad absorption spectra and high molar absorptivities. This particular QD property enables efficient excitation of multiple fluorophores using a single light source. Yet, this broad absorption spectrum produces narrow emission spectra due to transitions from a limited number of high energy to low energy levels, which emit very specific photon energies ($h\nu$). Thus, a light source with a wavelength shorter than the emission wavelength can lead to multiple excitations (and emissions) because of its broad absorption band. These properties that QDs exhibit, broad excitation spectra and narrow emission spectra [57], make them suitable for multiplexed imaging [58,59]. Figure 2 names some optical properties of QDs.

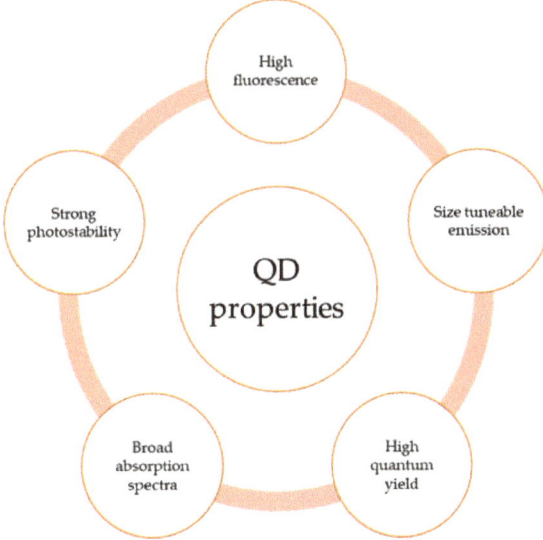

Figure 2. Diagram showing some optical properties of QDs.

Unlike organic dyes (1–5 ns), the decay rates of the excited state are slower in QDs. For example, after excitation, most QDs exhibit a relatively long fluorescence lifetime–10 to 50 ns–which is advantageous in differentiating QD signals from background fluorescence and attaining more sensitive detection. Thus, time-gated imaging can eliminate background autofluorescence. They also exhibit low photodegradation rates, which is often challenging for organic fluorophores.

Unlike organic fluorophores, which, when exposed to light, bleach after a few seconds of continuous exposure, QDs are quite photostable. Photostability is important in most fluorescence applications. This lack of photobleaching allows continuous or long-term monitoring of slow biological processes [59]. QDs can withstand hours of repeated excitation and fluorescence cycles with high brightness levels and photobleaching thresholds. It has been observed that QDs are more photostable than "stable" organic dyes such as Alexa488 [60], and thus offer several advantages in diagnostic applications [61,62].

As mentioned, the size- and chemically-tunable properties are advantageous in selecting an emission wavelength suitable to a specific experiment. For example, the emission wavelength of cadmium sulfide (CdS) and zinc selenide (ZnSe) dots can be tuned from blue to near-ultraviolet light. Similarly, cadmium selenide (CdSe) QDs of different sizes emit light across the visible spectrum. For far-infrared and near-infrared emissions, indium phosphide (InP) and indium arsenide (InAs) QDs can be used [63]. Table 1 lists the emission ranges for some common QDs.

Table 1. Size of QDs and their emission wavelengths. Reprinted with permission from [64].

Quantum Dots	Size Range (Diameter nm)	Emission Range (nm)
Cadmium sulfide (CdS)	2.8–5.4	410–460
Cadmium telluride (CdTe)	3.1–9.1	520–750
Cadmium selenide (CdSe)	2–8	480–680
CdTe/CdSe	4–9.2	650–840
Indium phosphide (InP)	2.5–4.5	610–710
Indium arsenide (InAs)	3.2–6	860–1270
Lead selenide (PbSe)	3.2–4.1	1110–1310
1-Dodecanethiol silver sulfide ((Dt)-Ag_2S)	5.4–10	1000–1300

These unique optical properties of QDs make them highly appealing to a wide array of research and diagnostic applications in diagnostic bioimaging, drug delivery, and more.

3. Synthesis of QDs

QDs must be carefully synthesized to meet specific optical requirements. Their synthesis can be divided into two general categories, the top-down method and the bottom-up approach [56].

3.1. Top-Down Approach

In the top-down approach, QDs are formed by the ablation of bulk semiconductor materials. This includes processes such as electron beam lithography, reactive ion etching, and focused ion beam. These processes synthesize QDs with diameters of around 30 nm. However, these processes have limitations, such as incorporating impurities during synthesis.

3.1.1. Electron Beam Lithography (EBL)

In electron beam lithography (EBL), the surface of a resist (electron-sensitive material) is patterned by scanning with a focused beam of electrons. The resist is made of a polymeric compound, which can either be a negative resist (i.e., long-chain polymer) or a positive resist (i.e., short-chain polymer). The solubility of the resist is altered by the electron beam, allowing the selective removal of either exposed regions or non-exposed regions of the resist when immersed in a solvent (called a developer). If the resist becomes soluble when immersed, it is a positive resist; if it becomes insoluble (i.e., unexposed parts removed), it is a negative resist. The purpose is to fabricate very small structures in the resist whose pattern can then be transferred to the substrate by etching. Although this technique can design patterns directly with sub-10 nm resolution, it is slow and expensive [65,66]. Nandwana et al. [67] reported direct patterning of QD nanostructures using EBL. In this example, functionalized CdSe/ZnS QDs were deposited onto a gold-coated silicon substrate, followed by direct patterning using EBL in the QD film. The QD film was washed using toluene, which removed the unexposed QDs, leaving the exposed areas anchored to the substrate due to the electron beam. QDs were observed to retain their optical properties after cross-linking. Similarly, Palankar et al. [68] reported using EBL to generate QD micropattern arrays. The QDs fabricated were reported to retain their fluorescence and bio-affinity during lithography.

3.1.2. Reactive Ion Etching

In dry etching, an etching chamber is used, where a reactive gas species is introduced, and plasma is formed by applying radio frequency energy by which the gas molecules are broken into reactive fragments. These high-energy species collide with the surface, reacting to form a volatile reaction product. Thus, the surface is slowly etched away. The surface can be protected from etching with a mask pattern. This process is also referred to as reactive ion etching [66,69,70]. Site- and dimension-controlled indium gallium nitride (InGaN) QDs were fabricated by Lee et al. [71]. The QDs were disk-shaped and integrated into a nanoscale pillar. They utilized inductively coupled plasma reactive ion etching to fabricate arrays of nanopillars with different densities and nanopillar diameters from InGaN/GaN. They observed single nanopillars that exhibited strong and distinct photoluminescence at room temperature. The advantages of this process include reducing the amount of etchants used, easy disposal, and eliminating the need to use dangerous liquid etchants. However, the drawback of this process is that it is both time-consuming and expensive, as it requires very specialized equipment [66].

3.1.3. Focused Ion Beam (FIB)

QDs can be fabricated with exceedingly high lateral precision through the focused ion beam technique. The semiconductor substrate's surface is sputtered using highly focused beams from a source of metal ions (Au/Si, Ga). The size, shape, and inter-particle distances of the QDs depend on the ion beam size. Furthermore, it has been reported that a beam with a minimum diameter of 8–20 nm allows QDs to be etched to <100 nm [70]. Choi et al. [72] used focused ion beam luminescence quenching (FIB-LQ) to enhance the single photon purity of the site-controlled QD emission. Optical quality was retained while the SNR of the QD improved, and at increased temperatures, single photon properties were maintained due to the improved signal-to-noise ratio (SNR). In a similar study, Zhang et al. [73] combined focused ion beam (FIB) patterning and self-assembly quantum dots to produce regular QD arrays. High resolution and high flexibility are the advantages of this process. However, the technique is slow and utilizes expensive equipment.

3.2. Bottom-Up Approach

In the bottom-up approach, small units are assembled (precipitated) into the desired structure's shape and size. This process involves nucleation, growth, and chemical decomposition [74]. QDs are synthesized with different techniques, which are further classified into wet-chemical and vapor-phase methods. Wet-chemical methods processes include sol–gel, and microemulsion, while vapor-phase methods processes include molecular beam epitaxy, physical vapor deposition, and sputtering [70]. In wet-chemical methods, conventional precipitation methods are followed by measured control of single solution parameters or a mixture of solutions. The process of precipitation always involves both nanoparticle nucleation and limited growth. Nucleation can involve homogenous, heterogenous, or secondary nucleation. QDs of the desired size, shape, and composition can be acquired by varying factors such as stabilizers, temperature, electrostatic double-layer thickness, and precursor concentration [70,75]. More details are given below.

3.2.1. Wet Chemical Methods
Sol–Gel

Sol–gel methods are commonly used to synthesize QDs [76,77]. The technique prepares a sol (a solution or suspension) of a metal precursor salt (acetates or nitrates, alkoxides) in a base or acidic medium. The process has three steps: hydrolysis, condensation (formation of sol), and growth (formation of gel). In brief, inside the solvent medium, the metal precursor hydrolyzes and condenses, thereby forming a sol, which then grows or polymerizes, forming a network (gel). This process can be used to prepare thin films, fibers, microspheres, etc. The advantages of this process for QD formation include good control of composition, better control of structure, incorporation of nanosized materials, and no use of special or expensive equipment. However, the process is slow, complex, and may involve toxic solvents [78]. QDs of semiconductor types II–VI and IV–VI zinc oxide, cadmium sulfide, and lead sulfide (ZnO, CdS, PbS) have been synthesized using this method [76,77,79]. For example, mixing solutions of Zn-acetate with alcohol and sodium hydroxide, followed by controlled aging in air, produced zinc oxide (ZnO) QDs [76]. Titanium dioxide (TiO_2) QDs were synthesized by Javed et al. [80] using the sol–gel reflux condensation method. They reported the QD to have an average 5–7 nm crystallite size, which offers a large surface area and exhibits photocatalytic properties. In another study by Jiang et al. [81], zinc selenide (ZnSe) QDs embedded in silicon oxide (SiO_2) thin films were synthesized using the sol–gel process. The synthesis was done with H_2SeO_4 as a source for selenium and $Zn\,(Ac)_2 \cdot H_2O$ as a source for zinc. One advantage of this approach to making $ZnSe/SiO_2$ thin films is a reduction in the amount of selenium volatilization. The sol–gel process was reported to be cost-effective and simple [80–82].

Microemulsion Process

A useful method for synthesizing QDs at room temperature is the microemulsion process. Two microemulsions of an aqueous phase in oil are prepared, each having a single chemical component of the semiconductor. While mixing slowly at room temperature, the water droplets collide and merge, thereby creating a mixture that forms QDs inside the very small water droplet. The process can also be done using an oil-in-water emulsion with the oil phase containing the semiconductor components. The use of alcohol instead of water has also been employed. In the reverse microemulsion process, water is dispersed into oil (immiscible liquid) and stirred vigorously in the presence of a surfactant to form extremely small emulsion droplets. The variation of the water-to-surfactant molar ratio controls the size of the water droplet, which in turn affects the size of the resulting QD [44,70,83]. The reverse micelle method has been used to prepare II–VI core and core/shell QDs. Shakur [84] synthesized zinc sulfide (ZnS) QDs by the reverse micelle method using polyvinyl pyrrolidone as a surfactant and produced a size of 2.1 nm. In another study, Karanikolos et al. [85] synthesized luminescent zinc selenide (ZnSe) QDs using a microemulsion process. The synthesized QDs were reported to exhibit excellent photostability and size-dependent luminescence. Cadmium sulfide (CdS) and CdS/ZnS semiconductor QDs were synthesized by the reverse micelle method in a study by Lien et al. [86]. Sodium bis (2-ethylhexyl) sulfosuccinate (AOT) was used as a surfactant. The synthesized QD had a diameter of ~2.5 to 4 nm, which was dependent on the surfactant concentration. In addition, the core/shell nanocrystal structure was reported to have excellent luminescence and photostability. This process is said to be cost-effective, easy to handle/control by modifying parameters such as the ratio of water to surfactant, inexpensive, highly reproducible, and displays good monodispersity [87–89]. However, this process has limitations, such as low yield and the need for large amounts of surfactant, which could result in the incorporation of impurities and presents difficulty in separating the surfactant from the final QDs [90].

3.2.2. Vapor-Phase Method

Vapor-phase methods to produce QDs involve QDs deposited in an atom-by-atom process, as described below.

Molecular Beam Epitaxy (MBE)

Molecular beam epitaxy (MBE) is one of the vapor-phase methods used under ultra-high vacuum conditions (~10^{-10} Torr). It involves the deposition of overlayers to grow elemental compound semiconductor materials of nanostructures on a heated substrate [91]. The process forms a beam of atoms or molecules from the evaporation of an apertured source. The beams can be formed from solids (Ga and As to form GaAs QDs) or a combination of solids and gases (e.g., PH_3 or tri-ethyl gallium). This method uses the large lattice mismatch to self-assemble QDs from II–VI semiconductors and III–V semiconductors [70]. During the process, a reflection high-energy electron diffraction gun is used to monitor the growth of the crystals. Although it is expensive and requires complex equipment, heating the material is slow and controlled, and the process does not involve a slow chemical reaction, resulting in a reduced amount of defects [66]. Brault et al. [92] used molecular beam epitaxy to grow $Al_yGa_{1-y}N$ QDs on $Al_xGa_{1-x}N$ (0001) for light-emitting-diode applications.

Physical Vapor Deposition (PVD)

Physical vapor deposition requires a high vacuum ($\leq 10^{-6}$ Torr) to retain a good vapor flow. A material is sublimated inside the vacuum by thermal evaporation, thereby condensing the substrate from the vapor. Techniques such as resistive heating, electron beam heating, and laser ablation have been used to evaporate the material. The quality of the films produced and their physical characteristics are influenced by the rate of deposition, pressure, substrate temperature, and distance between source and substrate. These factors control the creation of QDs from the thin films deposited [70]. As an example, niobium pentoxide (Nb_2O_5) QDs were grown by Dhawan et al. using PVD [93]. This process does not require expensive chemical reagents, the coatings by PVD have excellent adhesion, and the process allows the deposition of different types of materials. However, the equipment employed is complex and expensive [94,95].

Sputtering

The sputtering process produces nanostructures by bombarding a surface with high-energy particles (e.g., via gas or plasma). It is an effective technique for developing thin films of nanomaterials. During the process, high-energy gaseous ions bombard the semiconductor surface (target material), causing the physical expulsion of atoms or molecules from the surface, depending on the incident gaseous ion energy [96,97]. This technique is also referred to as ion sputtering and is commonly performed in an evacuated chamber. The process is done in different ways, such as radio frequency and magnetron sputtering [94,98,99]. Cadmium selenide (CdSe) QDs were synthesized by Dahi et al. [100] using radio frequency magnetron sputtering. The synthesized QDs had an average size of less than 10 nm in diameter using a radio frequency power of 14 W and a deposition duration of 7.5 min. It is noteworthy that increasing either RF power or deposition time (or both) increased the CdSe QD size. The advantages of this process are reduced surface contamination, no required solvents, and facile tuning of the size, shape, and optical density through careful control of pressure, temperature, and deposition time. However, a drawback of this process is the redeposition of unwanted atoms, which may contaminate the QDs.

3.3. Other Syntheses

QDs are also produced using hydrothermal synthesis. This is a one-pot synthesis by which inorganic salts are crystallized from aqueous solution by regulating temperature and pressure. In this technique, the temperature can be raised very high due to the pressure containment in the autoclave. This results in partial chemical decomposition and promotes molecular collisions, causing the formation of QD. By changing the pressure, temperature, reactants, and aging time, different QD sizes and shapes can be attained [70]. This method of preparing QDs gives excellent photostability and high quantum yield. The process is efficient, timesaving, and more convenient. However, a significant disadvantage of this process is the need for expensive autoclaves [101]. Shen et al. [102] developed nitrogen-doped carbon QDs (N-CQDs) by the hydrothermal synthesis of glucose and phenylenediamine. The synthesized N-CQDS were reported to have good photostability, water solubility, and low toxicity. They were also reported to be excellent fluorescent probes for Fe^{3+} and CrO_4^{2-} in addition to serving as cell imaging reagents for Hela cells. Likewise, QDs can be fabricated using the solvothermal method, which is similar to the hydrothermal except that organic solvents with high boiling points are used instead of water [103,104]. Luo et al. [105] synthesized multiple color emission iron disulfide (FeS_2) QDs by the solvothermal method. Temperature, time, and the reactant ratio were varied to make QDs with blue, green, yellow, and red fluorescence. The blue emission of the QDs was used as a fluorescent responsive signal and the yellow emission was used as a reference signal to construct a molecular imprinting radiometric sensor used for the visual detection of aconitine. The process was reported to be simple and low in cost.

The microwave-assisted synthesis of QDs is a rapid heating method that shortens reaction time and improves production yield. In this method, fewer solvents are used, and tiny particles with a narrow size distribution can be created [106,107]. Cadmium selenide (CdSe) QDs were synthesized by Abolghasemi et al. [108] using the microwave-assisted method. The QDs were synthesized in an N-methyl-2-pyrrolidone solvent with a microwave irradiation power of 900 W. It was reported that this method showed easy control of the size and band gap energy of the QDs, resulting in controllable emission from photoluminescence spectroscopy. The performance of the QDs was tested in photovoltaic solar cells, where results showed that the QDs are suitable sensitizers.

Recently, an ultrasonic technique was employed to synthesize QDs. This method utilizes ultrasound, which causes acoustic cavitation. This involves the formation, development, and implosive collapse of bubbles in a liquid, which produces high pressure and high energy [109,110]. Graphene QDs (GQDs) were synthesized by Zhu et al. [111] from graphene oxide (GO) by ultrasonication in $KMnO_4$ for 4 h. High-resolution transmission electron microscopy (HR-TEM) revealed that the GQDs had an average of 3.0 nm lateral diameter with a narrow size distribution. The GQDs were reported to be uniform and of high crystallinity. These QDs were used in an alkaline phosphate (ALP) activity assay. In another study by Chen et al. [112], perovskite QDs were synthesized using ultrasonic synthesis. This synthesis method was reported to produce smaller particle sizes with a more uniform particle-size distribution. They also used this method to prepare different chemical compositions of $CH_3NH_3PbX_3$ QDs that could tune emission wavelengths, thus providing a wider range of pure colors. Table 2 catalogs the various types of QDs and their synthesis.

Table 2. The different synthesis techniques used to fabricate QDs.

Synthesis Methods	QDs Fabricated	Properties	Refs.
Electron beam lithography	QD nanostructures	Optical properties retained after cross-linking	[67]
	QD microarrays	Fluorescence Bioaffinity	[68]
Reactive ion etching	Indium gallium nitride (InGaN) QDs	Strong and distinct photoluminescence signal	[71]
Sol-gel	Titanium dioxide (TiO_2) QDs	large surface area photocatalytic properties	[80]
	Zinc selenide (ZnSe) QDs embedded in Silicon dioxide (SiO_2)	-	[81]
	Cadmium sulfide (CdS) and Ni-doped CdS	Highly crystalline	[113]
	Zinc oxide (ZnO)@polymer core/shell	Quantum yield above 50%	[114]
	Zinc oxide (ZnO) QD	High photoluminescence quantum yield	[115]
Microemulsion (reverse micelle)	Zinc sulfide (ZnS) QDs	Pure nanocrystal Quantum confinement effect Photoluminescence peak at 365 nm	[84]
	Cadmium sulfide/Zinc sulfide (CdS/ZnS) semiconductor QDs	Excellent luminescence and photostability	[86]
	Cadmium selenide@Zinc sulfide (CdSe@ZnS) within monodisperse silica	Good monodispersity High luminescence	[89]
Microemulsion (gas contacting technique)	Zinc selenide (ZnSe) QDs	Excellent photostability and size-dependent luminescence	[85]
Microemulsion method + ultrasonic waves (sono-microemulsion method)	Cadmium sulfide (CdS)	Narrow size distribution High crystallinity and purity	[116]
Physical vapor deposition	Niobium pentoxide (Nb_2O_5) QDs	Quantum confinement effect	[93]

Table 2. Cont.

Synthesis Methods	QDs Fabricated	Properties	Refs.
RF magnetron sputtering	Cadmium selenide (CdSe) QDs	Optical properties	[100]
Solvothermal	Zinc Oxide (ZO) QDs	Small size Pure, high crystallinity and surface area	[117]
	Graphene QDs (GQDs)	11.4% photoluminescence quantum yield High stability Biocompatibility Low toxicity	[118]
Hydrothermal	Nitrogen- and sulfur-doped carbon QDs (N, S-doped CQDs)	Small Spherical Green emission Fluorescence quantum yield (10.35%)	[119]
	Nitrogen-doped carbon QDs (N-CQDs)	Low toxicity Good photostability Good water dispersibility	[102]
	Silicon QDs	Strong photoluminescence High pH stability	[120]
	Tin oxide/Tin sulfide in reduced bovine serum albumin (SnO_2/SnS_2 @r-BSA2)	Specific selectivity Long term stability Enhanced reproducibility	[121]
	Nitrogen-doped Graphene QDs (N-GQDs)	High quantum yield Long-term fluorescence stability High sensitivity and specificity	[122,123]
Molecular beam epitaxy	Indium arsenide gallium arsenide core/shell (InAs/GaAs) QDs	Strong photoluminescence intensity High structural properties	[124]

4. Surface Functionalization of QDs

QDs have been widely used in various applications such as bioimaging, drug delivery, and diagnostics [125–130]. This has only been possible due to functionalizing their surfaces, thereby enhancing biocompatibility, uptake, stability, and reducing biological toxicity [131]. After synthesis, QDs are generally hydrophobic, which could produce a cytotoxic effect on cells or reduce their uptake efficiency, limiting their use in clinical practice. Hence, the surfaces of QDs need to be altered for prospective diagnostic and therapeutic applications by making them hydrophilic, and by attaching various chemical groups and targeting molecules [132–134]. This can be achieved by coating or conjugating the surface of the QDs with molecular ligands, growing silica, or applying other coatings to the QDs, such as with amphiphilic polymers [135–138]. The next sections present general descriptions of methods for surface modification.

4.1. Ligand Exchange

This process involves exchanging hydrophobic ligands such as trioctylphosphine oxide (TOPO), trioctylphosphine (TOP), and hexadecyl amine (HDA) on the QD surface with hydrophilic ligands to promote the formation of stable suspensions in water [139]. The most common approach for ligand exchange is the use of thiols (-SH), such as mercaptoacetic acid (MAA), mercaptopropionic acid (MPA), mercaptoundecanoic acid (MUA), and dihydrolipoic acid (DHLA) as anchoring groups, all of which present carboxyl (-COOH) groups as hydrophilic and ionized groups to enhance hydrogen bonding with water. Furthermore, at the proper pH (pH 5 to 12), ionic groups provide charge repulsion between particles. The attachment of hydrophilic polymers such as PEG can enhance the solubility range of QDs by steric repulsion [51,139,140].

The as-synthesized QDs are reported to have a small hydrodynamic size, which is useful in fluorescence resonance energy transfer (FRET) experiments [141]. However, after the process, there is a decrease in fluorescence quantum yield.

In other studies, the multidentate ligands were used as sensing probes to detect bovine serum albumin (BSA) protein in aqueous media [142]. Similarly, Chen et al. [143] reported the ligand exchange of oleate-capped ZB-CdSe with oleylamine, resulting in a significant decrease in photoluminescence quantum yield (PLQY). In another study [144], a method was optimized to overcome the issue of the reduced fluorescence and stability of silver telluride (Ag_2Te) QDs. Tributylphosphine (TBP) was added during synthesis, which was used as a precursor (TBP-Te) to form a high fluorescent Ag_2Te core. The rapid injection of TBP-Te precursor in hot solvent resulted in a PLQY of up to 6.51%. This was then followed by phase transfer of NIR-II Ag_2Te QDs via direct ligand exchange of hydrophobic Ag_2Te surface ligands with ligands of the thiol family (e.g., glutathione (GSH), DL-cystine, dithiothreitol (DTT), dihydrolipoic acid (DHLA), DHLA-EA, cysteamine, and thiol-containing PEG). It was observed that the hydrophilic thiol ligands promoted the water solubility of QDs and that only ligands composed of free thiol groups were suitable for this technique. Moreover, the QDs were reported to retain a PLQY of nearly 5% as well as exhibiting good biocompatibility. PEGylated Ag_2Te QDs were used for "second" near-infrared (NIR-II) imaging in mice. Unlike near-infrared (NIR) imaging with emission wavelengths between 700–900 nm, which is reported to produce substantial background signal and affect the quality of images [145], the NIR-II window encompasses emission wavelengths between 1000–1700 nm. Thus, it exhibits excellent penetration capacity and high-resolution fluorescence imaging in the living body. Real-time imaging in mice showed high brightness in abdominal vessels, sacral lymph nodes, hindlimb arterial vessels, and tumor vessels [144].

4.2. Surface Silanization

This coating process produces a silica shell around the QDs. It is an effective process for the modification of hydroxyl-rich material surfaces. This technique initially deposits hydroxyl groups by ligand exchange of the surface hydrophobic groups with a thiol-derived silane ligand (e.g., mercaptopropyltris (methyloxy)silane (MPS)) to place silanol groups on the surface. This is followed by further silica shell growth, where other silanes can be added on the outer surface to modify the surface charge or provide reactive functional sites. Aminopropylsilanes (APS), phosphosilanes, and polyethylene glycol (PEG)-silane are the most frequently used silanes [138,140]. Due to the silica thickness, the aqueous stability, size, biocompatibility, and fluorescence of the QDs are enhanced after being covered with a silica layer [146]. The layer also serves as a platform for further coating processes due to the silane shell end terminal groups exposing either their thiol, phosphate, or methyl terminal ends for subsequent reactions [147]. The advantage of this process is that the silica shells are highly crosslinked, thereby stabilizing the silanized QDs [147]. Furthermore, this is a preferred approach because the QDs can be made more biocompatible, less toxic, and chemically inert. The presence of silica increases the photostability of QDs by preventing surface oxidation [148,149], which makes them useful for applications such as drug and gene delivery, therapy, and bioimaging [150,151].

For example, silica coating is reported to suppress photoluminescence bleaching through the reduction in photochemical oxidation of cadmium selenide (CdSe) surfaces [152]. Similarly, encapsulation in silica was reported to prevent the loss of Cd^{2+} ions [153]. However, the silica shell is reported to increase the hydrodynamic size. Ham et al. [154] fabricated SiO_2@InP QDs@SiO_2 NPs by encapsulating multiple indium phosphide/zinc sulfide (InP/ZnS) QDs onto silica templates and coating silica shells over them. The fabricated QDs were reported to exhibit hydrophilic properties due to the surface silica shell. The NPs were further applied in detecting tumors where the fluorescence signal was notably detected in the tumor. Goftman et al. synthesized silica-coated cadmium selenide (CdSe)

QDs by the reverse microemulsion method. The silica-capped QDs were reported to have high stability and initial brightness [155].

4.3. Amphiphilic Ligands

In this approach, the hydrophobic surfactants trioctylphosphine, trioctylphosphine oxide, and hexadecylamine (TOP/TOPO/HDA) are preserved on the surface of the QDs. They are coated or encapsulated with crosslinked amphiphilic polymers containing hydrophobic and hydrophilic segments. The synthesized QDs are hydrophobic, and upon encapsulation with an amphiphilic polymer, an attraction is formed between the hydrophobic alkyl chains and the hydrophobic components of surfactants on the surface of the QDs. In contrast, the hydrophilic component (carboxylic acid or polyethylene glycol chains) provides dispersibility in aqueous solution and chemical functionality. During the coating process, the amphiphilic polymers are hypothesized to provide additional stability to the QDs through crosslinking reactions [51,138,156]. Some amphiphilic polymers include poly (acrylic acid), phospholipids, and maleic anhydride copolymers [156].

Yoon et al. [157] fabricated CdSe@ZnS/ZnS core/shell QDs encapsulated with an amphiphilic polymer (i.e., poly(styrene-co-maleic anhydride) PSMA). The amphiphilic polymer (PSMA) served as a crosslinker for the matrix polymer between the maleic anhydride of QDs and the diamines of PDMS within a ring-opening reaction. This produced a highly transparent polymer at low curing temperature with enhanced compatibility between QDs and a polydimethylsiloxane (PDMS) matrix and also improved dispersion of QDs. The encapsulated QDs were also reported to preserve photoluminescence intensity as a result of using this encapsulation method. They further fabricated a light-emitting diode, which was observed to have excellent luminous efficacy.

Starch-g-poly(acrylic acid)/ZnSe-QDs hydrogel was fabricated by Abdolahi et al. [158]. The QDs were fabricated to serve as an effective adsorbent and photocatalyst. In another study, Speranskaya et al. [159] synthesized hydrophobic cadmium selenide (CdSe)-based QDs. The QDs were hydrophilized by coating with amphiphilic polymers (i.e., maleic anhydride-based polymers and Jeffamines). The polymer-coated QDs were reported to retain up to 90% of their initial brightness. Carolina and Wolfgang [160] synthesized pyridyl-modified amphiphilic polymeric ligands (Py-PMA) in order to overcome the limitations of QDs coated with amphiphilic polymers, such as a decrease in photoluminescence quantum yield and diffusion of small molecules causing oxidation. Poly (isobutylene-alt-maleic anhydride) backbone was used for synthesis with pyridyl and alkyl end groups. The synthesized polymer-coated QDs were reported to preserve photoluminescence quantum yield and exhibit good colloidal stability in water.

4.4. Microsphere Coating

The microsphere coating of QDs is of great interest in biological applications [161]. The formation of composite nanostructures in which micro-composite nanostructures are assembled from QDs by an encapsulant component that can serve as a glue, a scaffold, or a matrix has been developed by researchers. Different techniques have been used for encapsulating QDs into microspheres. These include dispersing synthesized microspheres, placing QDs in a solvent or non-solvent mixture [162], and electrostatic bondage of QD to the microsphere surface [163]. The reverse microemulsion method [164] and emulsion polymerization [165] are encapsulation techniques. For example, a uniform magnetic/fluorescent microsphere was synthesized by Li et al. [164] using the Pickering emulsion polymerization method. The authors synthesized QD-encoded magnetic microbeads that were closely covered with a Pickering structure containing many silica nanoparticles. This was done using a microfluidic device that produced homogenous microbeads by forming Pickering emulsion droplets. The oil-in-water emulsion (O/W) droplets fabricated contained the oil phase (i.e., Fe_3O_4 NPs and QDs along with PSMA polymer were dispersed in toluene) and the water phase (silica NPs dispersed in deionized water), with the silica NPs accumulated at the interface (i.e., the oil and water interface). Thus, the silica NPs served as stabilizers. They reported the successful synthesis of a CdSe/ZnS core/shell along with a Fe_3O_4 nanoparticle encapsulated in a magnetic fluorescence microsphere (MFM microsphere). The microspheres were observed to be highly homogenous in shape, to have a high surface area, and to be well dispersed. Moreover, they also exhibited excellent fluorescent stability under room temperature. Hence, they further tested the microspheres to detect tumor markers (CEA, CA199, CA125) in a single sample. Results showed the detection limits achieved to be 0.027 ng/mL, 1.09 KU/L, and 1.48 KU/L for CEA, CA199, and CA125, respectively. The microspheres exhibited excellent detection performance.

Zhao et al. [166] synthesized bismuth oxybromide/carbon quantum dots (BiOBr/CQDs) microspheres using the solvothermal method followed by the hydrothermal process. The synthesized microsphere QDs were reported to exhibit excellent photoactivity under visible light irradiation due to exceptional electron transfer, and the CQDs exhibited increased light harvesting capacity in addition to stability and enhanced visible-light absorption ability. Moreover, QD-based sensors have been observed to agglomerate, leading to self-absorption and non-radiative deactivation. Hence, to overcome this issue, microsphere-QD-sensor platforms are being utilized. For instance, Khan et al. [167] developed a fluorescent sensor platform for heavy metal sensing. The authors used non-toxic fluorescent zinc oxide ZnO–QDs that were conjugated with carboxymethyl cellulose (CMC) polymer (ZCM) for the synthesis of microspheres for sensing heavy metal (cationic metal ions, e.g., Pb^{2+}, Hg^{2+}, Fe^{3+}, Cr^{6+}, Cu^{2+}, Ni^{2+}, Mn^{2+}). To differentiate these metal ions, a fluorescence turn-off response was adopted. Their results showed that the developed sensor had an affinity towards the different heavy metal ions and excellent photostability. In addition to detecting the heavy metals, the sensor could also quantify them with an accuracy of 5%. However, only Fe^{3+}, Cr^{6+}, and Cu^{2+}, among the seven metals, showed high sensitivity toward the sensor system. Table 3 presents several examples of functionalization of the surface of QDs.

Table 3. A summary of surface functionalization of QDs (showing the advantages and disadvantages of the four main techniques).

Surface Modification Techniques	Advantages	Disadvantages	Refs.
Ligand exchange	Ease of processing Small QD size	Degradation of QD photophysical properties in an aqueous environment (i.e., reduced PLQY) QD core is susceptible to oxidation	[51,168–170]
Surface silanization	Improves biocompatibility Highly cross-linked ligand molecules End terminal groups allow further coating through the exposure of the terminal ends (e.g., thiol). Control of silica shell thickness encourages fine-tuning of QD response to light. Improves PLQY of QDs Improves photochemical stability	Large hydrodynamic size Aggregation of QDs in aqueous solution	[171–173]
Amphiphilic ligands	More chemically stable Increased colloidal stability Good biocompatibility and strong, stable fluorescence signals	Size enlargementSurface defects	[138,174,175]
Microsphere coating	Improve QD stability High fluorescence Can mask QD toxicity effectively	The formation of a uniform microsphere is hindered. Reduced PLQY Encapsulation of high concentrations of QDs results in QD aggregation	[167,176,177]

This table shows that each of these methods has its advantages. However, the final choice depends on the specific application and the requirements. For instance, the ligand exchange process decreased the photoluminescence quantum yield (PLQY). Hence, direct encapsulation of QDs with silica shell resolves the issue of reduced luminescence yields. This layer of silica on the QD is reported to provide enhanced aqueous stability and fluorescence by the silica's thickness [146]. Yet, it was reported that coating with silica shell yields larger QDs due to the difficulty in controlling the silica thickness [173]. Moreover, the encapsulation of QDs with an amphiphilic polymer also preserves quantum yield (QY) even after surface modification.

Regarding microspheres, they are reported to provide hydrophobic protection. This is because some QDs are hydrophobic in nature and not biologically useful. Thus, QDs are functionalized or coated to make them water-dispersible and enhance their biocompatibility. However, it was reported that the size of the photoluminescence (PL) microsphere determined QD stability, with a larger PL microsphere observed to give more hydrophobic protection of the interiors of QDs compared to smaller PL microspheres [178]. In other words, for every possible application, the prerequisite is to properly functionalize the surface of the QDs accordingly while ensuring they do not lose their physicochemical properties, which are enhanced in aqueous media. Figure 3 illustrates some surface functionalization approaches of a multifunctional QD.

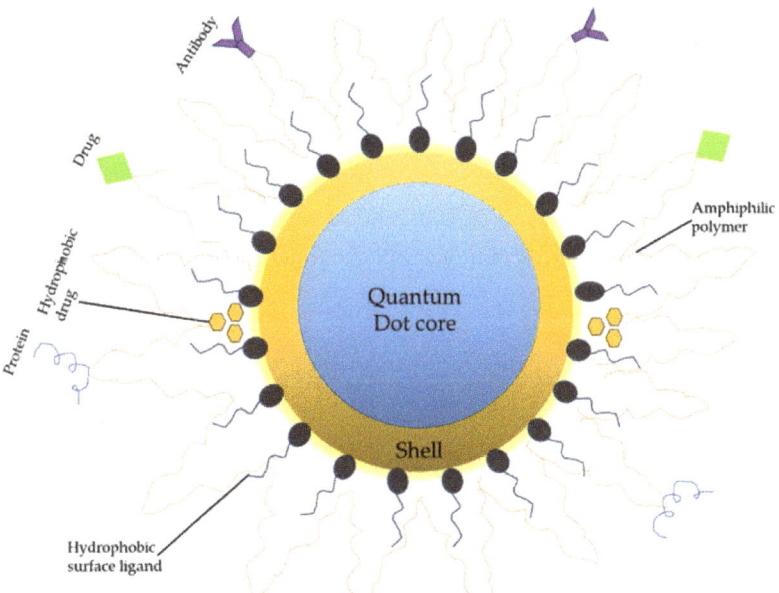

Figure 3. Surface functionalization of QD core/shell. The surface coating (e.g., amphiphilic polymer coating) enables antibodies, drugs, proteins, and other compounds to be linked with it. Hydrophobic drugs can also be integrated between the hydrophobic core and amphiphilic layer.

5. Application of QDs

5.1. QDs for In Vitro Tumor Imaging

One of the most important applications of QDs in recent research has been to produce in vitro fluorescent images of cancerous cells. The unique properties of QDs make them preferable to traditional fluorescence organic dyes. A schematic representation of QDs for in vitro tumor imaging is shown in Figure 4.

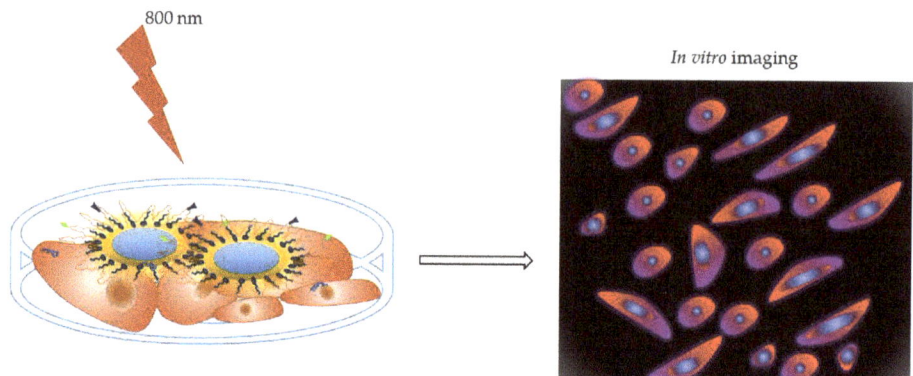

Figure 4. Schematic representation of QDs for in vitro tumor imaging.

Nitrogen-doped carbon QDs (N-CQDs) were synthesized hydrothermally by Wu et al. [179] using tetraphenyl porphyrin and its metal complex (Pd or Pt) as a precursor. As a result of the strong photoluminescence (PL) exhibited by the CQDs, they were investigated as imaging probes for living cells. HeLa cells treated with CQDs (0.2 mg/mL) exhibited blue, green, and red fluorescence at excitation wavelengths of 405 nm, 458 nm, and 514 nm. Fluorescence images showed CQDs to be mainly dispersed in the cell cytoplasm, and the nucleus showed weak emission signals. These experiments supported that CQDs enter into cells via endocytosis.

Near-infrared (NIR) emitting CdHgTe/CdS/CdZnS QDs were synthesized by Liu et al. [180]. The QDs were coated with N-acetyl-L-cysteine (NAC), 3-mercaptopropionic acid (MPA), and thioglycolic acid (TGA) thiol ligands. HeLa cells were stained with these QDs and exposed to continuous UV excitation. In vitro studies showed that after 20 min of irradiation, stained HeLa cells produced red emission. Fluorescence images revealed that after 40 min, NAC-tagged CdHgTe/CdS/CdZnS QD-stained cells showed high photostability in the intracellular environment compared to TGA- and MPA-capped QDs. This success was attributed to the NAC thiol capping of the QDs preventing degradation.

Near-infrared (NIR) CdTe/CdS was synthesized in an aqueous solution with 3-mercaptopropionic acid (MPA) as a stabilizer. These QDs were employed to monitor the change in Cu^{2+} concentration in living cells. HeLa cells were incubated with the synthesized QDs (5 μg/mL), followed by adding Cu^{2+} (30 μM) before fluorescence imaging. A bright fluorescence signal from the cells at 700–800 nm showed efficient uptake of CdTe/CdS. However, when HeLa cells were treated with 30 μM of Cu^{2+} before incubation with the QDs, significant fluorescence quenching (~90%) was observed. This observation was attributed to the aggregation of QDs mediated by the competitive binding between MPA and the Cu^{2+} in the solution. Overall, they reported the nanosensor to exhibit high selectivity, excellent photostability, and rapid response [181]. Fluorescence images generated during this study are shown in Figure 5.

Shi et al. synthesized molybdenum disulfide (MoS_2) QDs with Na_2MoO_4 as the molybdenum source and $2H_2O\cdot GSH$ as the sulfur source using hydrothermal synthesis [182]. The reaction conditions (i.e., precursor, precursor ratio, ratio, reaction time, and temperature) were optimized to improve the photoluminescence quantum yield (PLQY). These MoS_2 QDs were then used for fluorescence imaging. The in vitro studies reported glutathione–molybdenum disulfide (GSH-MoS_2) to be biocompatible after SW480 cells were exposed to the QDs (from 0 to 1.5 μM Mo). They reported that blue fluorescence was observed in the SW480 cells cytoplasm.

In another study, blue-fluorescent nitrogen-doped graphene quantum dots (N-GQDs) were produced by Tao et al. [183]. The QDs were synthesized using hydrothermal synthesis from citric acid and diethylamine, and the binding sites were highlighted. The doping with nitrogen element resulted in ample amide II bonds (this provides a structure for integrating HA with N-GQDs) and enough binding sites to conjugate hyaluronic acid (HA). In order to recognize the breast cancer cells (MCF-7 cells), the N-GQDs were conjugated to HA through an amide bond. It was reported that the formation of amide bonds was more conducive under alkaline conditions. In addition, MCF-7 cells exhibited stronger fluorescence as a result of combining HA-conjugated N-GQDs (HA-N-GQDs) with CD44 over-expressed on the MCF-7 cells surface. Their results showed the good cytocompatibility, low toxicity, and high fluorescence of HA-N-GQDs.

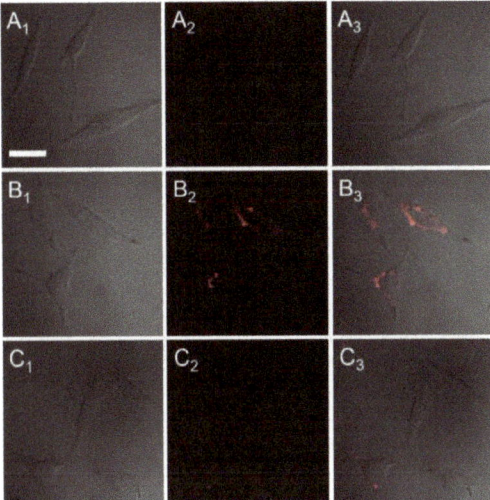

Figure 5. Confocal fluorescence images of HeLa cells (**A**) before and (**B**) after mixing with CdTe/CdS QDs at 5 μg/mL and (**C**) 30 μM of Cu^{2+} was then added to (**B**) to monitor concentration change of Cu^{2+} Showing as (1) brightfield images, (2) fluorescence images (700–800 nm filter), and (3) merging of (1) and (2). (Scale bar 30 μm. Reprinted with permission from [181]).

5.2. QDs for In Vivo Tumor Imaging

The excellent fluorescent signals and multiplex capabilities of QDs make them a promising tool for cancer bioimaging, specifically in vivo. Researchers have reported many examples of using QDs to image tumors in vivo. A schematic representation is shown in Figure 6.

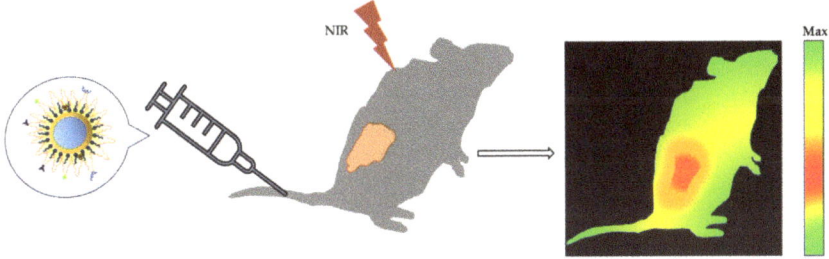

Figure 6. Schematic representation of QD injected into a tumor-bearing mouse for in vivo tumor imaging.

For instance, Zhu et al. [184] developed near-infrared (NIR) fluorescent silver selenide (Ag_2Se) QDs tagged with Cetuximab for targeted imaging and cancer therapy. The multifunctional nanoprobe was reported to display fluorescent contrast at the tumor site, and 24 h post-injection, the fluorescence was still easily detected at the tumor site, unlike with Ag_2Se QDs alone. Their results showed that this nanoprobe significantly inhibited tumor growth, and the survival rate of nude mice with orthotopic tongue cancer improved from 0% to 57.1%. This platform was claimed to have successfully targeted orthotopic tongue cancer.

Sulfonic-graphene QDs were used by Yao et al. [185] to target tumor cells in vivo. They showed that the sulfonic-GQDs had successfully penetrated the plasma membrane into tumor cells without modifying any bio-ligand, which they attributed to high interstitial fluid pressure. They also reported fluorescence of the sulfonic-GQDs at an excitation of 470 nm in tumor-bearing mice post-injection. Rapid accumulation of sulfonic-GQDs at the tumor site occurred 0.5 h after injection and was cleared 24 h later. This research demonstrated sulfonic-GQDs' ability to target nuclei of tumor cells in vivo with a low distribution in normal tissues.

In another study, Wu et al. [186] developed a novel strategy against tumor cells. They modified near-infrared fluorescent indium phosphide (InP) QDs using a vascular endothelial growth factor receptor 2 (anti-$VEGFR_2$) monoclonal antibody and attached miR-92a inhibitor to $VEGFR_2$-InP QDs. The miR-92a is said to enhance the expression of tumor suppressor p63. Their results showed that the functionalized InP nanocomposite showed an enhanced NIR fluorescence intensity at the tumor site, which had accumulated via enhanced permeability and retention effect, thereby targeting tumor angiogenic cells. Moreover, using nude mice inoculated with k562 cells, they investigated the suppression of tumor growth in vivo. They observed the functionalized InP nanocomposite to significantly inhibit tumor growth compared to InP QDs or miR-92a, which showed moderate suppression. Overall, the developed system may provide a new and promising chemotherapy strategy against tumor cells.

Fluorescent silver indium sulfide/zinc sulfide (Ag-In-S/ZnS (AIS/ZnS)) QDs with red emission were synthesized by Sun et al. [187] and then dispersed with poly(vinylpyrrolidone) (PVP) for imaging of tumor drainage lymph nodes. The synthesized QDs were subcutaneously injected in nude mice, and a bright red fluorescence was observed, suggesting that AIS/ZnS QDs are excellent fluorescent probes for in vivo imaging. To image sentinel lymph nodes, AIS/ZnS QDs were intradermally injected into the extremities of nude mice, and the QDs were observed to migrate to sentinel lymph nodes. Furthermore, within 10 min of intratumoral injection in mice bearing H460 tumors, AIS/ZnS QDs were observed to stain tumor drainage lymph nodes with bright red fluorescence. However, after 10 min, only weak fluorescence was observed in the tumor drainage lymph node.

Triple-negative breast cancer (TNBC) is known to develop rapidly and is associated with recurrence and metastasis. The efficacy of chemotherapy is reported to be poor, with the survival rate of patients affected being less than 30%. Hence, Zhao et al. [188] designed and constructed biomimetic black phosphorus QDs (BBPQDs) coated with cancer cell membranes for tumor-targeted photothermal therapy (PTT) and anti-PD-L1 mediated immunotherapy. The stability of the BBPQDs after encapsulating with cancer cell membrane exhibited active targeting and enrichment ability in tumors. Subsequently, Cy5.5-labelled BBPQDs were intravenously injected into BALB/c mice bearing 4T1 tumors to investigate tumor targeting and tissue distribution. The BBPQDs were reported to exhibit significant fluorescence intensity post-injection compared to Cy5.5-labeled BPQDs. Moreover, after 72 h, the BBPQDs showed good tumor targeting, high aggregation, and good retention at the tumor site. The BBPQDs exhibited excellent photothermal properties and could kill tumors directly and induce dendritic cell maturation and the activation of T cells. BBPQD-mediated PTT and αPD-L1 combined inhibited tumor recurrence and metastasis through the immune memory effect.

Stable fluorescent CQDs were synthesized by Huang et al. [189] under photobleaching treatment. The synthesized CQDs were reported to have a quantum yield (QY) of ~13% at an excitation of 365 nm, proving them to be viable in bioimaging mice with Smmc-7721 tumor cells. The CQDs were intravenously injected (0.2 μg/mL). Optical images of the distribution of the CQDs were obtained at different time points. The study reported detecting fluorescence signal 5 min post-injection, and CQDs accumulated at the tumor site after 3 h. Complete accumulation of the CQDs was reported to occur at 12 h. The CQDs appeared to exhibit good biocompatibility and could be used for a prolonged imaging period. Results also showed that CQDs accumulated in the tumor, kidney, and liver. However, no fluorescence signal was detected in the heart, lungs, and spleen. In addition, the CQDs were reported to exhibit excellent bioimaging performance, low cytotoxicity, and antioxidant activity.

Although the unique optical properties of QDs make them an attractive fluorescent probe, specifically in bioimaging, the potential toxicity of QDs, such as those containing toxic heavy metals, has limited their applications. Hence, Yaghini et al. [190] developed a heavy-metal-free biocompatible and good photoluminescence quantum yield (PLQY) Indium-based QD (bio CFQD® NP) for imaging in vivo. These metal-free QDs were investigated for in vivo axillary lymphatic mapping applications. Twenty-four hours post-injection of the QDs in the paw of rats, the QDs were observed to accumulate mainly in the regional lymph nodes with negligible accumulation in the spleen and liver while exhibiting stable photoluminescence. Their low intrinsic toxicity makes them attractive for in vivo tumor imaging.

5.3. QDs for Drug Delivery

QDs are just one example of the numerous nanoparticles (NPs) that have been widely investigated for drug delivery applications. Reports show that antitumor efficacy is increased while systemic side effect is reduced, which is attributed to effective nanoparticle entrapment of anti-cancer drugs and control of distribution in cells and in tissue. The use of nanoparticles as drug delivery agents has been reported to overcome the limitations posed by traditional cancer therapies, including but not limited to overcoming multidrug resistance, lack of specificity, and cytotoxicity. Their specific advantages, such as enhanced stability, reduced toxicity, precise targeting, and biocompatibility, promote the use of NPs as nanocarriers in cancer therapy [191–193].

Moreover, these nanocarriers have been found to facilitate the administrative routes and enhance the biodistribution of drugs [194]. They act as drug vehicles and can target tumor cells or tissues while shielding the drug during transport [192]. The delivery of drugs to the site occurs actively, i.e., a drug delivery system (DDS) is coupled with peptides and antibodies anchored with lipids or receptors at the target site, or passively, i.e., the drug is transported via self-assembled nanostructured material and released at the target site [195].

Nanoparticles, in general, are excellent nanocarriers for targeted drug delivery. They serve as potential candidates due to their biocompatibility, controlled drug release, prolonged circulation time, and accumulation at the tumor site due to enhanced permeability and retention (EPR) effect [196–198]. Table 4 lists some common nanoparticles used for drug delivery, along with their advantages and disadvantages.

Table 4. Advantages and disadvantages of organic and inorganic NPs used for drug delivery.

Organic Nanoparticles	Advantages	Disadvantages	Refs.
Liposomes	Enhances drug solubility Reduces drug toxicity	Decreased stability	[199,200]
Micelles	Improves circulation time Protects aqueous drug cargo	Lack of targeting moieties	[201,202]
Polymer NP (Chitosan)	Increase drug residence time in the bloodstream	Initial burst release results in loss of drug efficiency	[203,204]
Dendrimers	The hydrophobic core allows insoluble anti-tumor drugs to be absorbed and provides smooth delivery. The hydrophilic part increases stability and limits the particles' interaction with serum proteins.	Rapid clearance of reticuloendothelial system	[205,206]

Table 4. Cont.

Organic Nanoparticles	Advantages	Disadvantages	Refs.
Inorganic Nanoparticles			
Silver NPs	Enhances PTX distribution in tumor microenvironment	Release of silver ions in cytosol	[207,208]
Gold NPs	Enhances photothermal therapy Easily functionalized	Low tissue clearance	[209–211]
Mesoporous silica NPs	Controlled drug release	Slow biodegradation	[204,212]
Magnetic NPs (iron oxide)	Precise targeting of cancer cells Release of PTX under external magnetic field	Removal by macrophages	[213–215]
Quantum Dots	Improves the bioavailability of the drug	Leaching of heavy metals	[216,217]

The use of QD nanoparticles for targeted drug delivery is of great interest due to their unique properties, including their distinctive optical characteristics due to their quantum confinement effects. QDs are also an excellent choice because of their intrinsic fluorescence and unique properties to serve as a multifunctional nanosystem. This includes their ability to aid in targeted drug delivery and improve the bioavailability and stability of drugs by prolonging the circulation time in vivo and improving distribution [216].

The use of QDs for drug delivery requires the modification of their surface with target ligands (e.g., thioglycolic acid, polyethylene glycol (PEG), antibodies, DNA, biotin, or peptides) [218,219]. Some surface modifications enable the drug molecules to bind to the QDs through covalent bonds or electrostatic binding, which forms nano-drug carriers and then makes fluorescent tags of drug molecules in cells and live animals [220]. Hence, QDs can act as drug carriers as well as fluorescent probes to trace drug distribution in vivo [221]. However, the size of the QD should be considered because excretion from the body is important. Moreover, the drugs can be loaded into a polymer NP system containing either hydrophilic QDs or hydrophobic QDs, depending on the polymer particle type used for encapsulation. This is followed by delivery at the desired site, where the polymer particle releases the drug via degradation at low pH or diffuses out of the polymer [222]. Figure 7 shows the development of molybdenum disulfide (MoS_2) QDs for tumor fluorescence imaging, tumor targeting, and chemo/photodynamic therapy (PDT).

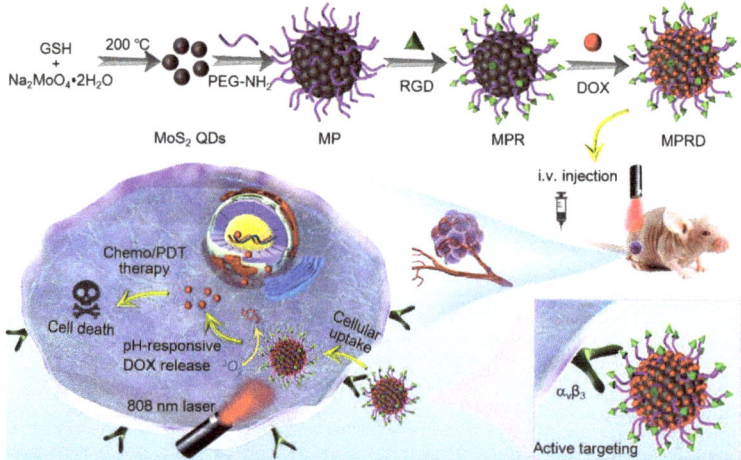

Figure 7. Illustration of synthesized PEGylated MoS_2 conjugated with arginine-glycine-aspartic acid (RGD) peptide to form MPR (i.e., novel nanocarrier) and then loaded with doxorubicin (DOX)

to form MPRD. MPRD exhibits tumor-targeting ability, pH-responsive drug release, and synergistic chemo/PDT performance under near-infrared (NIR) laser irradiation(grey circles: synthesized MoS_2 QDs, red circles: Dox, green triangles: RGD). Reprinted with permission from [223].

Table 5 shows in vitro and in vivo targeted drug delivery using QDs.

Table 5. In vitro and in vivo targeted drug delivery using QDs.

QDs Used In Vitro	Drug	Cell Line	Ref.
Iron oxide carbon QDs encapsulated in chitosan (Fe_2O_3/CQDs/Chitosan)	Curcumin	MCF-7 cells	[224]
Transferrin (TF)-conjugated Carbon QDs	Doxorubicin	MCF-7 cells	[225]
Graphene oxide QDs conjugated with glucosamine and boric acid (GOQDs-GlcN-BA)	Doxorubicin	MCF-7 cells	[226]
Magnesium nitride (Mg/N) doped carbon QDs (CQDs)	Epirubicin (EPI)	4T1 and MCF-7 cells	[227]
Nitrogen-doped Graphene QDs (N-GQDs)	Methotrexate (MTX)	MCF-7 human breast cancer cells	[228]
PEGylated molybdenum disulfide QDs (PEG-MoS_2 QDs)	Doxorubicin	U251 cells	[229]
Zinc oxide adipic dihydrazide heparin (ZnO-ADH-Hep)	Paclitaxel	A549 cells	[230]
Cadmium-sulfide-modified chitosan (CdS@CTS)	Sesamol	MCF-7 cell	[231]
PEGylated Silver graphene QDs (Ag-GQDs)	Doxorubicin	HeLa and DU145 cells	[232]
Magnetic carbon triazine dendrimer reacted with graphene QDs (Fe_3O_4@C@TD GQDs) microsphere	Doxorubicin	A549 cell	[233]
QDs used in vivo			
Graphene QDs	Doxorubicin	MCF-7 cells	[234]
Silver sulfide (Ag_2S) QDs conjugated with chitosan	Doxorubicin	HeLa cells	[235]
Manganese doped zinc sulfide (Mn-ZnS) QDs conjugated with folic acid (FA)	5-fluorouracil (5-FU)	4T1 breast cancer cells	[236]
PEGylated silver sulfide Ag_2S QDs	Doxorubicin	MDA-MB-231 human breast tumor cells	[237]
Graphene QD (GQD)-modified magnetic chitosan Fe_3O_4@CS	Doxorubicin	Hepatocellular carcinoma	[238]
Red-emissive carbon QDs (CQDs)	Doxorubicin	HeLa cells	[239]
Black phosphorus QDs (BPQDs) encapsulated in platelet-osteosarcoma hybrid membrane (OPM)	Doxorubicin	Osteosarcoma	[240]
Nitrogen-doped carbon QDs conjugated with folic acid (FA)	Doxorubicin	4T1 and MCF-7 cells	[241]
PEGylated molybdenum disulfide (MoS_2) QDs conjugated with arginylglycylaspartic acid (RGD) peptide	Doxorubicin	HepG2 cells	[223]
Polyethyleneimine (PEI)-conjugated graphene QDs (GQDs)	Doxorubicin	HCT116 cells	[242]

The anti-tumor drug Adriamycin was loaded into a drug delivery system (DDS) developed by Hao et al. [243] through covalent interactions and the formation of Zn^{2+}-DOX. The lanthanum-doped zinc oxide (La-ZnO) QDs were modified with hyaluronic acid (HA). This enables them to bind specifically to receptor CD44. In addition, the developed system was PEGylated to stabilize it under physiological conditions. Their results showed that an anti-tumor effect and dual fluorescence enhancement were achieved due to lanthanum doping.

Similarly, Cai et al. [55] used covalent interactions and the formulation of a zinc doxorubicin (Zn^{2+}-Dox) chelate complex to load Doxorubicin to hyaluronic-functionalized PEGylated zinc oxide (HA-ZnO-PEG). They reported that the system exhibited an acidic pH response, which triggered targeted drug release in tumors.

A polylactic acid (PLA) polymer matrix has been used for drug encapsulation as it provides sustained and controlled drug release. Gautam et al. [244] conjugated Gefitinib to polyethylene glycol graphene QDs (PEG-GQDs) and encapsulated the QDs in polylactic acid (PLA) microsphere for cancer therapy. They aimed to use the developed system for controlled drug (Gefitinib) delivery. They reported drug release to be around 65% after 48 h at an acidic pH (pH = 4.5). This was attributed to destabilized electrostatic interaction. At basic pH (pH = 7.4), drug release was observed to be slower. They suggested that their prepared system using PLA microspheres could be an excellent candidate for cell imaging and drug delivery. Figure 8 illustrates the in vitro release of Gefitinib-loaded microspheres.

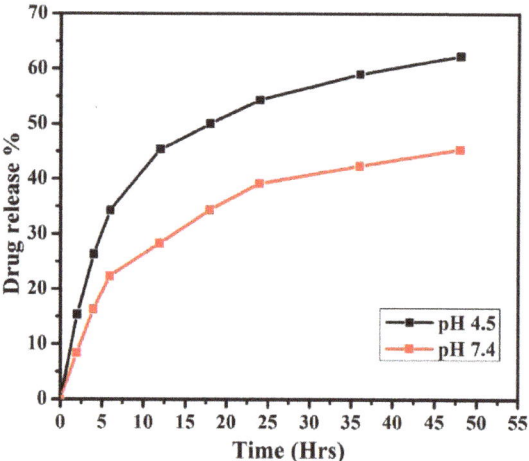

Figure 8. In vitro release of drug (Gefitinib)-loaded microspheres at pH 4.5 and 7.4. Reprinted with permission from [244].

Furthermore, Wei et al. [245] evaluated using QDs as an effective tool for microenvironment-targeted drug delivery. Using chemical oxidation and a covalent reaction, Pt-loaded and polyethylene glycol (PEG)-modified graphene QDs (GQDs) were developed as a drug delivery system. The Pt-loaded and PEG-GQDs were developed to overcome hypoxia-induced chemoresistance in oral squamous cell carcinoma. The accumulation of Pt within oral squamous cell carcinoma (OSCC) cells was significantly enhanced using polyethylene glycol–graphene QDs-Pt (GPt) in normoxia and hypoxia. The GPt was observed 2 h after incubation in the cytoplasm and in the nucleus 5–8 h after incubation. After 24 h, GPt luminescence was further enhanced, indicating that GQDs can transfer Pt and are potential platforms for nucleus-targeted drug delivery. The in vivo studies reported that GPt inhibited tumor growth.

In another study, graphene QDs (GQDs) were incorporated into carboxymethyl cellulose (CMC) hydrogels to design a hydrogel nanocomposite film loaded with doxorubicin as a drug model. They reported drug release to inversely depend on the concentration of GQD (i.e., release % of DOX from CMC/GQD decreases with increasing GQD concentration) even as the pH was varied. In addition, increasing GQD concentration resulted in increased drug loading capacity, showing that GQDs incorporated in CMC films resulted in pH sensitivity and the prolonged release of the therapeutic agent [246]. Olerile et al. [247] developed paclitaxel (PTX) and CdTe@CdS@ZnS QDs co-loaded in nanostructure lipid carriers (NLC). Their experiments showed that the encapsulation efficiency of PTX was 80.70 ± 2.11% and the drug loading was 4.68 ± 0.04%. In addition, the rate of tumor suppression was reported to be 77.85%. Their results showed that the co-loaded NLC could also detect H22 tumors, revealing some potential for bioimaging.

Zhao et al. [248] also used paclitaxel (PTX) as a model drug. They synthesized manganese-doped zinc selenide zinc sulfide (ZnSe:Mn/ZnS) core/shell, and the anti-cancer drug (PTX) was co-loaded into hybrid silica nanocapsules conjugated with folate. Folic acid (FA) conjugation was performed via an esterification reaction between FA carboxylic groups and animated F127 amino groups. The PTX solubility (0.1 μg/mL) was reported to be enhanced 630 times, improving the loading amount to 62.99 μg/mL. Their reports showed sustained release of PTX across 12 h. Overall, the developed hybrid nanocapsules showed the efficacy of anti-cancer drug loading and sustained release. Figure 9 illustrates the process of FA conjugation.

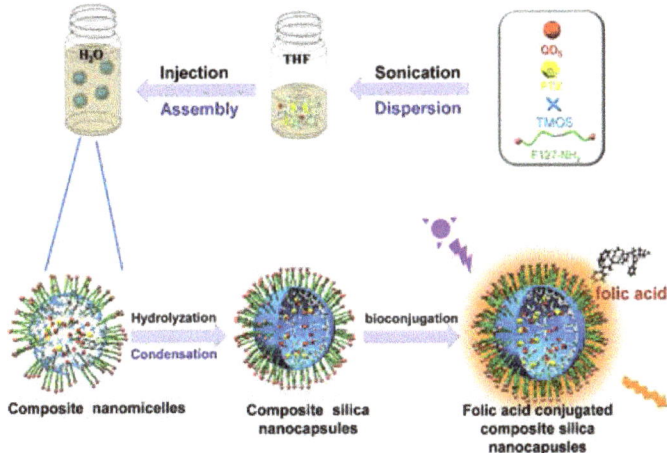

Figure 9. Schematic diagram of FA-conjugated hybrid silica nanocapsules. Reprinted with permission from [248].

Demir Duman et al. [249] evaluated the use of near-infrared-emitting silver sulfide (Ag_2S) QDs. The Ag_2S QDs surfaces were coated with PEG, functionalized with Cetuximab (Cet) antibodies to target and reveal tumor cells, and loaded with the 5-fluorouracil (5FU) anti-cancer drug. The QDs were developed for targeted NIR imaging and treatment of lung cancer via low and high epidermal growth factor receptors (EGFR). The Cet-conjugated QDs delivered 5FU effectively and selectively to A549 cells and provided exceptionally enhanced cell death associated with apoptosis. They suggested their novel system would significantly overcome drug resistance compared to the treatment of 5FU alone.

Yang et al. [250] developed GQDs loaded into hollow mesoporous silica nanoparticles (HMSN cavity) (GQDs@hMSN-PEG NPs). The singlet oxygen ($^1O_2$) generating capacity of the GQDs was not affected after hMSN loading. The developed GQDs@hMSN-PEG NPs were reported to exhibit excellent absorption and emission properties. The drug loading capacity was measured and the NPs were found to carry significant amounts of DOX. They further demonstrated drug delivery feasibility on mice bearing 4T1 tumors by injecting

GQDs@hMSN (DOX)-PEG, with results showing the feasibility of tumor-directed drug delivery. Table 6 summarizes some other applications relative to cancer involving QDs.

Table 6. Other recent applications of QDs relative to cancer [226,251–268].

QDs Utilized	Application	Target Cells
Carbon QDs (CQDs)	Drug delivery	Breast cancer cell line
Carbon QDs (CQDs)	Drug delivery	Breast MCF-7 cancer cells
Graphene QDs (GQDs)	Drug delivery	U251 glioma cells
Near-infrared (NIR) copper indium sulfide zinc sulfide core/shell (CuInS$_2$/ZnS) QDs	In vivo	RR1022 Cancer cell
Alloyed Zinc copper indium sulfide (ZCIS) QDs	In vitro	HER2-positive SKBR3 cancer cells
Molybdenum disulfide (MoS$_2$) QDs-MXene	Electrochemiluminescence (ECL) sensor for detection	Gastric cancer cell exosome
Zinc oxide (ZnO) QDs	Drug delivery	HepG2 cells
Molybdenum disulfide (MoS$_2$) QDs	Photodynamic therapy Drug delivery	HeLa and HepG2 cells
Manganese-doped molybdenum disulfide (Mn-MoS$_2$) QDs	In vivo MR imaging Fluorescence labeling	786-O Renal carcinoma cells
Titanium-ligand-coordinated black phosphorus QDs (TiL4@BPQDs)	In vivo Photoacoustic Imaging	MCF-7 cancer cells
Graphene QDs (GQDs)	Photothermal therapy	MDA-MB-231
Folic-acid-conjugated carbon QDs (FA-CQDs)	Fluorescence imaging	MCF-7 cells and ovarian cancer (HeLa)
Copper indium sulfide zinc sulfide core/shell (CuInS/ZnS) QDs	Sensor probe for targeted imaging	BEL-7402 cancer cells
Titanium nitride (Ti$_2$N) QDs	Photoacoustic (PA) imaging-guided photothermal therapy (PTT) in near-infrared (NIR-I/II) biowindows	293T, 4T1 and U87 cancer cells
Cadmium telluride cadmium sulfide (CdTe/CdS) core–shell QDs	Fluorescence imaging	MDA-MB-231/MDR
Zinc oxide (ZnO) QDs	Drug delivery	MCF-7
Cadmium selenide telluride zinc sulfide (CdSeTe/ZnS) QDs	Photothermal therapy	Hepatoma cells Huh7
Graphene QDs (GQDs)	Drug delivery	MCF-7 cells
Near-infrared (NIR) silver selenide (Ag$_2$Se) QDs	In vivo tumor imaging	MCF-7 human breast cancer cells and SW1990 pancreatic cancer cells

6. Cytotoxicity

The cytotoxicity of many QDs is a major deterrent to using QDs in widespread biomedical imaging and therapy. Despite their promising potential in various applications due to their optoelectronic properties, the toxicity of QDs limits their use to in vitro or animal studies. The toxicity of QDs is attributed to their chemical compositions containing heavy metal ions such as cadmium and indium [269]. In addition, their environmental conditions and physicochemical structure contribute to toxin availability (e.g., size, concentration, capping material, mechanical stability, etc.) [270–272]. For instance, the cardiotoxicity of cadmium selenide zinc sulfide (CdSe/ZnS) QDs was investigated by Li et al. [273]. A significant amount of cadmium (Cd) was detected in the hearts of mice bearing CdSe/ZnS QDs. Their results showed the accumulation of CdSe/ZnS QDs in the heart in addition to the incomplete QD excretion of up to 42 days.

In another study, the toxicity of copper indium disulfide zinc sulfide (CuInS$_2$/ZnS) core/shell QDs was investigated in vivo. Ninety days after injection, indium was detected in the kidney, heart, brain, and testis. In another study, CuInS$_2$/ZnS QDs were reported to accumulate in the liver and spleen [274].

Furthermore, QD toxicity results from the generation of reactive oxygen species (e.g., free radicals and the creation of singlet oxygen) [275], which could damage DNA.

Near-infrared (NIR) QDs have also been reported to present a health risk. For instance, Zhang et al. [276] reported that lead sulfide/cadmium sulfide (PbS/CdS) QDs (0.7%) remained in mice after 1 month. The QDs were observed in the liver, spleen, lungs, kidneys, stomach, and gut and distributed to other body parts. The toxicity and accumulation of QDs in off-target tissues is an issue that must be addressed.

Conversely, researchers have reported that the coating of QDs or the surface functionalized QDs reduced the leaching of ions [277], thereby reducing acute toxicity. For instance, Murase et al. [278] synthesized cadmium selenide/zinc sulfide (CdSe/ZnS) QDs encapsulated in highly emitting silica capsules by the sol–gel method. At a shell thickness of 15 nm, the release was suppressed effectively compared to a shell thickness of 10 nm. They further reported leakage suppression at a temperature of 40 °C. Their results revealed that the silica capsules were non-toxic to cells. There is still the need to consider the effective surface coating of QDs because a better-protecting shell is less likely to leach heavy metals; however, at the same time, the size of QDs is increased after encapsulation, which might hinder their use in some applications. Even if capping can effectively minimize toxic ion release and preclude acute toxicity, the long-term buildup of capped QDs must be addressed for clinical translation to be approved. Consequently, additional investigation is warranted to develop improved methods for synthesizing QDs that mitigate or eradicate their toxic properties.

7. Conclusions

The use of nanoparticles in the fight against cancer has been researched extensively. Nanoparticles possess several characteristics required to overcome the limitations of conventional cancer management strategies, thus providing a platform for early detection and treatment. Quantum dots are the latest nanoparticles to exhibit unique properties that could impact how cancer is diagnosed and treated. These features include their small tuneable size, stable photoluminescence, large surface-to-volume ratio, and potential biocompatibility. QDs have been extensively applied for in vitro and in vivo tumor imaging and, more specifically, integrated with therapeutic agents for targeted drug delivery in vivo. The flexibility to bioconjugate or modify the surface of QDs according to the needed application qualifies QDs to be potential candidates as multifunctional systems. Many studies have shown that drug encapsulation in QDs increased drug delivery efficacy. More importantly, surface-modified QDs show promise as a great platform that could simultaneously deliver loaded drugs and provide real-time imaging of the biodistribution of the drug at tumor sites in vitro and in vivo.

Furthermore, studies have revealed that QDs subjected to surface modification serve as fluorescent markers and can inhibit tumor growth substantially or directly induce tumor cell death when combined with the requisite receptors or ligands. While the toxicity issues associated with QDs containing heavy metals like cadmium have been acknowledged, their tendency to accumulate in bodily organs due to their overall size hinders some of their potential use in human in vivo imaging and drug delivery applications. Hence, the development of heavy-metal-free QDs is extensively studied for possible clinical applications [279]. While acknowledging the need to minimize QD dimensions and appropriately capping them to mitigate toxicity, all while considering the specific application needs, it is important to note that QDs have demonstrated novel and useful promise in cancer imaging and treatment. Without a doubt, persistently utilizing the QD platform for cancer-related biological research will lead to a noteworthy breakthrough that has the potential to reshape the current research landscape.

Author Contributions: Original draft writing and visualization: A.H.; reviewing and editing: W.G.P. and G.A.H. All authors have read and agreed to the published version of the manuscript.

Funding: This research was funded by the Dana Gas Endowed Chair for Chemical Engineering, the American University of Sharjah Faculty Research Grants (FRG20-L-E48, FRG22-C-E08), the Sheikh Hamdan Award for Medical Sciences MRG/18/2020, and the Friends of Cancer Patients (FoCP).

Data Availability Statement: Not applicable.

Acknowledgments: The authors would like to acknowledge the financial support of the American University of Sharjah Faculty Research Grants, the Al-Jalila Foundation (AJF 2015555), the Al Qasimi Foundation, the Patient's Friends Committee-Sharjah, the Biosciences and Bioengineering Research Institute (BBRI18-CEN-11), GCC Co-Fund Program (IRF17-003) the Takamul program (POC-00028-18), the Technology Innovation Pioneer (TIP) Healthcare Awards, the Sheikh Hamdan Award for Medical Sciences MRG/18/2020, the Friends of Cancer Patients (FoCP), and the Dana Gas Endowed Chair for Chemical Engineering. The work in this paper was supported, in part, by the Open Access Program from the American University of Sharjah. This paper represents the opinions of the author(s) and does not mean to represent the position or opinions of the American University of Sharjah.

Conflicts of Interest: The authors declare no conflict of interest.

References

1. Institute, N.C. What-Is-Cancer. 2015. Available online: http://www.cancer.gov/cancertopics/what-is-cancer (accessed on 18 June 2023).
2. Wu, S.; Zhu, W.; Thompson, P.; Hannun, Y.A. Evaluating intrinsic and non-intrinsic cancer risk factors. *Nat. Commun.* **2018**, *9*, 3490. [CrossRef] [PubMed]
3. Anand, P.; Kunnumakara, A.B.; Sundaram, C.; Harikumar, K.B.; Tharakan, S.T.; Lai, O.S.; Sung, B.; Aggarwal, B.B. Cancer is a Preventable Disease that Requires Major Lifestyle Changes. *Pharm. Res.* **2008**, *25*, 2097–2116. [CrossRef] [PubMed]
4. WHO. Cancer. 2022, p. 1. Available online: http://www.who.int/mediacentre/factsheets/fs297/en/ (accessed on 18 June 2023).
5. Ferlay, J.; Colombet, M.; Soerjomataram, I.; Parkin, D.M.; Piñeros, M.; Znaor, A.; Bray, F. Cancer statistics for the year 2020: An overview. *Int. J. Cancer* **2021**, *149*, 778–789. [CrossRef] [PubMed]
6. Pucci, C.; Martinelli, C.; Ciofani, G. Innovative approaches for cancer treatment: Current perspectives and new challenges. *Ecancermedicalscience* **2019**, *13*, 961. [CrossRef] [PubMed]
7. Sung, H.; Ferlay, J.; Siegel, R.L.; Laversanne, M.; Soerjomataram, I.; Jemal, A.; Bray, F. Global Cancer Statistics 2020: GLOBOCAN Estimates of Incidence and Mortality Worldwide for 36 Cancers in 185 Countries. *CA Cancer J. Clin.* **2021**, *71*, 209–249. [CrossRef]
8. Wang, X.; Yang, L.; Chen, Z.; Shin, D.M. Application of Nanotechnology in Cancer Therapy and Imaging. *CA A Cancer J. Clin.* **2008**, *58*, 97–110. [CrossRef]
9. Shakeri-Zadeh, A.; Zareyi, H.; Sheervalilou, R.; Laurent, S.; Ghaznavi, H.; Samadian, H. Gold nanoparticle-mediated bubbles in cancer nanotechnology. *J. Control. Release* **2020**, *330*, 49–60. [CrossRef]
10. Sharma, A.; Saini, A.K.; Kumar, N.; Tejwan, N.; Singh, T.A.; Thakur, V.K.; Das, J. Methods of preparation of metal-doped and hybrid tungsten oxide nanoparticles for anticancer, antibacterial, and biosensing applications. *Surf. Interfaces* **2021**, *28*, 101641. [CrossRef]
11. Almanghadim, H.G.; Nourollahzadeh, Z.; Khademi, N.S.; Tezerjani, M.D.; Sehrig, F.Z.; Estelami, N.; Shirvaliloo, M.; Sheervalilou, R.; Sargazi, S. Application of nanoparticles in cancer therapy with an emphasis on cell cycle. *Cell Biol. Int.* **2021**, *45*, 1989–1998. [CrossRef]
12. Liu, J.; Chen, Q.; Feng, L.; Liu, Z. Nanomedicine for tumor microenvironment modulation and cancer treatment enhancement. *Nano Today* **2018**, *21*, 55–73. [CrossRef]
13. Sheervalilou, R.; Shirvaliloo, M.; Sargazi, S.; Ghaznavi, H.; Shakeri-Zadeh, A. Recent advances in iron oxide nanoparticles for brain cancer theranostics: From in vitro to clinical applications. *Expert Opin. Drug Deliv.* **2021**, *18*, 949–977. [CrossRef] [PubMed]
14. Irajirad, R.; Ahmadi, A.; Najafabad, B.K.; Abed, Z.; Sheervalilou, R.; Khoei, S.; Shiran, M.B.; Ghaznavi, H.; Shakeri-Zadeh, A. Combined thermo-chemotherapy of cancer using 1 MHz ultrasound waves and a cisplatin-loaded sonosensitizing nanoplatform: An in vivo study. *Cancer Chemother. Pharmacol.* **2019**, *84*, 1315–1321. [CrossRef] [PubMed]
15. Dadwal, A.; Baldi, A.; Kumar Narang, R. Nanoparticles as carriers for drug delivery in cancer. *Artif. Cells Nanomed. Biotechnol.* **2018**, *46*, 295–305. [CrossRef] [PubMed]
16. Duncan, R. Polymer conjugates as anticancer nanomedicines. *Nat. Rev. Cancer* **2006**, *6*, 688–701. [CrossRef] [PubMed]
17. Ferrari, M. Cancer nanotechnology: Opportunities and challenges. *Nat. Rev. Cancer* **2005**, *5*, 161–171. [CrossRef] [PubMed]
18. Takáč, P.; Michalková, R.; Čižmáriková, M.; Bedlovičová, Z.; Balážová, Ľ.; Takáčová, G. The Role of Silver Nanoparticles in the Diagnosis and Treatment of Cancer: Are There Any Perspectives for the Future? *Life* **2023**, *13*, 466. [CrossRef] [PubMed]
19. Rahdar, A.; Hajinezhad, M.R.; Sargazi, S.; Zaboli, M.; Barani, M.; Baino, F.; Bilal, M.; Sanchooli, E. Biochemical, Ameliorative and Cytotoxic Effects of Newly Synthesized Curcumin Microemulsions: Evidence from In Vitro and In Vivo Studies. *Nanomaterials* **2021**, *11*, 817. [CrossRef]
20. Barani, M.; Rahdar, A.; Sargazi, S.; Amiri, M.S.; Sharma, P.K.; Bhalla, N. Nanotechnology for inflammatory bowel disease management: Detection, imaging and treatment. *Sens. Bio-Sens. Res.* **2021**, *32*, 100417. [CrossRef]
21. Arshad, R.; Barani, M.; Rahdar, A.; Sargazi, S.; Cucchiarini, M.; Pandey, S.; Kang, M. Multi-Functionalized Nanomaterials and Nanoparticles for Diagnosis and Treatment of Retinoblastoma. *Biosensors* **2021**, *11*, 97. [CrossRef]
22. Zarrabi, A.; Zarepour, A.; Khosravi, A.; Alimohammadi, Z.; Thakur, V.K. Synthesis of Curcumin Loaded Smart pH-Responsive Stealth Liposome as a Novel Nanocarrier for Cancer Treatment. *Fibers* **2021**, *9*, 19. [CrossRef]

23. Burz, C.; Pop, V.-V.; Buiga, R.; Daniel, S.; Samasca, G.; Aldea, C.; Lupan, I. Circulating tumor cells in clinical research and monitoring patients with colorectal cancer. *Oncotarget* **2018**, *9*, 24561–24571. [CrossRef] [PubMed]
24. Chen, T.; Ren, L.; Liu, X.; Zhou, M.; Li, L.; Xu, J.; Zhu, X. DNA Nanotechnology for Cancer Diagnosis and Therapy. *Int. J. Mol. Sci.* **2018**, *19*, 1671. [CrossRef] [PubMed]
25. Dessale, M.; Mengistu, G.; Mengist, H.M. Nanotechnology: A Promising Approach for Cancer Diagnosis, Therapeutics and Theragnosis. *Int. J. Nanomed.* **2022**, *17*, 3735–3749. [CrossRef]
26. Saleh, T.A. Nanomaterials: Classification, properties, and environmental toxicities. *Environ. Technol. Innov.* **2020**, *20*, 101067. [CrossRef]
27. Kagan, C.R.; Murray, C.B. Charge transport in strongly coupled quantum dot solids. *Nat. Nanotechnol.* **2015**, *10*, 1013–1026. [CrossRef]
28. Buzea, C.; Pacheco, I. Nanomaterials and their classification. In *EMR/ESR/EPR Spectroscopy for Characterization of Nanomaterials*; Springer: Berlin/Heidelberg, Germany, 2017; pp. 3–45.
29. Rizwan, M.; Shoukat, A.; Ayub, A.; Razzaq, B.; Tahir, M.B. Chapter 3-Types and classification of nanomaterials. In *Nanomaterials: Synthesis, Characterization, Hazards and Safety*; Elsevier: Amsterdam, The Netherlands, 2021; pp. 31–54.
30. Ali, A.A.; Abuwatfa, W.H.; Al-Sayah, M.H.; Husseini, G.A. Gold-Nanoparticle Hybrid Nanostructures for Multimodal Cancer Therapy. *Nanomaterials* **2022**, *12*, 3706. [CrossRef]
31. Chen, W.; Goldys, E.M.; Deng, W. Light-induced liposomes for cancer therapeutics. *Prog. Lipid Res.* **2020**, *79*, 101052. [CrossRef]
32. Gomes, H.I.O.; Martins, C.S.M.; Prior, J.A.V. Silver Nanoparticles as Carriers of Anticancer Drugs for Efficient Target Treatment of Cancer Cells. *Nanomaterials* **2021**, *11*, 964. [CrossRef]
33. Tang, L.; Xiao, Q.; Mei, Y.; He, S.; Zhang, Z.; Wang, R.; Wang, W. Insights on functionalized carbon nanotubes for cancer theranostics. *J. Nanobiotechnol.* **2021**, *19*, 1–28. [CrossRef]
34. Ekimov, A.; Efros, A.; Onushchenko, A. Quantum size effect in semiconductor microcrystals. *Solid State Commun.* **1985**, *56*, 921–924. [CrossRef]
35. Remya, V.R.; Prajitha, V.; George, J.S.; Jibin, K.P.; Thomas, S. Chapter 7-Quantum dots: A brief introduction. In *Micro and Nano Technologies*; Elsevier: Amsterdam, The Netherlands, 2021; pp. 181–196.
36. Alivisatos, P. The use of nanocrystals in biological detection. *Nat. Biotechnol.* **2003**, *22*, 47–52. [CrossRef] [PubMed]
37. Bentolila, L.A.; Ebenstein, Y.; Weiss, S. Quantum Dots for In Vivo Small-Animal Imaging. *J. Nucl. Med.* **2009**, *50*, 493–496. [CrossRef] [PubMed]
38. Zeng, Z.; Xiao, F.-X.; Phan, H.; Chen, S.; Yu, Z.; Wang, R.; Nguyen, T.-Q.; Yang Tan, T.T. Unraveling the cooperative synergy of zero-dimensional graphene quantum dots and metal nanocrystals enabled by layer-by-layer assembly. *J. Mater. Chem. A* **2018**, *6*, 1700–1713. [CrossRef]
39. Ji, X.; Peng, F.; Zhong, Y.; Su, Y.; He, Y. Fluorescent quantum dots: Synthesis, biomedical optical imaging, and biosafety assessment. *Colloids Surf. B:Biointerfaces* **2014**, *124*, 132–139. [CrossRef]
40. Shamsi, J.; Dang, Z.; Bianchini, P.; Canale, C.; Di Stasio, F.; Brescia, R.; Prato, M.; Manna, L. Colloidal Synthesis of Quantum Confined Single Crystal CsPbBr$_3$ Nanosheets with Lateral Size Control up to the Micrometer Range. *J. Am. Chem. Soc.* **2016**, *138*, 7240–7243. [CrossRef] [PubMed]
41. Segets, D. Analysis of Particle Size Distributions of Quantum Dots: From Theory to Application. *KONA Powder Part. J.* **2016**, *33*, 48–62. [CrossRef]
42. Jamieson, T.; Bakhshi, R.; Petrova, D.; Pocock, R.; Imani, M.; Seifalian, A.M. Biological applications of quantum dots. *Biomaterials* **2007**, *28*, 4717–4732. [CrossRef]
43. Reimann, S.M.; Manninen, M. Electronic structure of quantum dots. *Rev. Mod. Phys.* **2002**, *74*, 1283–1342. [CrossRef]
44. Maxwell, T.; Nogueira Campos, M.G.; Smith, S.; Doomra, M.; Thwin, Z.; Santra, S. Chapter 15-Quantum Dots. In *Micro and Nano Technologies*; Elsevier: Amsterdam, The Netherlands, 2020; pp. 243–265.
45. Fomenko, V.; Nesbitt, D.J. Solution Control of Radiative and Nonradiative Lifetimes: A Novel Contribution to Quantum Dot Blinking Suppression. *Nano Lett.* **2007**, *8*, 287–293. [CrossRef]
46. Ornes, S. Quantum dots. *Proc. Natl. Acad. Sci. USA* **2016**, *113*, 2796–2797. [CrossRef]
47. Sumanth Kumar, D.; Jai Kumar, B.; Mahesh, H.M. Chapter 3-Quantum Nanostructures (QDs): An Overview. In *Micro and Nano Technologies*; Woodhead Publishing: Sawston, UK, 2018; pp. 59–88.
48. Yoffe, A.D. Semiconductor quantum dots and related systems: Electronic, optical, luminescence and related properties of low dimensional systems. *Adv. Phys.* **2001**, *50*, 1–208. [CrossRef]
49. Hong, N.H. Chapter 1-Introduction to Nanomaterials: Basic Properties, Synthesis, and Characterization. In *Micro and Nano Technologies*; Elsevier: Amsterdam, The Netherlands, 2019; pp. 1–19.
50. Jin, T.; Tiwari, D.K.; Tanaka, S.-I.; Inouye, Y.; Yoshizawa, K.; Watanabe, T.M. Antibody–ProteinA conjugated quantum dots for multiplexed imaging of surface receptors in living cells. *Mol. Biosyst.* **2010**, *6*, 2325–2331. [CrossRef] [PubMed]
51. Liang, Z.; Khawar, M.B.; Liang, J.; Sun, H. Bio-Conjugated Quantum Dots for Cancer Research: Detection and Imaging. *Front. Oncol.* **2021**, *11*, 749970. [CrossRef] [PubMed]
52. Misra, K.P.; Misra, R.D.K. ZnO-Based Quantum Dots for Biosensing, Cancer Imaging and Therapy: An Overview. *Biomed. Mater. Devices* **2022**, *11*, 749970. [CrossRef]

53. Brunetti, J.; Riolo, G.; Gentile, M.; Bernini, A.; Paccagnini, E.; Falciani, C.; Lozzi, L.; Scali, S.; Depau, L.; Pini, A.; et al. Near-infrared quantum dots labelled with a tumor selective tetrabranched peptide for in vivo imaging. *J. Nanobiotechnol.* **2018**, *16*, 1–10. [CrossRef]
54. Ranjbar-Navazi, Z.; Eskandani, M.; Johari-Ahar, M.; Nemati, A.; Akbari, H.; Davaran, S.; Omidi, Y. Doxorubicin-conjugated D-glucosamine- and folate- bi-functionalised InP/ZnS quantum dots for cancer cells imaging and therapy. *J. Drug Target.* **2017**, *26*, 267–277. [CrossRef]
55. Cai, X.; Luo, Y.; Zhang, W.; Du, D.; Lin, Y. pH-Sensitive ZnO Quantum Dots–Doxorubicin Nanoparticles for Lung Cancer Targeted Drug Delivery. *ACS Appl. Mater. Interfaces* **2016**, *8*, 22442–22450. [CrossRef]
56. Mohamed, W.A.A.; El-Gawad, H.A.; Mekkey, S.; Galal, H.; Handal, H.; Mousa, H.; Labib, A. Quantum dots synthetization and future prospect applications. *Nanotechnol. Rev.* **2021**, *10*, 1926–1940. [CrossRef]
57. Kara, H.E.Ş. *Quantum Dots for Pharmaceutical and Biomedical Analysis*; Zafar, N.E.E.-E.S.E.-F., Ed.; IntechOpen: Rijeka, Croatia, 2017; p. 8.
58. Han, M.; Gao, X.; Su, J.Z.; Nie, S. Quantum-dot-tagged microbeads for multiplexed optical coding of biomolecules. *Nat. Biotechnol.* **2001**, *19*, 631–635. [CrossRef]
59. Bailey, R.E.; Smith, A.M.; Nie, S. Quantum dots in biology and medicine. *Phys. E:Low-Dimens. Syst. Nanostructures* **2004**, *25*, 1–12. [CrossRef]
60. Sutherland, A.J. Quantum dots as luminescent probes in biological systems. *Curr. Opin. Solid State Mater. Sci.* **2002**, *6*, 365–370. [CrossRef]
61. Ha, Y.; Jung, H.S.; Jeong, S.; Kim, H.-M.; Kim, T.H.; Cha, M.G.; Kang, E.J.; Pham, X.-H.; Jeong, D.H.; Jun, B.-H. Fabrication of Remarkably Bright QD Densely-Embedded Silica Nanoparticle. *Bull. Korean Chem. Soc.* **2019**, *40*, 9–13. [CrossRef]
62. Resch-Genger, U.; Grabolle, M.; Cavaliere-Jaricot, S.; Nitschke, R.; Nann, T. Quantum dots versus organic dyes as fluorescent labels. *Nat. Methods* **2008**, *5*, 763–775. [CrossRef] [PubMed]
63. Chan, W.C.; Maxwell, D.J.; Gao, X.; Bailey, R.E.; Han, M.; Nie, S. Luminescent quantum dots for multiplexed biological detection and imaging. *Curr. Opin. Biotechnol.* **2002**, *13*, 40–46. [CrossRef] [PubMed]
64. Reshma, V.; Mohanan, P. Quantum dots: Applications and safety consequences. *J. Lumin* **2018**, *205*, 287–298. [CrossRef]
65. Tennant, D.; Bleier, A. Electron Beam Lithography of Nanostructures. *Handb. Nanofabrication* **2010**, *4*, 121–148. [CrossRef]
66. Nagpal, R.; Gusain, M. Chapter 25-Synthesis methods of quantum dots. In *Woodhead Publishing Series in Electronic and Optical Materials*; Al-Douri, Y.B.T.-G., Ed.; Woodhead Publishing: Sawston, UK, 2022; pp. 599–630.
67. Nandwana, V.; Subramani, C.; Yeh, Y.-C.; Yang, B.; Dickert, S.; Barnes, M.D.; Tuominen, M.T.; Rotello, V.M. Direct patterning of quantum dot nanostructures via electron beam lithography. *J. Mater. Chem.* **2011**, *21*, 16859–16862. [CrossRef]
68. Palankar, R.; Medvedev, N.; Rong, A.; Delcea, M. Fabrication of Quantum Dot Microarrays Using Electron Beam Lithography for Applications in Analyte Sensing and Cellular Dynamics. *ACS Nano* **2013**, *7*, 4617–4628. [CrossRef]
69. Valizadeh, A.; Mikaeili, H.; Samiei, M.; Farkhani, S.M.; Zarghami, N.; Kouhi, M.; Akbarzadeh, A.; Davaran, S. Quantum dots: Synthesis, bioapplications, and toxicity. *Nanoscale Res. Lett.* **2012**, *7*, 480. [CrossRef]
70. Bera, D.; Qian, L.; Tseng, T.-K.; Holloway, P.H. Quantum Dots and Their Multimodal Applications: A Review. *Materials* **2010**, *3*, 2260–2345. [CrossRef]
71. Lee, L.K.; Ku, P. Fabrication of site-controlled InGaN quantum dots using reactive-ion etching. *Phys. Status Solidi C* **2011**, *9*, 609–612. [CrossRef]
72. Choi, M.; Jun, S.; Woo, K.Y.; Song, H.G.; Yeo, H.-S.; Choi, S.; Park, D.; Park, C.-H.; Cho, Y.-H. Nanoscale Focus Pinspot for High-Purity Quantum Emitters via Focused-Ion-Beam-Induced Luminescence Quenching. *ACS Nano* **2021**, *15*, 11317–11325. [CrossRef]
73. Zhang, H.; Ross, I.M.; Walther, T. Study of site controlled quantum dot formation on focused ion beam patterned GaAs substrate. *J. Physics: Conf. Ser.* **2013**, *471*, 012047. [CrossRef]
74. Lee, J.; Yang, J.; Kwon, S.G.; Hyeon, T. Nonclassical nucleation and growth of inorganic nanoparticles. *Nat. Rev. Mater.* **2016**, *1*, 16034. [CrossRef]
75. Aftab, S.; Shah, A.; Erkmen, C.; Kurbanoglu, S.; Uslu, B. Chapter 1-Quantum dots: Synthesis and characterizations. In *Micro and Nano Technologies*; Elsevier: Amsterdam, The Netherlands, 2021; pp. 1–35.
76. Bang, J.; Yang, H.; Holloway, P.H. Enhanced and stable green emission of ZnO nanoparticles by surface segregation of Mg. *Nanotechnology* **2006**, *17*, 973–978. [CrossRef] [PubMed]
77. Bera, D.; Qian, L.; Sabui, S.; Santra, S.; Holloway, P.H. Photoluminescence of ZnO quantum dots produced by a sol–gel process. *Opt. Mater.* **2008**, *30*, 1233–1239. [CrossRef]
78. Rai, A.K.; Jat, K.K. Chapter 3-Sol-gel synthesis of quantum dots. In *Quantum Dots*; Elsevier: Amsterdam, The Netherlands, 2023; pp. 35–52.
79. Sashchiuk, A.; Lifshitz, E.; Reisfeld, R.; Saraidarov, T.; Zelner, M.; Willenz, A. Optical and Conductivity Properties of PbS Nanocrystals in Amorphous Zirconia Sol-Gel Films. *J. Sol-Gel Sci. Technol.* **2002**, *24*, 31–38. [CrossRef]
80. Javed, S.; Islam, M.; Mujahid, M. Synthesis and characterization of TiO_2 quantum dots by sol gel reflux condensation method. *Ceram. Int.* **2018**, *45*, 2676–2679. [CrossRef]
81. Jiang, H.; Yao, X.; Che, J.; Wang, M.; Kong, F. Preparation of ZnSe quantum dots embedded in SiO_2 thin films by sol-gel process. *Ceram. Int.* **2004**, *30*, 1685–1689. [CrossRef]

82. Moghaddam, E.; Youzbashi, A.; Kazemzadeh, A.; Eshraghi, M. Preparation of surface-modified ZnO quantum dots through an ultrasound assisted sol–gel process. *Appl. Surf. Sci.* **2015**, *346*, 111–114. [CrossRef]
83. Malik, M.A.; Wani, M.Y.; Hashim, M.A. Microemulsion method: A novel route to synthesize organic and inorganic nanomaterials. *Arab. J. Chem.* **2010**, *5*, 397–417. [CrossRef]
84. Shakur, H.R. A detailed study of physical properties of ZnS quantum dots synthesized by reverse micelle method. *Phys. E Low-dimens. Syst. Nanostructures* **2011**, *44*, 641–646. [CrossRef]
85. Karanikolos, G.N.; Alexandridis, P.; Itskos, G.; Petrou, A.; Mountziaris, T.J. Synthesis and Size Control of Luminescent ZnSe Nanocrystals by a Microemulsion−Gas Contacting Technique. *Langmuir* **2004**, *20*, 550–553. [CrossRef] [PubMed]
86. Lien, V.T.K.; Ha, C.V.; Ha, L.T.; Dat, N.N. Optical properties of CdS and CdS/ZnS quantum dots synthesized by reverse micelle method. *J. Phys. Conf. Ser.* **2009**, *187*, 012028. [CrossRef]
87. Mohagheghpour, E.; Rabiee, M.; Moztarzadeh, F.; Tahriri, M.; Jafarbeglou, M.; Bizari, D.; Eslami, H. Controllable synthesis, characterization and optical properties of ZnS:Mn nanoparticles as a novel biosensor. *Mater. Sci. Eng. C* **2009**, *29*, 1842–1848. [CrossRef]
88. Hosseini, M.S.; Kamali, M. Synthesis and characterization of aspartic acid-capped CdS/ZnS quantum dots in reverse micelles and its application to Hg(II) determination. *J. Lumin* **2015**, *167*, 51–58. [CrossRef]
89. Darbandi, M.; Thomann, R.; Nann, T. Single Quantum Dots in Silica Spheres by Microemulsion Synthesis. *Chem. Mater.* **2005**, *17*, 5720–5725. [CrossRef]
90. Saran, A.D.; Bellare, J.R. Green engineering for large-scale synthesis of water-soluble and bio-taggable CdSe and CdSe–CdS quantum dots from microemulsion by double-capping. *Colloids Surf. A Physicochem. Eng. Asp.* **2010**, *369*, 165–175. [CrossRef]
91. Arthur, J.R. Molecular beam epitaxy. *Surf. Sci.* **2002**, *500*, 189–217. [CrossRef]
92. Brault, J.; Matta, S.; Ngo, T.-H.; Al Khalfioui, M.; Valvin, P.; Leroux, M.; Damilano, B.; Korytov, M.; Brändli, V.; Vennéguès, P.; et al. Internal quantum efficiencies of AlGaN quantum dots grown by molecular beam epitaxy and emitting in the UVA to UVC ranges. *J. Appl. Phys.* **2019**, *126*, 205701. [CrossRef]
93. Dhawan, S.; Dhawan, T.; Vedeshwar, A.G. Growth of Nb_2O_5 quantum dots by physical vapor deposition. *Mater. Lett.* **2014**, *126*, 32–35. [CrossRef]
94. Baptista, A.; Silva, F.J.G.; Porteiro, J.; Míguez, J.L.; Pinto, G. Sputtering Physical Vapour Deposition (PVD) Coatings: A Critical Review on Process Improvement and Market Trend Demands. *Coatings* **2018**, *8*, 402. [CrossRef]
95. Yap, Y.K.; Zhang, D. *Physical Vapor Deposition BT-Encyclopedia of Nanotechnology*; Bhushan, B., Ed.; Springer: Dordrecht, The Netherlands, 2014; pp. 1–8.
96. Son, H.H.; Seo, G.H.; Jeong, U.; Shin, D.Y.; Kim, S.J. Capillary wicking effect of a Cr-sputtered superhydrophilic surface on enhancement of pool boiling critical heat flux. *Int. J. Heat Mass Transf.* **2017**, *113*, 115–128. [CrossRef]
97. Wender, H.; Migowski, P.; Feil, A.F.; Teixeira, S.R.; Dupont, J. Sputtering deposition of nanoparticles onto liquid substrates: Recent advances and future trends. *Co-Ord. Chem. Rev.* **2013**, *257*, 2468–2483. [CrossRef]
98. Jeevanandam, J.; Balu, S.K.; Andra, S.; Danquah, M.K.; Vidyavathi, M.; Muthalagu, M. *Quantum Dots Synthesis and Application BT-Contemporary Nanomaterials in Material Engineering Applications*; Mubarak, N.M., Khalid, M., Walvekar, R., Numan, A., Eds.; Springer International Publishing: Cham, Germany, 2021; pp. 229–265.
99. Tiwari, P.K.; Sahu, M.; Kumar, G.; Ashourian, M. Pivotal Role of Quantum Dots in the Advancement of Healthcare Research. *Comput. Intell. Neurosci.* **2021**, *2021*, 1–9. [CrossRef] [PubMed]
100. Dahi, A.; Colson, P.; Jamin, C.; Cloots, R.; Lismont, M.; Dreesen, L. Radio-frequency magnetron sputtering: A versatile tool for CdSe quantum dots depositions with controlled properties. *J. Mater. Environ. Sci.* **2016**, *7*, 2277–2287.
101. Bhatt, J.P.; Godha, N. Chapter 2-Hydrothermal synthesis of quantum dots. In *Quantum Dots*; Elsevier: Amsterdam, The Netherlands, 2023; pp. 15–34.
102. Shen, T.-Y.; Jia, P.-Y.; Chen, D.-S.; Wang, L.-N. Hydrothermal synthesis of N-doped carbon quantum dots and their application in ion-detection and cell-imaging. *Spectrochim. Acta Part A Mol. Biomol. Spectrosc.* **2020**, *248*, 119282. [CrossRef]
103. Dalvand, P.; Mohammadi, M.R. Controlling morphology and structure of nanocrystalline cadmium sulfide (CdS) by tailoring solvothermal processing parameters. *J. Nanopart. Res.* **2011**, *13*, 3011–3018. [CrossRef]
104. Tian, R.; Zhong, S.; Wu, J.; Jiang, W.; Shen, Y.; Wang, T. Solvothermal method to prepare graphene quantum dots by hydrogen peroxide. *Opt. Mater.* **2016**, *60*, 204–208. [CrossRef]
105. Luo, K.; Chen, H.; Zhou, Q.; Yan, Z.; Su, Z.; Li, K. A facile one step solvothermal controllable synthesis of FeS_2 quantum dots with multiple color emission for the visual detection of aconitine. *Spectrochim. Acta Part A Mol. Biomol. Spectrosc.* **2020**, *240*, 118563. [CrossRef]
106. Bharti, D.; Bharati, A.V.; Wankhade, A.V. Synthesis, characterization and optical property investigation of CdS nanoparticles. *Luminescence* **2018**, *33*, 1445–1449. [CrossRef]
107. Khan, A.; Shkir, M.; Manthrammel, M.; Ganesh, V.; Yahia, I.; Ahmed, M.; El-Toni, A.M.; Aldalbahi, A.; Ghaithan, H.; AlFaify, S. Effect of Gd doping on structural, optical properties, photoluminescence and electrical characteristics of CdS nanoparticles for optoelectronics. *Ceram. Int.* **2019**, *45*, 10133–10141. [CrossRef]
108. Abolghasemi, R.; Rasuli, R.; Alizadeh, M. Microwave-assisted growth of high-quality CdSe quantum dots and its application as a sensitizer in photovoltaic cells. *Mater. Today Commun.* **2020**, *22*, 100827. [CrossRef]

109. Chen, W.; Lv, G.; Hu, W.; Li, D.; Chen, S.; Dai, Z. Synthesis and applications of graphene quantum dots: A review. *Nanotechnol. Rev.* **2018**, *7*, 157–185. [CrossRef]
110. Ghasempour, A.; Dehghan, H.; Ataee, M.; Chen, B.; Zhao, Z.; Sedighi, M.; Guo, X.; Shahbazi, M.-A. Cadmium Sulfide Nanoparticles: Preparation, Characterization, and Biomedical Applications. *Molecules* **2023**, *28*, 3857. [CrossRef]
111. Zhu, Y.; Wang, G.; Jiang, H.; Chen, L.; Zhang, X. One-step ultrasonic synthesis of graphene quantum dots with high quantum yield and their application in sensing alkaline phosphatase. *Chem. Commun.* **2014**, *51*, 948–951. [CrossRef]
112. Chen, L.-C.; Tseng, Z.-L.; Chen, S.-Y.; Yang, S. An ultrasonic synthesis method for high-luminance perovskite quantum dots. *Ceram. Int.* **2017**, *43*, 16032–16035. [CrossRef]
113. Mahdi, H.S.; Parveen, A.; Azam, A. Microstructural and Optical Properties of Ni doped CdS Nanoparticles Synthesized by Sol Gel route. *Mater. Today Proc.* **2018**, *5*, 20636–20640. [CrossRef]
114. Xiong, H.-M.; Xu, Y.; Ren, Q.-G.; Xia, Y.-Y. Stable Aqueous ZnO@Polymer Core−Shell Nanoparticles with Tunable Photoluminescence and Their Application in Cell Imaging. *J. Am. Chem. Soc.* **2008**, *130*, 7522–7523. [CrossRef]
115. Ye, Y. Photoluminescence property adjustment of ZnO quantum dots synthesized via sol–gel method. *J. Mater. Sci. Mater. Electron.* **2018**, *29*, 4967–4974. [CrossRef]
116. Entezari, M.H.; Ghows, N. Micro-emulsion under ultrasound facilitates the fast synthesis of quantum dots of CdS at low temperature. *Ultrason. Sonochem.* **2011**, *18*, 127–134. [CrossRef]
117. Mohamed, W.A.; Handal, H.T.; Ibrahim, I.A.; Galal, H.R.; Mousa, H.A.; Labib, A.A. Recycling for solar photocatalytic activity of Dianix blue dye and real industrial wastewater treatment process by zinc oxide quantum dots synthesized by solvothermal method. *J. Hazard. Mater.* **2020**, *404*, 123962. [CrossRef]
118. Zhu, S.; Zhang, J.; Qiao, C.; Tang, S.; Li, Y.; Yuan, W.; Li, B.; Tian, L.; Liu, F.; Hu, R.; et al. Strongly green-photoluminescent graphene quantum dots for bioimaging applications. *Chem. Commun.* **2011**, *47*, 6858–6860. [CrossRef]
119. Aladesuyi, O.A.; Oluwafemi, O.S. Synthesis of N, S co-doped carbon quantum dots (N,S-CQDs) for sensitive and selective determination of mercury (Hg^{2+}) in *Oreochromis niloctus* (Tilapia fish). *Inorg. Chem. Commun.* **2023**, *153*, 110843. [CrossRef]
120. Liao, B.; Wang, W.; Deng, X.; He, B.; Zeng, W.; Tang, Z.; Liu, Q. A facile one-step synthesis of fluorescent silicon quantum dots and their application for detecting Cu^{2+}. *RSC Adv.* **2016**, *6*, 14465–14467. [CrossRef]
121. Nathiya, D.; Gurunathan, K.; Wilson, J. Size controllable, pH triggered reduction of bovine serum albumin and its adsorption behavior with SnO2/SnS2 quantum dots for biosensing application. *Talanta* **2019**, *210*, 120671. [CrossRef]
122. Safardoust-Hojaghan, H.; Salavati-Niasari, M.; Amiri, O.; Hassanpour, M. Preparation of highly luminescent nitrogen doped graphene quantum dots and their application as a probe for detection of Staphylococcus aureus and E. coli. *J. Mol. Liq.* **2017**, *241*, 1114–1119. [CrossRef]
123. Su, J.; Zhang, X.; Tong, X.; Wang, X.; Yang, P.; Yao, F.; Guo, R.; Yuan, C. Preparation of graphene quantum dots with high quantum yield by a facile one-step method and applications for cell imaging. *Mater. Lett.* **2020**, *271*, 127806. [CrossRef]
124. Aouassa, M.; Franzò, G.; Assaf, E.; Sfaxi, L.; M'ghaieth, R.; Maaref, H. MBE growth of InAs/GaAs quantum dots on sintered porous silicon substrates with high optical quality in the 1.3 μm band. *J. Mater. Sci. Mater. Electron.* **2020**, *31*, 4605–4610. [CrossRef]
125. Ashokkumar, M.; Boopathyraja, A. Structural and optical properties of Mg doped ZnS quantum dots and biological applications. *Superlattices Microstruct.* **2018**, *113*, 236–243. [CrossRef]
126. Bruns, O.T.; Bischof, T.S.; Harris, D.K.; Franke, D.; Shi, Y.; Riedemann, L.; Bartelt, A.; Jaworski, F.B.; Carr, J.A.; Rowlands, C.J.; et al. Next-generation in vivo optical imaging with short-wave infrared quantum dots. *Nat. Biomed. Eng.* **2017**, *1*, 1–11. [CrossRef]
127. Choi, H.S.; Kim, Y.; Park, J.C.; Oh, M.H.; Jeon, D.Y.; Nam, Y.S. Highly luminescent, off-stoichiometric $Cu_xIn_yS_2$/ZnS quantum dots for near-infrared fluorescence bio-imaging. *RSC Adv.* **2015**, *5*, 43449–43455. [CrossRef]
128. Zhang, C.; Han, Y.; Lin, L.; Deng, N.; Chen, B.; Liu, Y. Development of Quantum Dots-Labeled Antibody Fluorescence Immunoassays for the Detection of Morphine. *J. Agric. Food Chem.* **2017**, *65*, 1290–1295. [CrossRef]
129. Xue, Q.; Zhang, H.; Zhu, M.; Pei, Z.; Li, H.; Wang, Z.; Huang, Y.; Deng, Q.; Zhou, J.; Du, S.; et al. Photoluminescent Ti_3C_2 MXene Quantum Dots for Multicolor Cellular Imaging. *Adv. Mater.* **2017**, *29*, 1604847. [CrossRef]
130. Matea, C.T.; Mocan, T.; Tabaran, F.; Pop, T.; Mosteanu, O.; Puia, C.; Iancu, C.; Mocan, L. Quantum dots in imaging, drug delivery and sensor applications. *Int. J. Nanomed.* **2017**, *12*, 5421–5431. [CrossRef]
131. Wei, N.; Li, L.; Zhang, H.; Wang, W.; Pan, C.; Qi, S.; Zhang, H.; Chen, H.; Chen, X. Characterization of the Ligand Exchange Reactions on CdSe/ZnS QDs by Capillary Electrophoresis. *Langmuir* **2019**, *35*, 4806–4812. [CrossRef]
132. Sperling, R.A.; Parak, W.J. Surface modification, functionalization and bioconjugation of colloidal inorganic nanoparticles. *Philos. Trans. R. Soc. A: Math. Phys. Eng. Sci.* **2010**, *368*, 1333–1383. [CrossRef]
133. Zhang, F.; Lees, E.; Amin, F.; RiveraGil, P.; Yang, F.; Mulvaney, P.; Parak, W.J. Polymer-Coated Nanoparticles: A Universal Tool for Biolabelling Experiments. *Small* **2011**, *7*, 3113–3127. [CrossRef]
134. Zhou, J.; Liu, Y.; Tang, J.; Tang, W. Surface ligands engineering of semiconductor quantum dots for chemosensory and biological applications. *Mater. Today* **2017**, *20*, 360–376. [CrossRef]
135. Lees, E.E.; Nguyen, T.-L.; Clayton, A.H.A.; Mulvaney, P. The Preparation of Colloidally Stable, Water-Soluble, Biocompatible, Semiconductor Nanocrystals with a Small Hydrodynamic Diameter. *ACS Nano* **2009**, *3*, 1121–1128. [CrossRef] [PubMed]
136. Heyne, B.; Arlt, K.; Geßner, A.; Richter, A.F.; Döblinger, M.; Feldmann, J.; Taubert, A.; Wedel, A. Mixed Mercaptocarboxylic Acid Shells Provide Stable Dispersions of InPZnS/ZnSe/ZnS Multishell Quantum Dots in Aqueous Media. *Nanomaterials* **2020**, *10*, 1858. [CrossRef] [PubMed]

137. Zhang, Y.; Clapp, A. Overview of Stabilizing Ligands for Biocompatible Quantum Dot Nanocrystals. *Sensors* **2011**, *11*, 11036–11055. [CrossRef] [PubMed]
138. Ma, Y.; Shen, H.; Zhang, M.; Zhang, Z. Quantum Dots (QDs) for Tumor Targeting Theranostics. In *Nanomaterials for Tumor Targeting Theranostics: A Proactive Clinical Perspective*; World Scientific: Singapore, 2016; pp. 85–141. [CrossRef]
139. Karakoti, A.S.; Shukla, R.; Shanker, R.; Singh, S. Surface functionalization of quantum dots for biological applications. *Adv. Colloid Interface Sci.* **2015**, *215*, 28–45. [CrossRef]
140. Wang, J.; Han, S.; Ke, D.; Wang, R. Semiconductor Quantum Dots Surface Modification for Potential Cancer Diagnostic and Therapeutic Applications. *J. Nanomater.* **2012**, *2012*, 129041. [CrossRef]
141. He, X.; Gao, L.; Ma, N. One-Step Instant Synthesis of Protein-Conjugated Quantum Dots at Room Temperature. *Sci. Rep.* **2013**, *3*, 2825. [CrossRef]
142. Sanjayan, C.; Jyothi, M.; Sakar, M.; Balakrishna, R.G. Multidentate ligand approach for conjugation of perovskite quantum dots to biomolecules. *J. Colloid Interface Sci.* **2021**, *603*, 758–770. [CrossRef]
143. Chen, O.; Yang, Y.; Wang, T.; Wu, H.; Niu, C.; Yang, J.; Cao, Y.C. Surface-Functionalization-Dependent Optical Properties of II–VI Semiconductor Nanocrystals. *J. Am. Chem. Soc.* **2011**, *133*, 17504–17512. [CrossRef]
144. Shi, X.-H.; Dai, Y.-Y.; Wang, L.; Wang, Z.-G.; Liu, S.-L. Water-Soluble High-Quality Ag_2Te Quantum Dots Prepared by Mutual Adaptation of Synthesis and Surface Modification for In Vivo Imaging. *ACS Appl. Bio Mater.* **2021**, *4*, 7692–7700. [CrossRef]
145. Gu, L.; Hall, D.J.; Qin, Z.; Anglin, E.; Joo, J.; Mooney, D.J.; Howell, S.B.; Sailor, M.J. In vivo time-gated fluorescence imaging with biodegradable luminescent porous silicon nanoparticles. *Nat. Commun.* **2013**, *4*, 2326. [CrossRef]
146. Serrano, I.C.; Vazquez-Vazquez, C.; Adams, A.M.; Stoica, G.; Correa-Duarte, M.A.; Palomares, E.; Alvarez-Puebla, R.A. The effect of the silica thickness on the enhanced emission in single particle quantum dots coated with gold nanoparticles. *RSC Adv.* **2013**, *3*, 10691–10695. [CrossRef]
147. Thanh, N.T.; Green, L.A. Functionalisation of nanoparticles for biomedical applications. *Nano Today* **2010**, *5*, 213–230. [CrossRef]
148. Li, Y.; Dai, C.; Wang, X.; Lv, W.; Zhou, H.; Zhao, G.; Li, L.; Sun, Y.; Wu, Y.; Zhao, M. A novel strategy to create bifunctional silica-protected quantum dot nanoprobe for fluorescence imaging. *Sens. Actuators B Chem.* **2019**, *282*, 27–35. [CrossRef]
149. Kambayashi, M.; Yamauchi, N.; Nakashima, K.; Hasegawa, M.; Hirayama, Y.; Suzuki, T.; Kobayashi, Y. Silica coating of indium phosphide nanoparticles by a sol–gel method and their photobleaching properties. *SN Appl. Sci.* **2019**, *1*, 1576. [CrossRef]
150. Knopp, D.; Tang, D.; Niessner, R. Review: Bioanalytical applications of biomolecule-functionalized nanometer-sized doped silica particles. *Anal. Chim. Acta* **2009**, *647*, 14–30. [CrossRef]
151. Jun, B.H.; Hwang, D.W.; Jung, H.S.; Jang, J.; Kim, H.; Kang, H.; Kang, T.; Kyeong, S.; Lee, H.; Jeong, D.H.; et al. Ultrasensitive, Biocompatible, Quantum-Dot-Embedded Silica Nanoparticles for Bioimaging. *Adv. Funct. Mater.* **2012**, *22*, 1843–1849. [CrossRef]
152. Correa-Duarte, M.A.; Giersig, M.; Liz-Marzán, L.M. Stabilization of CdS semiconductor nanoparticles against photodegradation by a silica coating procedure. *Chem. Phys. Lett.* **1998**, *286*, 497–501. [CrossRef]
153. Du, Y.; Yang, P.; Matras-Postolek, K.; Wang, J.; Che, Q.; Cao, Y.; Ma, Q. Low toxic and highly luminescent $CdSe/Cd_xZn_{1-x}S$ quantum dots with thin organic SiO_2 coating for application in cell imaging. *J. Nanopart. Res.* **2016**, *18*, 1–11. [CrossRef]
154. Ham, K.-M.; Kim, M.; Bock, S.; Kim, J.; Kim, W.; Jung, H.S.; An, J.; Song, H.; Kim, J.-W.; Kim, H.-M.; et al. Highly Bright Silica-Coated InP/ZnS Quantum Dot-Embedded Silica Nanoparticles as Biocompatible Nanoprobes. *Int. J. Mol. Sci.* **2022**, *23*, 10977. [CrossRef]
155. Goftman, V.V.; Aubert, T.; Ginste, D.V.; Van Deun, R.; Beloglazova, N.V.; Hens, Z.; De Saeger, S.; Goryacheva, I.Y. Synthesis, modification, bioconjugation of silica coated fluorescent quantum dots and their application for mycotoxin detection. *Biosens. Bioelectron.* **2016**, *79*, 476–481. [CrossRef]
156. Anderson, R.E.; Chan, W.C.W. Systematic Investigation of Preparing Biocompatible, Single, and Small ZnS-Capped CdSe Quantum Dots with Amphiphilic Polymers. *ACS Nano* **2008**, *2*, 1341–1352. [CrossRef]
157. Yoon, C.; Yang, K.P.; Kim, J.; Shin, K.; Lee, K. Fabrication of highly transparent and luminescent quantum dot/polymer nanocomposite for light emitting diode using amphiphilic polymer-modified quantum dots. *Chem. Eng. J.* **2019**, *382*, 122792. [CrossRef]
158. Abdolahi, G.; Dargahi, M.; Ghasemzadeh, H. Synthesis of starch-g-poly (acrylic acid)/ZnSe quantum dot nanocomposite hydrogel, for effective dye adsorption and photocatalytic degradation: Thermodynamic and kinetic studies. *Cellulose* **2020**, *27*, 6467–6483. [CrossRef]
159. Speranskaya, E.S.; Beloglazova, N.V.; Lenain, P.; De Saeger, S.; Wang, Z.; Zhang, S.; Hens, Z.; Knopp, D.; Niessner, R.; Potapkin, D.V.; et al. Polymer-coated fluorescent CdSe-based quantum dots for application in immunoassay. *Biosens. Bioelectron.* **2014**, *53*, 225–231. [CrossRef] [PubMed]
160. Carrillo-Carrion, C.; Parak, W.J. Design of pyridyl-modified amphiphilic polymeric ligands: Towards better passivation of water-soluble colloidal quantum dots for improved optical performance. *J. Colloid Interface Sci.* **2016**, *478*, 88–96. [CrossRef] [PubMed]
161. Wang, Y.; Yu, L.; Kong, X.; Sun, L. Application of nanodiagnostics in point-of-care tests for infectious diseases. *Int. J. Nanomed.* **2017**, *12*, 4789–4803. [CrossRef] [PubMed]
162. Wang, G.; Zhang, P.; Dou, H.; Li, W.; Sun, K.; He, X.; Han, J.; Xiao, L.; Li, Y. Efficient Incorporation of Quantum Dots into Porous Microspheres through a Solvent-Evaporation Approach. *Langmuir* **2012**, *28*, 6141–6150. [CrossRef]

163. Kuznetsova, V.; Osipova, V.; Tkach, A.; Miropoltsev, M.; Kurshanov, D.; Sokolova, A.; Cherevkov, S.; Zakharov, V.; Fedorov, A.; Baranov, A.; et al. Lab-on-Microsphere—FRET-Based Multiplex Sensor Platform. *Nanomaterials* **2021**, *11*, 109. [CrossRef]
164. Li, Z.; Ma, H.; Guo, Y.; Fang, H.; Zhu, C.; Xue, J.; Wang, W.; Luo, G.; Sun, Y. Synthesis of uniform Pickering microspheres doped with quantum dot by microfluidic technology and its application in tumor marker. *Talanta* **2023**, *262*, 124495. [CrossRef]
165. Vaidya, S.V.; Couzis, A.; Maldarelli, C. Reduction in Aggregation and Energy Transfer of Quantum Dots Incorporated in Polystyrene Beads by Kinetic Entrapment due to Cross-Linking during Polymerization. *Langmuir* **2015**, *31*, 3167–3179. [CrossRef]
166. Zhao, C.; Li, W.; Liang, Y.; Tian, Y.; Zhang, Q. Synthesis of BiOBr/carbon quantum dots microspheres with enhanced photoactivity and photostability under visible light irradiation. *Appl. Catal. A Gen.* **2016**, *527*, 127–136. [CrossRef]
167. Khan, M.R.; Mitra, T.; Sahoo, D. Metal oxide QD based ultrasensitive microsphere fluorescent sensor for copper, chromium and iron ions in water. *RSC Adv.* **2020**, *10*, 9512–9524. [CrossRef] [PubMed]
168. Lin, W.; Niu, Y.; Meng, R.; Huang, L.; Cao, H.; Zhang, Z.; Qin, H.; Peng, X. Shell-thickness dependent optical properties of CdSe/CdS core/shell nanocrystals coated with thiol ligands. *Nano Res.* **2016**, *9*, 260–271. [CrossRef]
169. Ko, J.; Jeong, B.G.; Chang, J.H.; Joung, J.F.; Yoon, S.-Y.; Lee, D.C.; Park, S.; Huh, J.; Yang, H.; Bae, W.K.; et al. Chemically resistant and thermally stable quantum dots prepared by shell encapsulation with cross-linkable block copolymer ligands. *NPG Asia Mater.* **2020**, *12*, 19. [CrossRef]
170. Smith, A.M.; Duan, H.; Mohs, A.M.; Nie, S. Bioconjugated quantum dots for in vivo molecular and cellular imaging. *Adv. Drug Deliv. Rev.* **2008**, *60*, 1226–1240. [CrossRef] [PubMed]
171. Mulvaney, P.; Liz-Marzán, L.M.; Giersig, M.; Ung, T. Silica encapsulation of quantum dots and metal clusters. *J. Mater. Chem.* **2000**, *10*, 1259–1270. [CrossRef]
172. Aubert, T.; Soenen, S.J.; Wassmuth, D.; Cirillo, M.; Van Deun, R.; Braeckmans, K.; Hens, Z. Bright and Stable CdSe/CdS@SiO$_2$ Nanoparticles Suitable for Long-Term Cell Labeling. *ACS Appl. Mater. Interfaces* **2014**, *6*, 11714–11723. [CrossRef]
173. Pham, X.-H.; Park, S.-M.; Ham, K.-M.; Kyeong, S.; Son, B.S.; Kim, J.; Hahm, E.; Kim, Y.-H.; Bock, S.; Kim, W.; et al. Synthesis and Application of Silica-Coated Quantum Dots in Biomedicine. *Int. J. Mol. Sci.* **2021**, *22*, 10116. [CrossRef]
174. Cheng, R.; Li, F.; Zhang, J.; She, X.; Zhang, Y.; Shao, K.; Lin, Y.; Wang, C.-F.; Chen, S. Fabrication of amphiphilic quantum dots towards high-colour-quality light-emitting devices. *J. Mater. Chem. C* **2019**, *7*, 4244–4249. [CrossRef]
175. Li, C.; Ji, Y.; Wang, C.; Liang, S.; Pan, F.; Zhang, C.; Chen, F.; Fu, H.; Wang, K.; Cui, D. BRCAA1 antibody- and Her2 antibody-conjugated amphiphilic polymer engineered CdSe/ZnS quantum dots for targeted imaging of gastric cancer. *Nanoscale Res. Lett.* **2014**, *9*, 244. [CrossRef]
176. Nie, Q.; Tan, W.B.; Zhang, Y. Synthesis and characterization of monodisperse chitosan nanoparticles with embedded quantum dots. *Nanotechnology* **2005**, *17*, 140–144. [CrossRef]
177. Sheng, W.; Kim, S.; Lee, J.; Kim, S.-W.; Jensen, K.; Bawendi, M.G. In-Situ Encapsulation of Quantum Dots into Polymer Microspheres. *Langmuir* **2006**, *22*, 3782–3790. [CrossRef] [PubMed]
178. Zhou, C.; Yuan, H.; Shen, H.; Guo, Y.; Li, X.; Liu, D.; Xu, L.; Ma, L.; Li, L.S. Synthesis of size-tunable photoluminescent aqueous CdSe/ZnS microspheres via a phase transfer method with amphiphilic oligomer and their application for detection of HCG antigen. *J. Mater. Chem.* **2011**, *21*, 7393–7400. [CrossRef]
179. Wu, F.; Su, H.; Wang, K.; Wong, W.-K.; Zhu, X. Facile synthesis of N-rich carbon quantum dots from porphyrins as efficient probes for bioimaging and biosensing in living cells. *Int. J. Nanomed.* **2017**, *12*, 7375–7391. [CrossRef]
180. Liu, X.; Zhou, P.; Liu, H.; Zhan, H.; Zhang, Q.; Zhao, Y.; Chen, Y. Design of bright near-infrared-emitting quantum dots capped with different stabilizing ligands for tumor targeting. *RSC Adv.* **2018**, *8*, 4221–4229. [CrossRef]
181. Tao, J.; Zeng, Q.; Wang, L. Near-infrared quantum dots based fluorescent assay of Cu2+ and in vitro cellular and in vivo imaging. *Sens. Actuators B Chem.* **2016**, *234*, 641–647. [CrossRef]
182. Shi, M.; Dong, L.; Zheng, S.; Hou, P.; Cai, L.; Zhao, M.; Zhang, X.; Wang, Q.; Li, J.; Xu, K. "Bottom-up" preparation of MoS$_2$ quantum dots for tumor imaging and their in vivo behavior study. *Biochem. Biophys. Res. Commun.* **2019**, *516*, 1090–1096. [CrossRef]
183. Tao, J.; Feng, S.; Liu, B.; Pan, J.; Li, C.; Zheng, Y. Hyaluronic acid conjugated nitrogen-doped graphene quantum dots for identification of human breast cancer cells. *Biomed. Mater.* **2021**, *16*, 055001. [CrossRef]
184. Zhu, C.-N.; Chen, G.; Tian, Z.-Q.; Wang, W.; Zhong, W.-Q.; Li, Z.; Zhang, Z.-L.; Pang, D.-W. Near-Infrared Fluorescent Ag$_2$Se-Cetuximab Nanoprobes for Targeted Imaging and Therapy of Cancer. *Small* **2016**, *13*, 1602309. [CrossRef]
185. Yao, C.; Tu, Y.; Ding, L.; Li, C.; Wang, J.; Fang, H.; Huang, Y.; Zhang, K.; Lu, Q.; Wu, M.; et al. Tumor Cell-Specific Nuclear Targeting of Functionalized Graphene Quantum Dots In Vivo. *Bioconjug. Chem.* **2017**, *28*, 2608–2619. [CrossRef]
186. Wu, Y.-Z.; Sun, J.; Zhang, Y.; Pu, M.; Zhang, G.; He, N.; Zeng, X. Effective Integration of Targeted Tumor Imaging and Therapy Using Functionalized InP QDs with VEGFR2 Monoclonal Antibody and miR-92a Inhibitor. *ACS Appl. Mater. Interfaces* **2017**, *9*, 13068–13078. [CrossRef]
187. Sun, X.; Shi, M.; Zhang, C.; Yuan, J.; Yin, M.; Du, S.; Yu, S.; Ouyang, B.; Xue, F.; Yang, S.-T. Fluorescent Ag–In–S/ZnS Quantum Dots for Tumor Drainage Lymph Node Imaging In Vivo. *ACS Appl. Nano Mater.* **2021**, *4*, 1029–1037. [CrossRef]
188. Zhao, P.; Xu, Y.; Ji, W.; Zhou, S.; Li, L.; Qiu, L.; Qian, Z.; Wang, X.; Zhang, H. Biomimetic black phosphorus quantum dots-based photothermal therapy combined with anti-PD-L1 treatment inhibits recurrence and metastasis in triple-negative breast cancer. *J. Nanobiotechnol.* **2021**, *19*, 181. [CrossRef] [PubMed]

189. Huang, C.; Dong, H.; Su, Y.; Wu, Y.; Narron, R.; Yong, Q. Synthesis of Carbon Quantum Dot Nanoparticles Derived from Byproducts in Bio-Refinery Process for Cell Imaging and In Vivo Bioimaging. *Nanomaterials* **2019**, *9*, 387. [CrossRef] [PubMed]
190. Yaghini, E.; Turner, H.D.; Le Marois, A.M.; Suhling, K.; Naasani, I.; MacRobert, A.J. In vivo biodistribution studies and ex vivo lymph node imaging using heavy metal-free quantum dots. *Biomaterials* **2016**, *104*, 182–191. [CrossRef]
191. Shah, A.; Aftab, S.; Nisar, J.; Ashiq, M.N.; Iftikhar, F.J. Nanocarriers for targeted drug delivery. *J. Drug Deliv. Sci. Technol.* **2021**, *62*, 102426. [CrossRef]
192. Brigger, I.; Dubernet, C.; Couvreur, P. Nanoparticles in cancer therapy and diagnosis. *Adv. Drug Deliv. Rev.* **2012**, *64*, 24–36. [CrossRef]
193. Hu, Q.; Sun, W.; Wang, C.; Gu, Z. Recent advances of cocktail chemotherapy by combination drug delivery systems. *Adv. Drug Deliv. Rev.* **2015**, *98*, 19–34. [CrossRef]
194. Mirza, A.Z.; Siddiqui, F.A. Nanomedicine and drug delivery: A mini review. *Int. Nano Lett.* **2014**, *4*, 94. [CrossRef]
195. Lu, H.; Wang, J.; Wang, T.; Zhong, J.; Bao, Y.; Hao, H. Recent Progress on Nanostructures for Drug Delivery Applications. *J. Nanomater.* **2016**, *2016*, 5762431. [CrossRef]
196. Grigoletto, A.; Maso, K.; Mero, A.; Rosato, A.; Schiavon, O.; Pasut, G. Drug and protein delivery by polymer conjugation. *J. Drug Deliv. Sci. Technol.* **2016**, *32*, 132–141. [CrossRef]
197. Matai, I.; Sachdev, A.; Gopinath, P. Self-Assembled Hybrids of Fluorescent Carbon Dots and PAMAM Dendrimers for Epirubicin Delivery and Intracellular Imaging. *ACS Appl. Mater. Interfaces* **2015**, *7*, 11423–11435. [CrossRef] [PubMed]
198. Fang, J.; Islam, W.; Maeda, H. Exploiting the dynamics of the EPR effect and strategies to improve the therapeutic effects of nanomedicines by using EPR effect enhancers. *Adv. Drug Deliv. Rev.* **2020**, *157*, 142–160. [CrossRef] [PubMed]
199. Wang, Z.; Li, J.; Lin, G.; He, Z.; Wang, Y. Metal complex-based liposomes: Applications and prospects in cancer diagnostics and therapeutics. *J. Control. Release Off. J. Control. Release* **2022**, *348*, 1066–1088. [CrossRef] [PubMed]
200. Yuan, Z.; Gottsacker, C.; He, X.; Waterkotte, T.; Park, Y.C. Repetitive drug delivery using Light-Activated liposomes for potential antimicrobial therapies. *Adv. Drug Deliv. Rev.* **2022**, *187*, 114395. [CrossRef] [PubMed]
201. Mitchell, M.J.; Billingsley, M.M.; Haley, R.M.; Wechsler, M.E.; Peppas, N.A.; Langer, R. Engineering precision nanoparticles for drug delivery. *Nat. Rev. Drug Discov.* **2020**, *20*, 101–124. [CrossRef] [PubMed]
202. Kurawattimath, V.; Wilson, B.; Geetha, K.M. Nanoparticle-based drug delivery across the blood-brain barrier for treating malignant brain glioma. *OpenNano* **2023**, *10*, 100128. [CrossRef]
203. Hu, Q.; Luo, Y. Chitosan-based nanocarriers for encapsulation and delivery of curcumin: A review. *Int. J. Biol. Macromol.* **2021**, *179*, 125–135. [CrossRef]
204. Ying, N.; Liu, S.; Zhang, M.; Cheng, J.; Luo, L.; Jiang, J.; Shi, G.; Wu, S.; Ji, J.; Su, H.; et al. Nano delivery system for paclitaxel: Recent advances in cancer theranostics. *Colloids Surf. B Biointerfaces* **2023**, *228*, 113419. [CrossRef]
205. Cagel, M.; Tesan, F.C.; Bernabeu, E.; Salgueiro, M.J.; Zubillaga, M.B.; Moretton, M.A.; Chiappetta, D.A. Polymeric mixed micelles as nanomedicines: Achievements and perspectives. *Eur. J. Pharm. Biopharm.* **2017**, *113*, 211–228. [CrossRef]
206. Qiu, J.; Kong, L.; Cao, X.; Li, A.; Wei, P.; Wang, L.; Mignani, S.; Caminade, A.-M.; Majoral, J.-P.; Shi, X. Enhanced Delivery of Therapeutic siRNA into Glioblastoma Cells Using Dendrimer-Entrapped Gold Nanoparticles Conjugated with β-Cyclodextrin. *Nanomaterials* **2018**, *8*, 131. [CrossRef]
207. Zou, J.; Zhu, B.; Li, Y. Functionalization of Silver Nanoparticles Loaded with Paclitaxel-induced A549 Cells Apoptosis Through ROS-Mediated Signaling Pathways. *Curr. Top. Med. Chem.* **2020**, *20*, 89–98. [CrossRef] [PubMed]
208. Chen, J.; Fang, S.; Yang, L.; Ling, X.; Liao, J.; Zhou, X.; Li, M.; Zhong, W. Functionalized Silver Nanoparticles Enhance Therapeutic Effect of Paclitaxel for Prostate Cancer Therapy by Arresting the Cellular Cycle and Producing ROS. *Nano* **2021**, *16*, 2150126. [CrossRef]
209. Abdel-Rashid, R.S.; Omar, S.M.; Teiama, M.S.; Khairy, A.; Magdy, M.; Anis, B. Fabrication of Gold Nanoparticles in Absence of Surfactant as In Vitro Carrier of Plasmid DNA. *Int. J. Nanomed.* **2019**, *14*, 8399–8408. [CrossRef] [PubMed]
210. Kadkhoda, J.; Aghanejad, A.; Safari, B.; Barar, J.; Rasta, S.H.; Davaran, S. Aptamer-conjugated gold nanoparticles for targeted paclitaxel delivery and photothermal therapy in breast cancer. *J. Drug Deliv. Sci. Technol.* **2021**, *67*, 102954. [CrossRef]
211. Yang, W.; Liang, H.; Ma, S.; Wang, D.; Huang, J. Gold nanoparticle based photothermal therapy: Development and application for effective cancer treatment. *Sustain. Mater. Technol.* **2019**, *22*, e00109. [CrossRef]
212. Chen, Q.; Chen, Y.; Zhang, W.; Huang, Q.; Hu, M.; Peng, D.; Peng, C.; Wang, L.; Chen, W. Acidity and Glutathione Dual-Responsive Polydopamine-Coated Organic-Inorganic Hybrid Hollow Mesoporous Silica Nanoparticles for Controlled Drug Delivery. *ChemMedChem* **2020**, *15*, 1940–1946. [CrossRef]
213. Feng, Z.-Q.; Yan, K.; Li, J.; Xu, X.; Yuan, T.; Wang, T.; Zheng, J. Magnetic Janus particles as a multifunctional drug delivery system for paclitaxel in efficient cancer treatment. *Mater. Sci. Eng. C* **2019**, *104*, 110001. [CrossRef]
214. Kong, X.; Qi, Y.; Wang, X.; Jiang, R.; Wang, J.; Fang, Y.; Gao, J.; Hwang, K.C. Nanoparticle drug delivery systems and their applications as targeted therapies for triple negative breast cancer. *Prog. Mater. Sci.* **2023**, *134*, 101070. [CrossRef]
215. Arias, L.S.; Pessan, J.P.; Vieira, A.P.M.; de Lima, T.M.T.; Delbem, A.C.B.; Monteiro, D.R. Iron Oxide Nanoparticles for Biomedical Applications: A Perspective on Synthesis, Drugs, Antimicrobial Activity, and Toxicity. *Antibiotics* **2018**, *7*, 46. [CrossRef]
216. Zhao, M.-X.; Zhu, B.-J. The Research and Applications of Quantum Dots as Nano-Carriers for Targeted Drug Delivery and Cancer Therapy. *Nanoscale Res. Lett.* **2016**, *11*, 1–9. [CrossRef]

217. Abdelhamid, H.N. Chapter 13-Quantum dots hybrid systems for drug delivery. In *Woodhead Publishing Series in Biomaterials*; Woodhead Publishing: Sawston, UK, 2022; pp. 323–338.
218. Garcia-Cortes, M.; González-Iglesias, H.; Ruiz Encinar, J.; Costa-Fernández, J.M.; Coca-Prados, M.; Sanz-Medel, A. Sensitive targeted multiple protein quantification based on elemental detection of Quantum Dots. *Anal. Chim. Acta* **2015**, *879*, 77–84.
219. Banerjee, A.; Pons, T.; Lequeux, N.; Dubertret, B. Quantum dots–DNA bioconjugates: Synthesis to applications. *Interface Focus* **2016**, *6*, 20160064. [CrossRef]
220. Wolfbeis, O.S. An overview of nanoparticles commonly used in fluorescent bioimaging. *Chem. Soc. Rev.* **2015**, *44*, 4743–4768. [CrossRef] [PubMed]
221. Probst, C.E.; Zrazhevskiy, P.; Bagalkot, V.; Gao, X. Quantum dots as a platform for nanoparticle drug delivery vehicle design. *Adv. Drug Deliv. Rev.* **2013**, *65*, 703–718. [CrossRef] [PubMed]
222. Yong, K.-T.; Wang, Y.; Roy, I.; Rui, H.; Swihart, M.T.; Law, W.-C.; Kwak, S.K.; Ye, L.; Liu, J.; Mahajan, S.D.; et al. Preparation of Quantum Dot/Drug Nanoparticle Formulations for Traceable Targeted Delivery and Therapy. *Theranostics* **2012**, *2*, 681–694. [CrossRef] [PubMed]
223. Li, Y.; Dang, G.; Younis, M.R.; Cao, Y.; Wang, K.; Sun, X.; Zhang, W.; Zou, X.; Shen, H.; An, R.; et al. Peptide functionalized actively targeted MoS_2 nanospheres for fluorescence imaging-guided controllable pH-responsive drug delivery and collaborative chemo/photodynamic therapy. *J. Colloid Interface Sci.* **2023**, *639*, 302–313. [CrossRef]
224. Zoghi, M.; Pourmadadi, M.; Yazdian, F.; Nigjeh, M.N.; Rashedi, H.; Sahraeian, R. Synthesis and characterization of chitosan/carbon quantum dots/Fe2O3 nanocomposite comprising curcumin for targeted drug delivery in breast cancer therapy. *Int. J. Biol. Macromol.* **2023**, *249*, 125788. [CrossRef]
225. Mahani, M.; Pourrahmani-Sarbanani, M.; Yoosefian, M.; Divsar, F.; Mousavi, S.M.; Nomani, A. Doxorubicin delivery to breast cancer cells with transferrin-targeted carbon quantum dots: An in vitro and in silico study. *J. Drug Deliv. Sci. Technol.* **2021**, *62*, 102342. [CrossRef]
226. Mohammed-Ahmed, H.K.; Nakipoglu, M.; Tezcaner, A.; Keskin, D.; Evis, Z. Functionalization of graphene oxide quantum dots for anticancer drug delivery. *J. Drug Deliv. Sci. Technol.* **2023**, *80*, 104199. [CrossRef]
227. Ziaee, N.; Farhadian, N.; Abnous, K.; Matin, M.M.; Khoshnood, A.; Yaghoobi, E. Dual targeting of Mg/N doped-carbon quantum dots with folic and hyaluronic acid for targeted drug delivery and cell imaging. *BioMedicine* **2023**, *164*, 114971. [CrossRef]
228. Khodadadei, F.; Safarian, S.; Ghanbari, N. Methotrexate-loaded nitrogen-doped graphene quantum dots nanocarriers as an efficient anticancer drug delivery system. *Mater. Sci. Eng. C* **2017**, *79*, 280–285. [CrossRef] [PubMed]
229. Liu, L.; Jiang, H.; Dong, J.; Zhang, W.; Dang, G.; Yang, M.; Li, Y.; Chen, H.; Ji, H.; Dong, L. PEGylated MoS_2 quantum dots for traceable and pH-responsive chemotherapeutic drug delivery. *Colloids Surf. B Biointerfaces* **2020**, *185*, 110590. [CrossRef] [PubMed]
230. Xie, C.; Zhan, Y.; Wang, P.; Zhang, B.; Zhang, Y. Novel Surface Modification of ZnO QDs for Paclitaxel-Targeted Drug Delivery for Lung Cancer Treatment. *Dose-Response* **2020**, *18*, 1559325820926739. [CrossRef] [PubMed]
231. Abdelhamid, H.N.; El-Bery, H.M.; Metwally, A.A.; Elshazly, M.; Hathout, R.M. Synthesis of CdS-modified chitosan quantum dots for the drug delivery of Sesamol. *Carbohydr. Polym.* **2019**, *214*, 90–99. [CrossRef]
232. Habiba, K.; Encarnacion-Rosado, J.; Garcia-Pabon, K.; Villalobos-Santos, J.C.; Makarov, V.I.; Avalos, J.A.; Weiner, B.R.; Morell, G. Improving cytotoxicity against cancer cells by chemo-photodynamic combined modalities using silver-graphene quantum dots nanocomposites. *Int. J. Nanomed.* **2015**, *11*, 107–119. [CrossRef]
233. Karimi, S.; Namazi, H. Simple preparation of maltose-functionalized dendrimer/graphene quantum dots as a pH-sensitive biocompatible carrier for targeted delivery of doxorubicin. *Int. J. Biol. Macromol.* **2020**, *156*, 648–659. [CrossRef]
234. Sawy, A.M.; Barhoum, A.; Gaber, S.A.A.; El-Hallouty, S.M.; Shousha, W.G.; Maarouf, A.A.; Khalil, A.S. Insights of doxorubicin loaded graphene quantum dots: Synthesis, DFT drug interactions, and cytotoxicity. *Mater. Sci. Eng. C* **2021**, *122*, 111921. [CrossRef]
235. Tan, L.; Huang, R.; Li, X.; Liu, S.; Shen, Y.-M.; Shao, Z. Chitosan-based core-shell nanomaterials for pH-triggered release of anticancer drug and near-infrared bioimaging. *Carbohydr. Polym.* **2017**, *157*, 325–334. [CrossRef]
236. Bwatanglang, I.B.; Mohammad, F.; Yusof, N.A.; Abdullah, J.; Alitheen, N.B.; Hussein, M.Z.; Abu, N.; Mohammed, N.E.; Nordin, N.; Zamberi, N.R.; et al. In vivo tumor targeting and anti-tumor effects of 5-fluorouracil loaded, folic acid targeted quantum dot system. *J. Colloid Interface Sci.* **2016**, *480*, 146–158. [CrossRef]
237. Hu, F.; Li, C.; Zhang, Y.; Wang, M.; Wu, D.; Wang, Q. Real-time in vivo visualization of tumor therapy by a near-infrared-II Ag_2S quantum dot-based theranostic nanoplatform. *Nano Res.* **2015**, *8*, 1637–1647. [CrossRef]
238. Chen, L.; Hong, W.; Duan, S.; Li, Y.; Wang, J.; Zhu, J. Graphene quantum dots mediated magnetic chitosan drug delivery nanosystems for targeting synergistic photothermal-chemotherapy of hepatocellular carcinoma. *Cancer Biol. Ther.* **2022**, *23*, 281–293. [CrossRef]
239. Su, W.; Guo, R.; Yuan, F.; Li, Y.; Li, X.; Zhang, Y.; Zhou, S.; Fan, L. Red-Emissive Carbon Quantum Dots for Nuclear Drug Delivery in Cancer Stem Cells. *J. Phys. Chem. Lett.* **2020**, *11*, 1357–1363. [CrossRef] [PubMed]
240. Xu, Y.; Du, L.; Han, B.; Wang, Y.; Fei, J.; Xia, K.; Zhai, Y.; Yu, Z. Black phosphorus quantum dots camouflaged with platelet-osteosarcoma hybrid membrane and doxorubicin for combined therapy of osteosarcoma. *J. Nanobiotechnol.* **2023**, *21*, 1–19. [CrossRef] [PubMed]

241. Khoshnood, A.; Farhadian, N.; Abnous, K.; Matin, M.M.; Ziaee, N.; Yaghoobi, E. N doped-carbon quantum dots with ultra-high quantum yield photoluminescent property conjugated with folic acid for targeted drug delivery and bioimaging applications. *J. Photochem. Photobiol. A Chem.* **2023**, *444*, 114972. [CrossRef]
242. Lee, G.-Y.; Lo, P.-Y.; Cho, E.-C.; Zheng, J.-H.; Li, M.; Huang, J.-H.; Lee, K.-C. Integration of PEG and PEI with graphene quantum dots to fabricate pH-responsive nanostars for colon cancer suppression in vitro and in vivo. *FlatChem* **2021**, *31*, 100320. [CrossRef]
243. Hao, R.; Luo, S.; Wang, F.; Pan, X.; Yao, J.; Wu, J.; Fang, H.; Li, W. Enhancement of fluorescence and anti-tumor effect of ZnO QDs by La doping. *Front. Chem.* **2022**, *10*, 1042038. [CrossRef]
244. Gautam, A.; Pal, K. Gefitinib conjugated PEG passivated graphene quantum dots incorporated PLA microspheres for targeted anticancer drug delivery. *Heliyon* **2022**, *8*, e12512. [CrossRef]
245. Wei, Z.; Yin, X.; Cai, Y.; Xu, W.; Song, C.; Wang, Y.; Zhang, J.; Kang, A.; Wang, Z.; Han, W. Antitumor effect of a Pt-loaded nanocomposite based on graphene quantum dots combats hypoxia-induced chemoresistance of oral squamous cell carcinoma. *Int. J. Nanomed.* **2018**, *13*, 1505–1524. [CrossRef]
246. Javanbakht, S.; Namazi, H. Doxorubicin loaded carboxymethyl cellulose/graphene quantum dot nanocomposite hydrogel films as a potential anticancer drug delivery system. *Mater. Sci. Eng. C* **2018**, *87*, 50–59. [CrossRef]
247. Olerile, L.D.; Liu, Y.; Zhang, B.; Wang, T.; Mu, S.; Zhang, J.; Selotlegeng, L.; Zhang, N. Near-infrared mediated quantum dots and paclitaxel co-loaded nanostructured lipid carriers for cancer theragnostic. *Colloids Surf. B Biointerfaces* **2017**, *150*, 121–130. [CrossRef]
248. Zhao, T.; Liu, X.; Li, Y.; Zhang, M.; He, J.; Zhang, X.; Liu, H.; Wang, X.; Gu, H. Fluorescence and drug loading properties of ZnSe:Mn/ZnS-Paclitaxel/SiO2 nanocapsules templated by F127 micelles. *J. Colloid Interface Sci.* **2017**, *490*, 436–443. [CrossRef] [PubMed]
249. Duman, F.D.; Akkoc, Y.; Demirci, G.; Bavili, N.; Kiraz, A.; Gozuacik, D.; Acar, H.Y. Bypassing pro-survival and resistance mechanisms of autophagy in EGFR-positive lung cancer cells by targeted delivery of 5FU using theranostic Ag_2S quantum dots. *J. Mater. Chem. B* **2019**, *7*, 7363–7376. [CrossRef] [PubMed]
250. Yang, D.; Yao, X.; Dong, J.; Wang, N.; Du, Y.; Sun, S.; Gao, L.; Zhong, Y.; Qian, C.; Hong, H. Design and Investigation of Core/Shell GQDs/hMSN Nanoparticles as an Enhanced Drug Delivery Platform in Triple-Negative Breast Cancer. *Bioconjug. Chem.* **2018**, *29*, 2776–2785. [CrossRef] [PubMed]
251. Samimi, S.; Ardestani, M.S.; Dorkoosh, F.A. Preparation of carbon quantum dots- quinic acid for drug delivery of gemcitabine to breast cancer cells. *J. Drug Deliv. Sci. Technol.* **2020**, *61*, 102287. [CrossRef]
252. Sun, Z.; Zhao, Y.; Li, Z.; Cui, H.; Zhou, Y.; Li, W.; Tao, W.; Zhang, H.; Wang, H.; Chu, P.K.; et al. As an Efficient Contrast Agent for In Vivo Photoacoustic Imaging of Cancer. *Small* **2017**, *13*, 1602896. [CrossRef] [PubMed]
253. Kumawat, M.K.; Thakur, M.; Bahadur, R.; Kaku, T.; Prabhuraj, R.S.; Suchitta, A.; Srivastava, R. Preparation of graphene oxide-graphene quantum dots hybrid and its application in cancer theranostics. *Mater. Sci. Eng. C* **2019**, *103*, 109774. [CrossRef] [PubMed]
254. Saljoughi, H.; Khakbaz, F.; Mahani, M. Synthesis of folic acid conjugated photoluminescent carbon quantum dots with ultrahigh quantum yield for targeted cancer cell fluorescence imaging. *Photodiagnosis Photodyn. Ther.* **2020**, *30*, 101687. [CrossRef]
255. Zhang, F.; He, X.; Ma, P.; Sun, Y.; Wang, X.; Song, D. Rapid aqueous synthesis of CuInS/ZnS quantum dots as sensor probe for alkaline phosphatase detection and targeted imaging in cancer cells. *Talanta* **2018**, *189*, 411–417. [CrossRef]
256. Shao, J.; Zhang, J.; Jiang, C.; Lin, J.; Huang, P. Biodegradable titanium nitride MXene quantum dots for cancer phototheranostics in NIR-I/II biowindows. *Chem. Eng. J.* **2020**, *400*, 126009. [CrossRef]
257. Xu, N.; Piao, M.; Arkin, K.; Ren, L.; Zhang, J.; Hao, J.; Zheng, Y.; Shang, Q. Imaging of water soluble CdTe/CdS core-shell quantum dots in inhibiting multidrug resistance of cancer cells. *Talanta* **2019**, *201*, 309–316. [CrossRef]
258. Sobhani, Z.; Khalifeh, R.; Banizamani, M.; Rajabzadeh, M. Water-soluble ZnO quantum dots modified by polyglycerol: The pH-sensitive and targeted fluorescent probe for delivery of an anticancer drug. *J. Drug Deliv. Sci. Technol.* **2022**, *76*, 103452. [CrossRef]
259. Wang, J.; Su, X.; Zhao, P.; Gao, D.; Chen, R.; Wang, L. Cancer photothermal therapy based on near infrared fluorescent CdSeTe/ZnS quantum dots. *Anal. Methods* **2021**, *13*, 5509–5515. [CrossRef] [PubMed]
260. Ge, X.-L.; Huang, B.; Zhang, Z.-L.; Liu, X.; He, M.; Yu, Z.; Hu, B.; Cui, R.; Liang, X.-J.; Pang, D.-W. Glucose-functionalized near-infrared Ag_2Se quantum dots with renal excretion ability for long-term *in vivo* tumor imaging. *J. Mater. Chem. B* **2019**, *7*, 5782–5788. [CrossRef]
261. Li, X.; Vinothini, K.; Ramesh, T.; Rajan, M.; Ramu, A. Combined photodynamic-chemotherapy investigation of cancer cells using carbon quantum dot-based drug carrier system. *Drug Deliv.* **2020**, *27*, 791–804. [CrossRef] [PubMed]
262. Dong, J.; Wang, K.; Sun, L.; Sun, B.; Yang, M.; Chen, H.; Wang, Y.; Sun, J.; Dong, L. Application of graphene quantum dots for simultaneous fluorescence imaging and tumor-targeted drug delivery. *Sens. Actuators B Chem.* **2018**, *256*, 616–623. [CrossRef]
263. Kim, E.-M.; Lim, S.T.; Sohn, M.-H.; Jeong, H.-J. Facile synthesis of near-infrared CuInS2/ZnS quantum dots and glycol-chitosan coating for in vivo imaging. *J. Nanopart. Res.* **2017**, *19*, 251. [CrossRef]
264. Michalska, M.; Florczak, A.; Dams-Kozlowska, H.; Gapinski, J.; Jurga, S.; Schneider, R. Peptide-functionalized ZCIS QDs as fluorescent nanoprobe for targeted HER2-positive breast cancer cells imaging. *Acta Biomater.* **2016**, *35*, 293–304. [CrossRef]
265. Guo, Y.; Nie, Y.; Wang, P.; Li, Z.; Ma, Q. MoS2 QDs-MXene heterostructure-based ECL sensor for the detection of miRNA-135b in gastric cancer exosomes. *Talanta* **2023**, *259*, 124559. [CrossRef]

266. Wang, Y.; He, L.; Yu, B.; Chen, Y.; Shen, Y.; Cong, H. ZnO Quantum Dots Modified by pH-Activated Charge-Reversal Polymer for Tumor Targeted Drug Delivery. *Polymers* **2018**, *10*, 1272. [CrossRef]
267. Cao, Y.; Wang, K.; Zhu, P.; Zou, X.; Ma, G.; Zhang, W.; Wang, D.; Wan, J.; Ma, Y.; Sun, X.; et al. A near-infrared triggered upconversion/MoS$_2$ nanoplatform for tumour-targeted chemo-photodynamic combination therapy. *Colloids Surf. B Biointerfaces* **2022**, *213*, 112393. [CrossRef]
268. Zheng, S.; Zhang, M.; Bai, H.; He, M.; Dong, L.; Cai, L.; Zhao, M.; Wang, Q.; Xu, K.; Li, J. Preparation of AS1411 Aptamer Modified Mn-MoS$_2$ QDs for Targeted MR Imaging and Fluorescence Labelling of Renal Cell Carcinoma. *Int. J. Nanomed.* **2019**, *14*, 9513–9524. [CrossRef] [PubMed]
269. Badıllı, U.; Mollarasouli, F.; Bakirhan, N.K.; Ozkan, Y.; Ozkan, S.A. Role of quantum dots in pharmaceutical and biomedical analysis, and its application in drug delivery. *TrAC Trends Anal. Chem.* **2020**, *131*, 116013. [CrossRef]
270. Oh, E.; Liu, R.; Nel, A.; Gemill, K.B.; Bilal, M.; Cohen, Y.; Medintz, I.L. Meta-analysis of cellular toxicity for cadmium-containing quantum dots. *Nat. Nanotechnol.* **2016**, *11*, 479–486. [CrossRef] [PubMed]
271. Zheng, H.; Mortensen, L.; Ravichandran, S.; Bentley, K.; Delouise, L. Effect of Nanoparticle Surface Coating on Cell Toxicity and Mitochondria Uptake. *J. Biomed. Nanotechnol.* **2017**, *13*, 155–166. [CrossRef] [PubMed]
272. Manshian, B.B.; Martens, T.F.; Kantner, K.; Braeckmans, K.; De Smedt, S.C.; Demeester, J.; Jenkins, G.J.S.; Parak, W.J.; Pelaz, B.; Doak, S.H.; et al. The role of intracellular trafficking of CdSe/ZnS QDs on their consequent toxicity profile. *J. Nanobiotechnol.* **2017**, *15*, 1–14. [CrossRef]
273. Li, L.; Tian, J.; Wang, X.; Xu, G.; Jiang, W.; Yang, Z.; Liu, D.; Lin, G. Cardiotoxicity of Intravenously Administered CdSe/ZnS Quantum Dots in BALB/c Mice. *Front. Pharmacol.* **2019**, *10*, 1179. [CrossRef]
274. Zou, W.; Li, L.; Chen, Y.; Chen, T.; Yang, Z.; Wang, J.; Liu, D.; Lin, G.; Wang, X. In Vivo Toxicity Evaluation of PEGylated CuInS2/ZnS Quantum Dots in BALB/c Mice. *Front. Pharmacol.* **2019**, *10*, 437. [CrossRef]
275. Zhou, Y.; Sun, H.; Wang, F.; Ren, J.; Qu, X. How functional groups influence the ROS generation and cytotoxicity of graphene quantum dots. *Chem. Commun.* **2017**, *53*, 10588–10591. [CrossRef]
276. Zhang, M.; Yue, J.; Cui, R.; Ma, Z.; Wan, H.; Wang, F.; Zhu, S.; Zhou, Y.; Kuang, Y.; Zhong, Y.; et al. Bright quantum dots emitting at ~1600 nm in the NIR-IIb window for deep tissue fluorescence imaging. *Proc. Natl. Acad. Sci. USA* **2018**, *115*, 6590–6595. [CrossRef]
277. Hu, S.-H.; Gao, X. Stable Encapsulation of Quantum Dot Barcodes with Silica Shells. *Adv. Funct. Mater.* **2010**, *20*, 3721–3726. [CrossRef]
278. Murase, N.; Horie, M.; Sawai, T.; Kawasaki, K. Silica layer-dependent leakage of cadmium from CdSe/ZnS quantum dots and comparison of cytotoxicity with polymer-coated analogues. *J. Nanopart. Res.* **2019**, *21*, 10. [CrossRef]
279. Ko, N.R.; Nafiujjaman, M.; Lee, J.S.; Lim, H.-N.; Lee, Y.-K.; Kwon, I.K. Graphene quantum dot-based theranostic agents for active targeting of breast cancer. *RSC Adv.* **2017**, *7*, 11420–11427. [CrossRef]

Disclaimer/Publisher's Note: The statements, opinions and data contained in all publications are solely those of the individual author(s) and contributor(s) and not of MDPI and/or the editor(s). MDPI and/or the editor(s) disclaim responsibility for any injury to people or property resulting from any ideas, methods, instructions or products referred to in the content.

Article

Time-Dependent Size and Shape Evolution of Gold and Europium Nanoparticles from a Bioproducing Microorganism, a Cyanobacterium: A Digitally Supported High-Resolution Image Analysis

Melanie Fritz [1], Susanne Körsten [1], Xiaochen Chen [2], Guifang Yang [2], Yuancai Lv [2], Minghua Liu [2], Stefan Wehner [1] and Christian B. Fischer [1,3,*]

1 Department of Physics, University Koblenz-Landau, Universitätsstraße 1, D-56070 Koblenz, Germany
2 Fujian Provincial Engineering Research Center of Rural Waste Recycling Technology, College of Environment & Resources, Fuzhou University, Fuzhou 350116, China
3 Materials Science, Energy and Nano-Engineering Department, Mohammed VI Polytechnic University, Ben Guerir 43150, Morocco
* Correspondence: chrbfischer@uni-koblenz.de; Tel.: +49-26128-72345

Citation: Fritz, M.; Körsten, S.; Chen, X.; Yang, G.; Lv, Y.; Liu, M.; Wehner, S.; Fischer, C.B. Time-Dependent Size and Shape Evolution of Gold and Europium Nanoparticles from a Bioproducing Microorganism, a Cyanobacterium: A Digitally Supported High-Resolution Image Analysis. *Nanomaterials* 2023, 13, 130. https://doi.org/10.3390/nano13010130

Academic Editor: Thomas Dippong

Received: 4 December 2022
Revised: 23 December 2022
Accepted: 26 December 2022
Published: 27 December 2022

Copyright: © 2022 by the authors. Licensee MDPI, Basel, Switzerland. This article is an open access article distributed under the terms and conditions of the Creative Commons Attribution (CC BY) license (https://creativecommons.org/licenses/by/4.0/).

Abstract: Herein, the particle size distributions (PSDs) and shape analysis of in vivo bioproduced particles from aqueous Au^{3+} and Eu^{3+} solutions by the cyanobacterium *Anabaena* sp. are examined in detail at the nanoscale. Generally, biosynthesis is affected by numerous parameters. Therefore, it is challenging to find the key set points for generating tailored nanoparticles (NPs). PSDs and shape analysis of the Au and Eu-NPs were performed with *ImageJ* using high-resolution transmission electron microscopy (HR-TEM) images. As the HR-TEM image analysis reflects only a fraction of the detected NPs within the cells, additional PSDs of the complete cell were performed to determine the NP count and to evaluate the different accuracies. Furthermore, local PSDs were carried out at five randomly selected locations within a single cell to identify local hotspots or agglomerations. The PSDs show that particle size depends mainly on contact time, while the particle shape is hardly affected. The particles formed are distributed quite evenly within the cells. HR-PSDs for Au-NPs show an average equivalent circular diameter (ECD) of 8.4 nm (24 h) and 7.2 nm (51 h). In contrast, Eu-NPs preferably exhibit an average ECD of 10.6 nm (10 h) and 12.3 nm (244 h). Au-NPs are classified predominantly as "very round" with an average reciprocal aspect ratio (RAR) of ~0.9 and a Feret major axis ratio (FMR) of ~1.17. Eu-NPs mainly belong to the "rounded" class with a smaller RAR of ~0.6 and a FMR of ~1.3. These results show that an increase in contact time is not accompanied by an average particle growth for Au-NPs, but by a doubling of the particle number. *Anabaena* sp. is capable of biosorbing and bioreducing dissolved Au^{3+} and Eu^{3+} ions from aqueous solutions, generating nano-sized Au and Eu particles, respectively. Therefore, it is a low-cost, non-toxic and effective candidate for a rapid recovery of these sought-after metals via the bioproduction of NPs with defined sizes and shapes, providing a high potential for scale-up.

Keywords: cyanobacteria; nanoparticle size distribution; digital image processing; growth monitoring; shape classification

1. Introduction

Microbial in vitro or in vivo synthesized nanoparticles (NPs) are of great interest as they can be generated in an environmentally friendly and cost-effective manner. Especially in the field of biomedical applications, biogenic synthesis routes offer a sustainable and safe technique [1]. Bioproduction can be carried out under ambient temperatures, unlike conventional synthetic processes [2]. The microorganisms used as biological "nanofactories" are capable of accumulating gold (Au), rare earths (REs) and other dissolved elements from industrial waste or other anthropogenically contaminated sites. At the same time, they are able to satisfy the objectives of resource recovery and pollution reduction [3,4]. Eu ions accumulate in soil and water due to the gasoline industry, nuclear wastewater, or discarded equipment containing Eu dyes [5]. This harms the reproduction, nervous system and cell membranes of aquatic organisms and poses a potential threat to human health [6].

In particular, Au and RE recovery from mine waste is becoming increasingly attractive, as it is necessary to rehabilitate mine tailings using nature-based solutions and afterwards use them economically. The effort to discard the biomass contaminated with metals should be used profitably. For this purpose, the processing of plants to products is carried out, which is called farming for metals or agro-mining [7]. Rising mining costs also favor the biotechnological recycling of critical metals, which include Au and Eu. Biorecovery of Eu from primary (mineral deposits) and secondary (mining wastes) resources is of interest due to its scarcity and inherent luminescence properties [8]. Besides the recovery from mine tailings, industrial electronic waste, especially waste printed circuit boards [9], are attractive resources of trivalent actinides. E-waste, is one of the fastest increasing waste streams (>50 Mt in the year 2019, estimated 74 Mt in 2030), whose disposal and effective management is a global challenge according to the UN's Global E-waste Monitor 2020 [10].

Some studies are conducted to find suitable "green" sorbents (e.g., cell walls of *Stephanopyxis turris* and *Thalassiosira pseudonana*) to remediate Eu and model sorption reaction by using diffuse double layer model [11]. In sorption, different uptake mechanisms are possible: surface adsorption and incorporation/precipitation. The extent of adsorption is known to depend on solution pH, contact time, metal ion concentration and adsorbent dose [12].

To the best of the authors' knowledge, only few studies have been conducted on microbial Eu (nano-)particle synthesis concerning more the fact of demonstrating bioaccumulation and/or biosorption that can be used for biological recovery. Hence, Cadogan et al. successfully removed Eu^{3+} ions by *Arthrobacter* sp. by biomass and crab shell powder amongst others from aqueous solution through biosorption [5,12]. Serna et al. provided insights into the passive biosorption of Eu^{3+} binding properties to chemically distinct sites of different biomaterials using luminescence spectroscopy [13]. In contrast to the aforementioned methods, Kim et al. synthesized europium selenide (EuSe) NPs with recombinant *Escherichia coli* cells in vivo and showed the anti-cancer effect of these generated NPs [14]. Maleke et al. studied the reduction and intracellular bioaccumulation of Eu by a *Clostridium* strain and found it suitable for biorecovery of this critical metal [8].

In addition to intracellular uptake in vivo biosynthesis of NPs [15,16], there are methods in which stable crystalline Au particles are formed extracellularly in vitro using biomass extracts [17]. Commonly, in vitro methods for the quasi biosynthesis of (Au-)NPs by using biomass extracts are used to (extracellularly) generate particles.

The state of the art Au-NP biosynthesis is better established [18–21] than that for REs such as Eu. For example, Dahoumane et al. studied improvements of kinetics yield and colloidal stability of biogenic Au-NPs [21]. Furthermore, Castro et al. found that the initial pH value of the solution and the concentration of Au precursor influences the morphology and NP formation of crystalline gold nanowires using orange peel extract [22].

All these microbially synthesized metallic nanostructures clearly outperform chemical and physical processes by providing a better eco-balance, being faster and capable of being producing in larger scales [23]. However, a big issue of bioproduction is still the control of NPs with a certain size and shape [24]. Nevertheless, studies to analyze data on the diversity of shapes and sizes of NPs obtained by green synthesis are scare [25,26]. There are numerous variables, such as pH, metal ion concentration in the solution, adsorbent dose, contact time, etc., that affect microbial NP bioproduction [27]. Therefore, it is challenging to find the crucial points for the generation of specific NPs and optimize relevant parameters to examine the best practice [28].

Within our group, we already demonstrated the capability of biosorption and accumulation of precious and RE elements, such as Au, Eu and Sm, leading to well-defined nanosized particles [15,16,29,30]. Cyanobacteria of the genus *Anabaena* sp. are a species well suited for the intracellular uptake of precious metals, such as Au, Ag, transition metals (e.g., Cu, Al, etc.) and various REs [31,32].

Furthermore, they are good candidates to replace synthetic routes, which often require toxic solvents or produce undesirable waste products. They are easy to handle in contrast to some other microorganisms, why they have attracted increasing attention among scientists worldwide [33]. The kinetics and mechanisms of NP formation need to be understood before extending laboratory experiments to higher capacity processes [28]. If production could be accomplished by natural bioorganisms in effluents from treatment plants, this would be very beneficial. On the one hand, the algae could serve to minimize metal toxicity in the aqueous environment, and on the other hand, elemental metal would be recovered from the ions by biosorption through regular harvesting [34].

In this study, the influence of the contact time of *Anabaena* sp. with the metal-containing nutrient solution on particle sizes and shaping was investigated. High-resolution transmission electron microscopy images showing Au- and Eu-NPs formed in *Anabaena* sp. were analyzed using digital image processing. It was found that the particle growth time factor can affect particle size and number, but has less effect on particle shaping.

2. Materials and Methods

Biosorption experiments were performed according to our previous work [15,16]. Stock culture of *Anabaena* sp. (SAG 12.82, Algae Culture Collection (SAG) Göttingen, Germany) was transferred in a sterile 250 mL Erlenmeyer flask filled with 150 mL modified Bold's Basal Medium (BBM, pH 6.8). All biomass suspensions were kept at a low nitrate concentration of 50% in order to increase heterocyst growth. Incubation took place at 22 °C and pH 7.3 with a 12 h day–night cycle simulation (4200 K) under continuous mixing by an orbital shaker in a temperature-controlled incubator. Appropriate samples were separated in half, one was taken as a reference and the other was incubated with the respective salt BBM solutions of $HAuCl_4$ or $Eu(NO_3)_3 \cdot 6H_2O$, each 1×10^{-4} mol/L (ABCR GmbH, Karlsruhe, Germany). Aliquots of 2 mL of the biomass suspensions incubated with Au were taken after 24 h and 51 h, and of the samples mixed with Eu after 10 h and 244 h. Centrifugation (14,000 rpm, 15 min, 16,000× g) was used to extract biomass from the medium and the biomass pellet was prepared for TEM measurements. The supernatant was used for inductively coupled plasma mass spectrometry (ICP-MS) to prove the uptake of metal ions. More details can be taken from [15,16,30].

Prior to TEM-imaging, samples were fixed with glutaraldehyde and osmium tetroxide, dehydrated and embedded in epoxy resin, according to standard procedures. Ultrathin films (~60 nm) were cut with an Ultracut EM UC6 ultramicrotome (Leica Microsystems, Wetzlar, Germany) using a diamond knife (type ultra 35°, Diatome, Biel, Switzerland). Films were placed on pioloform-coated copper grids (Plano, Wetzlar, Germany), stained with uranyl acetate and lead citrate. The method follows established procedures [35] and has previously been described in more detail [15,16]. Nanoscale imaging was performed on a HT-7700 TEM 7700 (Hitachi, Tokyo, Japan) in high-resolution (HR) imaging mode. The system was operated at an accelerating voltage of 100 kV.

For digital image processing, *ImageJ* 1.53d (National Institute of Health, Rockville, MD, USA, freeware) with Java^{TM} version 1.8.0_112 (64-bit) under Windows 10 Pro edition was used to identify all pixels that contribute to recorded Au or Eu particles. A Fujitsu Siemens H19-1 monitor (resolution of 1280 × 1024, refresh rate of 60.020 Hz) was applied for the evaluation. The bit depth was 8-bit, the color format was RGB and the color space was SR (Standard Dynamic Range). The graphics card used was Intel(R) HD Graphics 620. TEM images were implemented as JPG (resolution 3296 × 2563 pixels, depth 24-bit), whereas the registered scale bar served as a known distance for the transfer to the particle units. To capture particles more accurately, images were cropped, smoothed and sharpened before individual thresholding took place. An appropriate size range was set for each image, to remove artefacts and cell compounds at the upper scale and scattered pixels originating from the background noise at the lower end. Particles touching the edges were excluded. Results were displayed as equivalent circular diameters (*ECD*, Equation (1)) in nm,

$$ECD = \sqrt{4 \cdot A / \pi} \tag{1}$$

assuming that the particles are perfectly spherical. Since this assumption does not correspond to reality and most particles are irregular in shape by nature, the Feret major axis ratio (*FMR*, Equation (2)) and the reciprocal aspect ratio (*RAR*, Equation (3)) were calculated from the Feret diameter D_f and the major a and minor axes b of the fitted ellipses for shape analysis.

$$FMR = D_f / a \tag{2}$$

$$RAR = b / a \tag{3}$$

The parameters used in Equations (1)–(3) were taken from Igathinathane's object identification strategy and are standard outputs generated by *ImageJ* to evaluate particles shape and size [36]. The exact procedure for particle size and shape analysis can be found in detail in the previous work of the authors [37].

3. Results and Discussion

3.1. Exemplary TEM Images with Nanoparticles

TEM measurements clearly show Au and Eu particles inside the cells of *Anabaena* sp. (Figure 1). After 24 and 51 h of in vivo contact time with the Au^{3+} containing solution (respectively, 10 and 244 h with the Eu^{3+} solution), cells of the living organism fully incorporated these ions, and generated nano-sized Au and Eu particles, respectively. Au-NPs were formed only in vegetative cells (Figure 1A–D), whereas Eu-NPs were generated exclusively in its heterocysts (Figure 1E–H). In high-resolution (HR) imaging, the Au-NPs (some indicated by blue arrows in Figure 1B,D) look more spherical and electron denser than the irregularly shaped Eu-NPs (red arrows in Figure 1F,H) at first glance. Therefore, the particle formation in different cells obviously affects the shape and size of the generated particles.

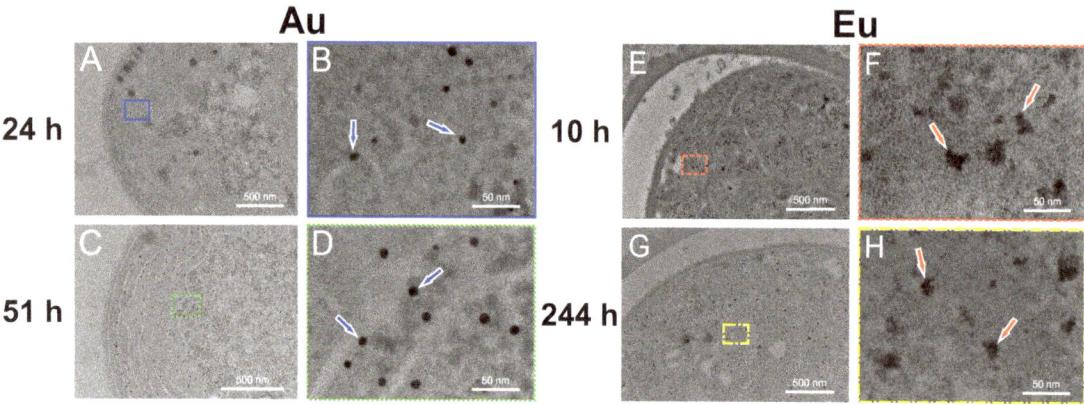

Figure 1. TEM images of vegetative cells of *Anabaena* sp. showing Au-NPs after 24 h (**A,B**) and 51 h (**C,D**) and its heterocysts with Eu-NPs after 10 h (**E,F**) and 244 h (**G,H**). Indicated are magnifications (**B**) (blue outline) of (**A,D**) (green dotted) of (**C,F**) (red dashed) of (**E**), and (**H**) (yellow dotted and dashed) of (**G**). Blue arrows mark some selected Au-NPs and red arrows Eu-NPs.

3.2. Particle Size Distributions (PSDs)

3.2.1. PSDs of all NPs in Complete Cells

The PSDs of the Au- and Eu-NPs shown in Figure 2 were determined from the TEM images provided in the Supporting Information Figure S1 (Supplementary Materials Section S1 Used Datasets) and the settings listed in Table 1 during digital image analysis.

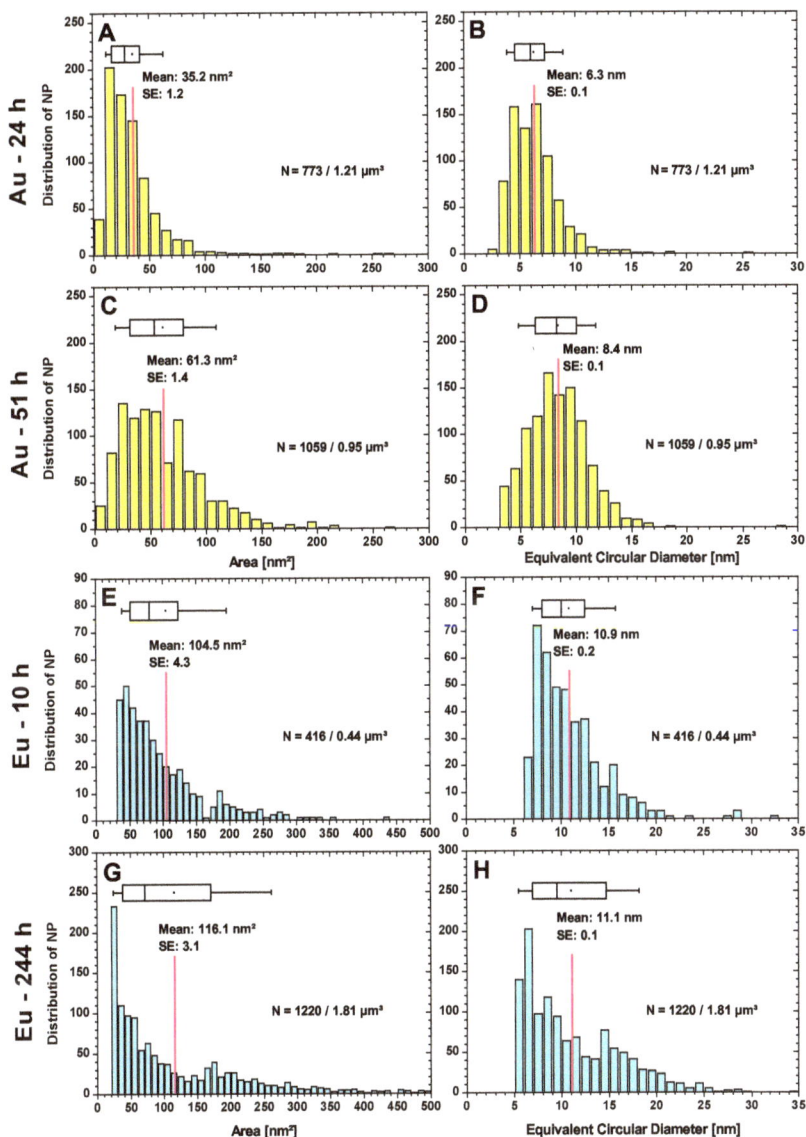

Figure 2. PSDs for Au-NPs after 24 h (**A,B**) and 51 h (**C,D**) and for Eu-NPs after 10 h (**E,F**) and 244 h (**G,H**) with the mean value and standard error (SE). The number of particles N per investigated cell volume and the average area or ECD (red lines) are indicated in the plots. The ECDs are calculated from the areas assuming round particles. The boxplots show the 25% and 75% percentiles, with the whiskers at 10% and 90%.

Table 1. Underlying TEM image settings for the determination of Au- and Eu-NP sizes in the complete vegetative cells (Au) and heterocysts (Eu), respectively (inf = infinity).

Ion Contact Time	Cell No.	Zoom	Range [px]	Scale Ratio [px/μm]	Threshold [%]	Counts
Au—24 h	1	×8.0 k	5–500	736	0.2	178
	2	×8.0 k	10–inf	732	0.2	30
	3	×15.0 k	10–inf	1374	0.2	194
	4	×10.0 k	10–300	917	0.2	371
Au—51 h	5	×8.0 k	5–inf	734	0.4	552
	6	×7.0 k	10–1000	639	0.5	507
Eu—10 h	7	×7.0 k	15–500	643	1.4	416
Eu—244 h	8	×5.0 k	5–100	458	0.5	742
	9	×7.0 k	20–400	639	1.3	258
	10	×6.0 k	40–300	496	1.4	220

The results of Figure 2 give an indication of the average NP-forming capacity of the investigated cell volumes and their PSDs in complete cells. The vegetative cells forming Au-NPs have a total volume of 1.21 μm^3 for cells isolated after 24 h of contact time and a volume of 0.95 μm^3 for cells extracted after 51 h. Each cell area was measured by digital image analysis and the cell volume was calculated by assuming the sliced ultrafilms have a thickness of 60 nm. In these cell volumes, 773 particles were detected after a growth time of 24 h and 1059 Au-NPs after 51 h. For Eu, a total of 416 NPs could be registered in a cell volume of 0.44 μm^3 (10 h) and 1220 of them in a volume of 1.81 μm^3 (244 h). To facilitate comparison, the values are converted to a standard volume of 1.00 μm^3. This results in the fact that with the same cell volume, about 639 particles would have formed with a growth time of 24 h and about 1115 Au-NPs with 51 h. The increase rate of particle number at slightly more than twice the time is about 57%. In the case of Eu, the situation is completely the other way round. Here, about 946 Eu-NPs would have formed at 10 h and only 674 at 244 h. It remains questionable whether this decrease of nearly 29% in the number of Eu-NPs is due to a partial agglomeration of the amorphous particles. Considering the range from approx. 300 nm^2, it can be seen that larger particles formed after 244 h compared to after 10 h. However, there is still a large number of small particles (<50 nm^2). The flatter distribution from about 100 nm^2 (244 h) and from about 200 nm^2 (10 h) indicates that, with increasing growth time, there is a higher tendency towards larger particles as well as a significantly higher number of small Eu-NPs. When analyzing images with *ImageJ*, manual adjustments must be made due to the different acquisition parameters. For example, the zoom varied for Au-NP evaluation from 7.0 k to 15.0 k, depending on the different cell sizes, which affects the selected scale and analysis range. In the analysis, it is important to place restrictions (a range), so that cellular components such as ribosomes or vacuoles with a similar or same gray scale to the Au-NPs are not included. Different measurement limits (lower/upper area limit in Table 1) result from the fact that the ranges have to be selected differently. For images No. 2 and No. 5 (not shown here, please refer to the Supplementary Materials Section S1 Used Datasets), an upward restriction was not necessary because the image section did not contain any artifacts. The limit is transferred by converting the particle size given in the area to ECD (lower/upper ECD limit in Table 1). The threshold varies slightly (0.19-0.21) due to the different brightness levels of the TEM images. The evaluations of cells No. 1-4 are summarized in Figure 2A,B and those of No. 5 and 6 in Figure 2C,D (for cell images see Supplementary Materials Section S1 Used Datasets).

Most Au-NPs (26.2%) in the size distribution (Figure 2A) belong to the size class 10–20 nm^2, followed by the second largest number (22.4%), which have an area of 20–30 nm^2. The average value is 35.2 nm^2, as there are still isolated particles (≤0–4) between 90 nm^2 and 300 nm^2. The majority of Au-NPs (83.2%) has a size between 10 and 50 nm^2 after 24 h, so the size range is much narrower than after 51 h (Figure 2C). Here, the average value is much higher at 61.3 nm^2 and the distribution ranges up to 150 nm^2 with a much higher number of particles. In contrast to 24 h, 87.3% of the particles are found in a size range of 10–100 nm^2, which is twice as large after 51 h of growth time. To get a better impression, the areas were converted to ECD, assuming spherical particles. After 24 h (Figure 2B), the Au-NPs have an average diameter of 6.3 nm, which expands to 8.4 nm after 51 h (Figure 2D). About 20.4% of the Au-NPs have an ECD between 4 and 5 nm, which is approximately equal to the number of NPs exhibiting an ECD between 6 and 7 nm (20.8%). The intermediate class (5–6 nm) has slightly fewer Au-NP with an ECD of 17.5%. Even after 51 h, the size distribution shows a local minimum at 8–9 nm. Among the adjacent classes, 15.7% of the Au-NPs belong to the 7–8 nm size class and 13.4% fall in the 8–9 nm range. Followed by 11.2% with a diameter between 6 and 7 nm and 10.8% between 10 and 11 nm. There are no Au-NP that have an area <9 nm^2 and only 734 of the 1059 have an area <25 nm^2 (or <3.4 nm and only a few <5.6 nm ECD) due to the restriction made in the particle evaluation for image No. 5 and No. 6. For the 24 h case, the measurement limit was also at an area of 9 nm^2 (or a diameter of 3.4 nm), although for some particles this limit had to be set higher to avoid counting false positives.

When looking at Figure 2F, it is noticeable that for the 416 Eu-NPs registered, the average ECD is 8.4 nm, with the dominant class here, between 8 and 10 nm, covering 44.0% of the particles. Conversely, the average ECD decreases by 1.2 nm to 7.2 nm at about a 25th growth time, and of the 1220 particles, 53.4% likely belong to the 5–10 nm (Figure 2H). This also confirms the increasing agglomeration over time.

Considering that the ECD is calculated assuming all NPs to be spherical, which is more or less the case depending on the particle shape, the unadjusted PSDs (Figure 2A,C,E,G) are also reported in area sizes for completeness, but are not discussed in detail.

3.2.2. High-Resolution (HR)-PSD of Cell Sections

The high-resolution (HR) images (No. 11–75) (Supplementary Materials Figures S2–S5) can only capture a portion of the total particle count (Au-NPs: ~20.7% for 24 h, ~29.7% for 51 h; Eu-NPs: ~15.1% for 10 h, ~12.2% for 244 h), but provide a more valuable analysis of particle size and shape. The results of the HR evaluation are presented in Figure 3 (see Supplementary Materials Table S1 for TEM image settings data for Au HR-PSDs in Figures S2 and S3, and Eu HR-PSDs in Figures S4 and S5). With the more precise size distribution (Figure 3A), the average area of the Au-NPs is larger, rises to 62.8 nm^2 instead of 35.2 nm^2 (Figure 2A) and the corresponding ECD increases from 6.3 nm (Figure 2B) to 8.4 nm (Figure 3B). The size classes are highest in the 60–70 nm^2 (11.9%), 70–80 nm^2 (11.9%), and 90–100 nm^2 (10.0%) range. The smaller particles with sizes between 0 and 20 nm^2 (17.5% of all particles) outweigh those between 20 and 50 nm^2 (6.3% for 20–30 nm^2, 6.3% for 30–40 nm^2 and 7.5% for 40–50 nm^2). The transfer to the ECD (Figure 3B) shows it more clearly: about half of all detected particles (45.6%) have a diameter of 8–11 nm. If the range is extended to 7–12 nm, this is already 63.1% of all particles. The count at 22–23 nm was caused by two closely spaced particles, which were scored as one particle. After a growth time of 51 h, the particles have an average area of 46.7 nm^2 (Figure 3C) instead of 61.3 nm^2 (Figure 2C) as in the complete cells.

Here, a relatively large number of Au-NPs have a small area of 0–30 nm^2. After 51 h, mainly smaller particles are detected, except for the class at 60–70 nm^2 with 12.7% total particle amount. There are no particles with an area larger than 170 nm^2 (or ECD of 15 nm). Compared to the shorter growth time, 51.0% more particles were registered for the same number of evaluated TEM images and, thus, approximately the same cell area. Again, there is a local minimum at an area of 30–60 nm^2, but this is higher than the areas from 80 nm^2, which is more pronounced here compared to 24 h particle growth time. The corresponding diameters (Figure 3D) show more clearly that most particles in the two fields are between 4 and 6 nm (21.7%) and 8 and 10 nm (28.0%). Including the intermediate classes, 68.8% of all particles belong to the size range 4–10 nm.

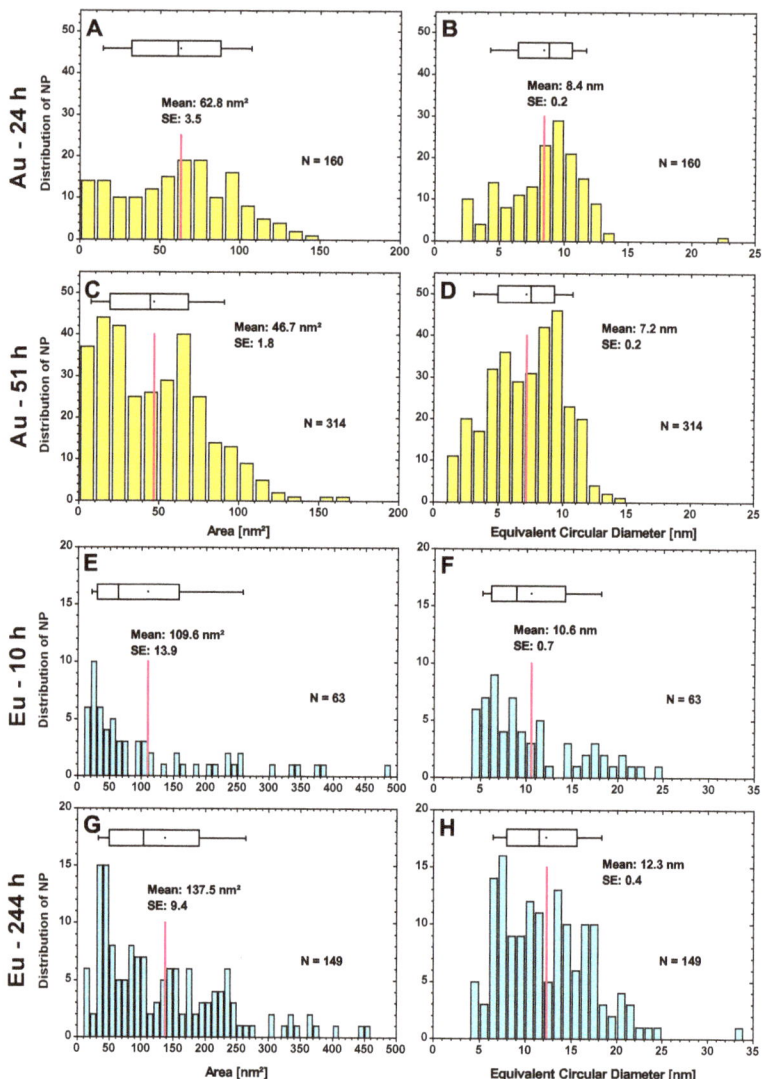

Figure 3. HR-PSD for Au-NPs after 24 h (**A**,**B**) and 51 h (**C**,**D**) and for Eu-NPs after 10 h (**E**,**F**) and 244 h (**G**,**H**) with the mean value and standard error (SE). The number of particles N counted in total for the same cell volume and the average area or particle diameter (red lines) are indicated in the plots. The ECDs are calculated from the areas assuming round particles. Again, the boxplots show the 25% and 75% percentiles, with the whiskers at 10% and 90%.

In the case of Eu, Figure 3F reveals that for the HR evaluation, most Eu-NPs are in the dominant classes between 5 and 7 nm with 25.4% of the detected 63 particles. The average ECD is calculated at 10.6 nm, which is quite higher than the Eu-NPs belonging to the predominant class. The reason for this is the high number of individual particles in the upper size classes, shifting the mean to higher values. In contrast to the Au-NPs with ECDs ranging up to 15 nm, the Eu-NPs reach ECDs of up to 25 nm, and for 244 h, even a particle with an ECD of 34 nm is recorded. This much larger particle is the result of a fusion of two particles. For the 149 detected Eu-NPs with a contact time of 244 h, the average ECD is 12.3 nm. Thus, there is a clear growth in particle size with time, of which most particles are found in the 6–8 nm, 10–12 nm, and 13–15 nm classes. In contrast to the PSDs of the complete cells, the average ECD decreased only minimally by 0.3 nm for a contact time of 10 h and increased by 1.2 nm for 244 h. However, by the much more accurate HR analysis, only 15.1% (10 h) and 12.2% (244 h) of all Eu-NPs within the cell could be covered. The discrepancy can be explained by the missing particles and the greater susceptibility to error due to the small magnification scale.

3.2.3. Local PSDs within an Exemplary Cell

To determine if the PSDs differed locally within the cells, five sites were randomly selected and local PSDs (Figure 4 and Supplementary Materials Figures S6–S8) were performed for them.

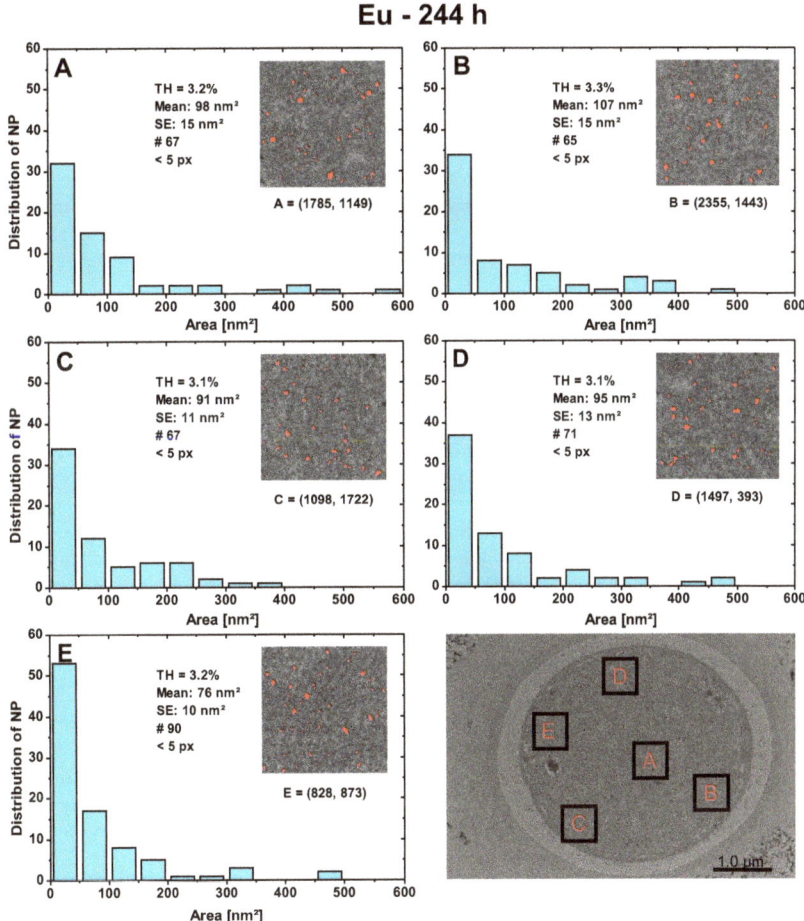

Figure 4. Local PSDs for Eu-NPs after 244 h growth time of five randomly selected 500 × 500 nm² areas (**A**–**E**) within the presented heterocyst No. 10. Indicated in the plots are the thresholds (TH), the mean values with the standard error of the mean (SE), the total number of particles detected (#), the lower measurement limit in pixels (px), and the local (x,y)-coordinates for each digital image analysis.

These should provide information on whether the particle number or size is uniformly distributed within the cells. All tests indicate that there are no local hotspots for either Au or Eu where NPs are formed more frequently or where particles are larger/smaller than elsewhere. Results for Au are summarized in the Supplementary Materials Section S2 (Supplementary Materials Figures S7 and S8) for the incubation time of 24 h and 51 h, as well as for Eu at 10 h. As an example, the local PSDs for the detected Eu-NPs with a contact time of 244 h (Figure 4) are discussed here. Since the locations within the cell have similar gray levels, the threshold values (TH) vary only slightly. It is important to set a lower limit for the particle size (<5 pixels), otherwise even single pixels of the adjusted grayscale value will be counted as particles. A maximum limit is not necessary if there are no artifacts. In case of artifacts, an upper value can be selected as a limit. In area E, the particle count of 90 is slightly higher than the average of ~67 for the other zones. The particle areas are largest for the classes 0–50 nm^2, with area E having 58.9% of all particles in this class. This is slightly higher than the other areas (A: 47.8%, B: 52.3%, C: 50.7%, D: 52.1%) with an average value of only 50.7%. Most of the largest particle areas are located in area A inside the cell, followed by area D, which is close to the cell membrane. Approximately equal numbers of particles in the different classes with an area size of 50 nm^2 and larger are found in area B. However, the differences between the areas are not meaningful enough to identify local hotspots or agglomeration. Similarly, no significant changes were observed in Eu-NPs among the five different zones at 10 h contact time (Supplementary Materials Figure S6). There is also no apparent preferential accumulation of Au-NPs during the 24 h or 51 h growth period at cellular constituents or local sites, indicating a greater particle number or size (Supplementary Materials Figures S7 and S8).

3.3. Shape Classification

Figure 5 reveals that the Au-NPs have an average RAR (red lines) of 0.88 for 24 h and 0.85 for 51 h, with most of the particles (%) being in the 0.85–0.95 class (Figure 5A,C). Additionally, the FMR with an average value of 1.17 and 1.16, respectively, indicates that the parameters are approximately the same for both times and the main weighting falls on the classes between 1.05 and 1.2 (Figure 5B,D). In the case of Eu, the particle shapes exhibit a more diversified spectrum (0.3–1.0) than those of Au-NPs, where the dominant classes are mainly found in the range of 0.7–1.0. The RAR is 0.63 for the detected 63 particles after an incubation time of 10 h and 0.65 for the 149 particles after 244 h. The small difference in the mean values at the different times suggests that the particle shapes are only imperceptibly affected by the change in growth time. A comparison of the dominant classes also does not reveal a strongly differentiated pattern. For the FMR, the distribution is dominant both times for the classes at 1.2–1.3. Again, a frequent particle shape is observed, with greater variance than for Au.

To obtain a more descriptive shape classification, a subdivision into the shape classes "very angular" to "very round" was made according to the RAR interval assignments shown in Table 2.

It can be seen that the majority of all Au-NPs are very rounded, 95.6% for 24 h and 88.9% for 51 h (Figure 6A,B). Increasing the time factor leads to an absence of sub-angular Au-NPs, but in return, less are very rounded. In the case of Eu-NPs, about 50% of the particles are just rounded for both incubation times. The remaining particles are mainly classified to very rounded (33.3% for 10 h and 36.9% for 244 h), followed by sub-rounded (14.3% for 10 h and 12.8% for 244 h). These results show that the Eu-NPs are round enough to justify the previous calculations of the ECD assuming spherical particles. For the shorter growth time (10 h), some Eu-NPs are also found in the sub-angular class, which disappear with increasing time. It appears that the initially more irregular amorphous Eu-NPs agglomerate into rounder particles over time. On the one hand, with increasing growth time, some of the crystalline Au-NPs become a little more irregular. On the other hand, there are less angular exotics. However, the changes are so minor that varying the incubation time does not significantly affect the overall shape of the particles.

Previous studies have shown that microbial synthesis and growth of metal NPs depend on the initial concentration of the respective metal ions, in addition to other factors such as adsorbent dose, pH value, temperature, and contact time. At the same time, this also affects the size, shape and agglomeration of the NPs [38–40]. In similar studies [19,41], higher initial concentrations of HAuCl$_4$ (0.5 and 1.0 mM, and 25 mg/L resp. 0.13 mM) at incubation times from 24 h up to several days resulted in more diverse shapes, e.g., spherical, triangular, hexagonal and irregular ones, with even larger NP sizes (30 nm up to 100 nm). However, in the case for *Anabaena laxa*, the higher concentrations of 0.5 and 1.0 mM also led to the death of the cells used within the 24 h [19]. Rösken et al. made a similar observation for *Anabaena* sp. with initial concentrations of 0.8 mM HAuCl$_4$, where cell viability ceased within a week [16]. In contrast, at lower concentrations of around 0.1 mM and slightly higher, there appears to be a preference for smaller and more spherical gold NPs [19,41].

Table 2. Interval assignments of the RAR values into six shape classes.

Class	Very Angular	Angular	Sub-Angular	Sub-Rounded	Rounded	Very Rounded
RAR value	0.12–0.17	0.17–0.25	0.25–0.35	0.35–0.49	0.49–0.70	0.70–1.00

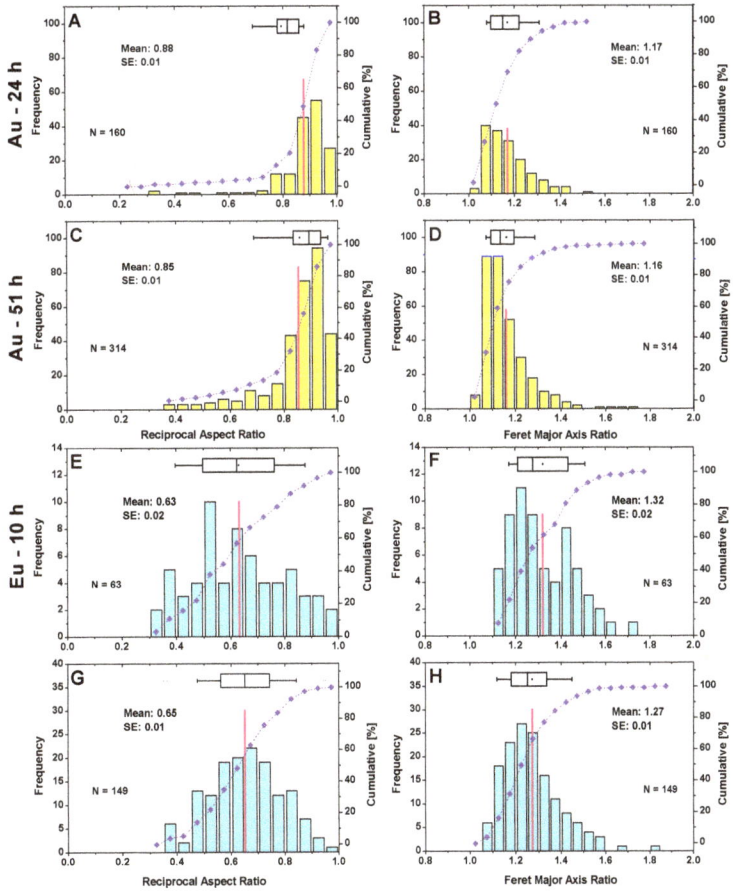

Figure 5. Shape classification for Au-NPs after 24 h (**A,B**) and 51 h (**C,D**) and for Eu after 10 h (**E,F**)

and 244 h (**G,H**) by using RAR and FMR as shape parameters (see Equations (2) and (3)). The total particle number N, mean values (red lines) with standard errors (SE), and the cumulative frequency in percentage (purple curves) are indicated in the plots. The boxplots show the 25% and 75% percentiles, with the whiskers at 10% and 90%.

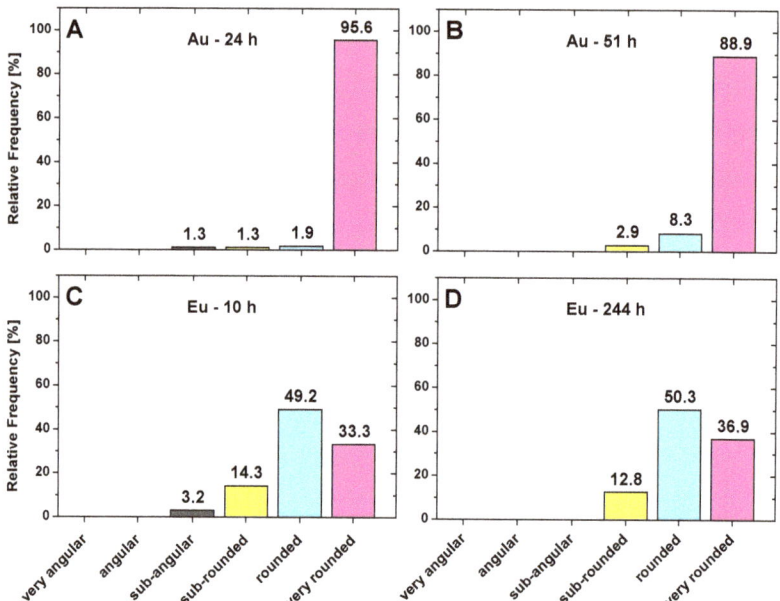

Figure 6. Shape classification into six different classes for Au-NPs after 24 h (**A**) and 51 h (**B**) growth time and for Eu-NPs after 10 h (**C**) and 244 h (**D**) with the indication of percentages.

In this biorecovery process, metal ions are adsorbed via the cell surface and subsequently reductively converted to NPs in the presence of enzymes synthesized by the microbes. The transformation occurs both intracellularly and extracellularly, and the metal nanoparticles are deposited accordingly, either inside or around the cell [38]. Studies by Chakraborty et al. [42] showed that *Cyanophyceae*, *Lyngbya majuscule* and *Spirulina subsalsa* actively absorb Au ions within 72 h of exposure, with absorption maxima greater than 96% and 86%, respectively. After processing the washing agent ethylenediaminetetraacetic acid, which chelates metal ions absorbed at the exterior of the cell, the intracellular fractions are obtained after appropriate workup. It turned out that for both organisms, gold in 20% amount for *Lyngbya majuscule* and 50% for *Spirulina subsalsa* could be determined at the cell surface, thus decreasing the intracellular yield and consequently the NPs production.

Therefore, the here used *Anabaena* sp. appears to effectively produce stable gold NPs intracellularly without agglomeration within a shorter time (Au-NPs within 24 h and for Eu within 10 h), but this would need to be verified by a more comparative study. For Eu-NPs, comparisons are not very meaningful since data in this field are still very sparse [43].

4. Conclusions

In this study, the influence of incubation time on the bioproduction of metallic gold (Au) and amorphous europium (Eu) nanoparticles (NP) in the cyanobacterium *Anabaena* sp. was investigated. Digital image processing was used to analyze HR-TEM images showing that Au- and Eu-NPs formed in the cyanobacterium. Special attention was drawn to the development of the size and shape classification of the nanoscale, bioproduced particles. Increasing the incubation time from 24 h to 51 h results in an almost double Au particle number. The average equivalent circular diameter (ECD) shrinks from 8.4 to 7.2 nm, with two predominant particle size classes (4–6 and 9–10 nm), instead of previously only one (8–10 nm). For the microbial formed amorphous Eu-NPs with a significantly higher growth time factor, an increasing average ECD from 10.6 to 12.3 nm, distributed in three classes (6–8, 10–12 and 13–15 nm), is found for 244 h compared to just the single class at 5–7 nm for 10 h. The factors leading to a decrease in particle productivity of almost 29% in the number of Eu-NPs with higher growth time are still questionable and have been attributed to partial agglomeration of the amorphous particles.

A comparison between five local particle size distributions (PSDs) within one cell compartment show no hotspots of preferred agglomerate zones.

Shape classification was performed by using the reciprocal aspect ratio (RAR) and Feret major axis ratio (FMR) as shape parameters. For the detected Au-NP, the RAR and FMR has no significant changes for the different growth times. The formed NPs are clearly classified in both cases as very rounded, 95.6% at 24 h and 88.9% at 51 h contact time. Compared to the Au-NPs, the shape of the Eu-NPs is less "rounded" and also shows no time-dependent shape effects, but has a broader shape spectrum. Therefore, half of the Eu-NP shapes are classified to be "rounded", followed by "very rounded" (~35%) and then "sub-rounded" (~13%). Thus, a higher growth time factor only very slightly enhanced the tendency toward more spherical particles.

Anabaena sp. was able to bioform NPs of very small size with a high tendency toward spherical shaped NPs within a very short time. For short growth time factors, no agglomeration and a relatively high productivity could be recorded. This makes these particles interesting for many applications. The initial salt concentrations used were lower, but resulted in better efficiencies in comparable studies.

The results showed only negligible time-dependent shape effects, but more obvious size effects. Since biosynthesis is more difficult to control and a variety of factors influence particle growth, further investigation is needed to obtain tailor-made particles.

Furthermore, it was found that cyanobacteria of the genus *Anabaena* spec. can be successfully used to extract raw materials from aqueous solutions containing trivalent Au or Eu ions by recovery means. Thus, *Anabaena* sp. is an appropriate and potential candidate for a cost-effective and non-toxic synthesis of the specific NP sizes and shapes presented in this work by just varying the growth time. A scale-up procedure is potentially feasible. Since crystalline Au-NPs are formed only in its vegetative cells and Eu-NPs only in its heterocysts, a separation process would be conceivable.

Supplementary Materials: The following supporting information can be downloaded at: https://www.mdpi.com/article/10.3390/nano13010130/s1, Section S1 Used Datasets: Figure S1 Vegetative cells of *Anabaena* sp. showing Au-NPs and respective heterocysts with Eu-NPs; Figure S2 HR-TEM images of Au-NPs after 24 h; Figure S3 HR-TEM images of Au-NPs after 54 h; Figure S4 HR-TEM images of Eu-NPs after 10 h; Figure S5 HR-TEM images of Eu-NPs after 244 h; Table S1 TEM image settings for the analysis of Au and Eu-NPs sizes; Section S2 Local particle size distributions (PSDs): Figure S6 Local PSDs for Eu-NPs of contact time 10 h; Figure S7 Local PSDs for Au-NPs of contact time 24 h; Figure S8 Local PSDs for Au-NPs of contact time 51 h.

Author Contributions: M.F.: conceptualization, data evaluation of HR-TEM, visualization, writing—original draft, review, and editing. S.K.: organization, project administration, review, and editing. X.C., G.Y. and Y.L.: TEM organization and measurements, data acquisition, project administration, review and editing. M.L. and S.W.: funding acquisition, supervision and overall guidance. C.B.F.: project administration, conceptualization, organization, visualization, writing, supervision, review, and editing. All authors have read and agreed to the published version of the manuscript.

Funding: This research was funded by the DAAD (PPP) Project ID: 57445545 and the CSC-DAAD program for the Project-Related Personal Exchange (grant agreement No. [2019]12039). All authors gratefully acknowledge the financial support.

Data Availability Statement: The raw data and processed data required to reproduce these findings are available on request of the authors.

Conflicts of Interest: The authors declare no conflict of interest.

References

1. Ghosh, S.; Ahmad, R.; Banerjee, K.; AlAjmi, M.F.; Rahman, S. Mechanistic Aspects of Microbe-Mediated Nanoparticle Synthesis. *Front. Microbiol.* **2021**, *12*, 1–12. [CrossRef] [PubMed]
2. Xie, J.; Ping, H.; Tan, T.; Lei, L.; Xie, H.; Yang, X.Y.; Fu, Z. Bioprocess-inspired fabrication of materials with new structures and functions. *Prog. Mater. Sci.* **2019**, *105*, 100571. [CrossRef]
3. Kratochvil, D.; Volesky, B. Advances in the biosorption of heavy metals. *Trends Biotechnol.* **1998**, *16*, 291–300. [CrossRef]
4. Vieira, R.H.S.F.; Volesky, B. Biosorption: A solution to pollution? *Int. Microbiol.* **2000**, *3*, 17–24. [CrossRef] [PubMed]
5. Cadogan, E.I.; Lee, C.H.; Popuri, S.R.; Lin, H.Y. Efficiencies of chitosan nanoparticles and crab shell particles in europium uptake from aqueous solutions through biosorption: Synthesis and characterization. *Int. Biodeterior. Biodegrad.* **2014**, *95*, 232–240. [CrossRef]
6. Gwenzi, W.; Makuvara, Z.; Marumure, J. Rare earth elements: Human exposure, risk factors, and health risks (Chapter 15). In *Emerging Contaminants in the Terrestrial-Aquatic-Atmosphere Continuum. Occurrence, Health Risks, and Mitigation*; Gwenzi, W., Ed.; Elsevier: Amsterdam, The Netherlands, 2022; pp. 273–290.
7. Jally, B.; Laubie, B.; Tang, Y.-T.; Simonnot, M.-O. Processing of Plants to Products: Gold, REEs and Other Elements. In *Agromining: Farming for Metals: Extracting Unconventional Resources Using Plants*, 2nd ed.; van der Ent, A., Baker, A.J.M., Echevarria, G., Simonnot, M.-O., Morel, J.L., Eds.; Springer International Publishing: Cham, Switzerland, 2021; pp. 63–74. [CrossRef]
8. Maleke, M.; Valverde, A.; Gomez-Arias, A.; Cason, E.D.; Vermeulen, J.G.; Coetsee-Hugo, L.; Swart, H.; van Heerden, E.; Castillo, J. Anaerobic reduction of europium by a Clostridium strain as a strategy for rare earth biorecovery. *Sci. Rep.* **2019**, *9*, 14339. [CrossRef] [PubMed]
9. Srivastava, R.R.; Ilyas, S.; Kim, H.; Choi, S.; Trinh, H.B.; Ghauri, M.A.; Ilyas, N. Biotechnological recycling of critical metals from waste printed circuit boards. *J. Chem. Technol. Biotechnol.* **2020**, *95*, 2796–2810. [CrossRef]
10. Forti, V.; Baldé, C.P.; Kuehr, R.; Bel, G. Surge in Global E-Waste, up 21 Per Cent in 5 Years. The Global E-waste Monitor 2020. 2020. Available online: https://api.globalewaste.org/publications/file/271/The-Global-E-waste-Monitor-2020-Quantities-flows-and-the-circular-economy-potential.zip (accessed on 1 December 2022).
11. Kammerlander, K.K.K.; Köhler, L.; Huittinen, N.; Bok, F.; Steudtner, R.; Oschatz, C.; Vogel, M.; Stumpf, T.; Brunner, E. Sorption of europium on diatom biosilica as model of a "green" sorbent for f-elements. *Appl. Geochem.* **2021**, *126*, 104823. [CrossRef]
12. Cadogan, E.I.; Lee, C.H.; Popuri, S.R. Facile synthesis of chitosan derivatives and *Arthrobacter* sp. biomass for the removal of europium(III) ions from aqueous solution through biosorption. *Int. Biodeterior. Biodegrad.* **2015**, *102*, 286–297. [CrossRef]
13. Serna, D.D.; Ortiz, M.; Rayson, G.D. Spectroscopic Comparison of Eu(III) Binding to Various Biosorbents. *Sep. Sci. Technol.* **2014**, *49*, 209–213. [CrossRef]
14. Kim, E.B.; Seo, J.M.; Kim, G.W.; Lee, S.Y.; Park, T.J. In vivo synthesis of europium selenide nanoparticles and related cytotoxicity evaluation of human cells. *Enzyme Microb. Technol.* **2016**, *95*, 201–208. [CrossRef] [PubMed]
15. Rösken, L.M.; Cappel, F.; Körsten, S.; Fischer, C.B.; Schönleber, A.; van Smaalen, S.; Geimer, S.; Beresko, C.; Ankerhold, G.; Wehner, S. Time-dependent growth of crystalline Au0-nanoparticles in cyanobacteria as self-reproducing bioreactors: 2. *Anabaena cylindrical*. *Beilstein J. Nanotechnol.* **2016**, *7*, 312–327. [CrossRef] [PubMed]
16. Rösken, L.M.; Körsten, S.; Fischer, C.B.; Schönleber, A.; van Smaalen, S.; Geimer, S.; Wehner, S. Time-dependent growth of crystalline Au0-nanoparticles in cyanobacteria as self-reproducing bioreactors: 1. *Anabaena* sp. *J. Nanopart. Res.* **2014**, *16*, 12–27. [CrossRef]
17. Naveed, S.; Li, C.; Lu, X.; Chen, S.; Yin, B. Technology Microalgal extracellular polymeric substances and their interactions with metalloids: A review. *Crit. Rev. Environ. Sci. Technol.* **2019**, *49*, 1769–1802. [CrossRef]
18. Dahoumane, S.A.; Yéprémian, C.; Djédiat, C.; Couté, A.; Fiévet, F.; Coradin, T.; Brayner, R. A global approach of the mechanism involved in the biosynthesis of gold colloids using micro-algae. *J. Nanopart. Res.* **2014**, *16*, 2607. [CrossRef]
19. Lenartowicz, M.; Marek, P.H.; Madura, I.D.; Lipok, J. Formation of Variously Shaped Gold Nanoparticles by *Anabaena laxa*. *J. Clust. Sci.* **2017**, *28*, 3035–3055. [CrossRef]
20. Khanna, P.; Kaur, A.; Goyal, D. Algae-based metallic nanoparticles: Synthesis, characterization and applications. *J. Microbiol. Methods.* **2019**, *163*, 105656. [CrossRef] [PubMed]
21. Dahoumane, S.A.; Yéprémian, C.; Djédiat, C.; Couté, A.; Fiévet, F.; Coradin, T.; Brayner, R. Improvement of kinetics, yield, and colloidal stability of biogenic gold nanoparticles using living cells of *Euglena gracilis* microalga. *J. Nanopart. Res.* **2016**, *18*, 79. [CrossRef]
22. Castro, L.; Blázquez, M.L.; González, F.; Muñoz, J.Á.; Ballester, A. Exploring the possibilities of biological fabrication of gold nanostructures using orange peel extract. *Metals* **2015**, *5*, 1609–1619. [CrossRef]
23. Ramrakhiani, L.; Ghosh, S. Metallic nanoparticle synthesised by biological route: Safer candidate for diverse applications. *IET Nanobiotechnol.* **2018**, *12*, 392–404. [CrossRef]
24. Jana, S. Advances in nanoscale alloys and intermetallics: Low temperature solution chemistry synthesis and application in catalysis. *Dalt. Trans.* **2015**, *44*, 18692–18717. [CrossRef] [PubMed]
25. Vetchinkina, E.; Loshchinina, E.; Kupryashina, M.; Burov, A.; Nikitina, V. Shape and Size Diversity of Gold, Silver, Selenium, and Silica Nanoparticles Prepared by Green Synthesis Using Fungi and Bacteria. *Ind. Eng. Chem. Res.* **2019**, *58*, 17207–17218. [CrossRef]

26. Klekotko, M.; Brach, K.; Olesiak-Banska, J.; Samoc, M.; Matczyszyn, K. Popcorn-shaped gold nanoparticles: Plant extract-mediated synthesis, characterization and multiphoton-excited luminescence properties. *Mater. Chem. Phys.* **2019**, *229*, 56–60. [CrossRef]
27. Anastopoulos, I.; Bhatnagar, A.; Lima, E.C. Adsorption of rare earth metals: A review of recent literature. *J. Mol. Liq.* **2016**, *221*, 954–962. [CrossRef]
28. Dahoumane, S.A.; Wujcik, E.K.; Jeffryes, C. Noble metal, oxide and chalcogenide-based nanomaterials from scalable phototrophic culture systems. *Enzyme Microb. Technol.* **2016**, *95*, 13–27. [CrossRef]
29. Rochert, A.S.; Rösken, L.M.; Fischer, C.B.; Schönleber, A.; Ecker, D.; van Smaalen, S.; Geimer, S.; Wehner, S. Bioselective synthesis of gold nanoparticles from diluted mixed Au, Ir, and Rh ion solution by Anabaena cylindrica. *J. Nanopart. Res.* **2017**, *19*, 355. [CrossRef]
30. Fischer, C.B.; Körsten, S.; Rösken, L.M.; Cappel, F.; Beresko, C.; Ankerhold, G.; Schönleber, A.; Geimer, S.; Ecker, D.; Wehner, S. Cyanobacterial promoted enrichment of rare earth elements europium, samarium and neodymium and intracellular europium particle formation. *RSC Adv.* **2019**, *9*, 32581–32593. [CrossRef]
31. Brayner, R.; Claude, Y.; Djediat, C.; Coradin, T.; Livage, J.; Fi, F.; Cout, A. Photosynthetic Microorganism-Mediated Synthesis of Akaganeite (β -FeOOH) Nanorods. *Langmuir* **2009**, *25*, 10062–10067. [CrossRef]
32. Nicolaisen, K.; Hahn, A.; Valdebenito, M.; Moslavac, S.; Samborski, A.; Maldener, I.; Wilken, C.; Valladares, A.; Flores, E.; Hantke, K.; et al. The interplay between siderophore secretion and coupled iron and copper transport in the heterocyst-forming cyanobacterium Anabaena sp. PCC 7120. *Biochim. Biophys. Acta-Biomembr.* **2010**, *1798*, 2131–2140. [CrossRef]
33. Dahoumane, S.A.; Mechouet, M.; Wijesekera, K.; Filipe, C.D.M.; Sicard, C.; Bazylinski, D.A.; Jeffryes, C. Algae-mediated biosynthesis of inorganic nanomaterials as a promising route in nanobiotechnology-a review. *Green Chem.* **2017**, *19*, 552–587. [CrossRef]
34. Henriques, B.; Morais, T.; Cardoso, C.E.D.; Freitas, R.; Viana, T.; Ferreira, N.; Fabre, E.; Pinheiro-Torres, J.; Pereira, E. Can the recycling of europium from contaminated waters be achieved through living macroalgae? Study on accumulation and toxicological impacts under realistic concentrations. *Sci. Total Environ.* **2021**, *786*, 147176. [CrossRef] [PubMed]
35. Reynolds, E.S. The use of lead citrate at high pH as an electron-opaque stain in electron microscopy. *J. Cell Biol.* **1963**, *17*, 208–212. [CrossRef] [PubMed]
36. Igathinathane, C.; Pordesimo, L.O.; Columbus, E.P.; Batchelor, W.D.; Methuku, S.R. Shape identification and particles size distribution from basic shape parameters using ImageJ. *Comput. Electron. Agric.* **2008**, *63*, 168–182. [CrossRef]
37. Fritz, M.; Körsten, S.; Chen, X.; Yang, G.; Lv, Y.; Liu, M.; Wehner, S.; Fischer, C.B. High-resolution particle size and shape analysis of the first Samarium nanoparticles biosynthesized from aqueous solutions via cyanobacteria Anabaena cylindrical. *NanoImpact* **2022**, *26*, 100398. [CrossRef]
38. Saravanan, A.; Kumar, P.S.; Karishma, S.; Vo, D.-V.N.; Jeevanantham, S.; Yaashikaa, P.R.; George, C.S. A review on biosynthesis of metal nanoparticles and its environmental applications. *Chemosphere* **2021**, *264*, 128580. [CrossRef]
39. Chaudhary, R.; Nawaz, K.; Khan, A.K.; Hano, C.; Abbasi, B.H.; Anjum, S. An overview of the algae-mediated biosynthesis of nanoparticles and their biomedical applications. *Biomolecules* **2020**, *10*, 1498. [CrossRef]
40. Koul, B.; Poonia, A.K.; Yadav, D.; Jin, J.O. Microbe-mediated biosynthesis of nanoparticles: Applications and future prospects. *Biomolecules* **2021**, *11*, 886. [CrossRef]
41. Roychoudhury, P.; Bhattacharya, A.; Dasgupta, A.; Pal, R. Biogenic synthesis of gold nanoparticle using fractioned cellular components from eukaryotic algae and cyanobacteria. *Phycol. Res.* **2016**, *64*, 133–140. [CrossRef]
42. Chakraborty, N.; Banerjee, A.; Lahiri, S.; Panda, A.; Gosh, A.N.; Pal, R. Biorecovery of gold using cyanobacteria and an eukaryotic alga with special reference to nanogold formation—A novel phenomenon. *J. Appl. Phycol.* **2009**, *21*, 145–152. [CrossRef]
43. Atalah, J.; Espina, G.; Blamey, L.; Muñoz-Ibacache, S.A.; Blamey, J.M. Advantages of using extremophilic bacteria for the biosynthesis of metallic nanoparticles and its potential for rare earth element recovery. *Front. Microbiol.* **2022**, *13*, 1–6. [CrossRef]

Disclaimer/Publisher's Note: The statements, opinions and data contained in all publications are solely those of the individual author(s) and contributor(s) and not of MDPI and/or the editor(s). MDPI and/or the editor(s) disclaim responsibility for any injury to people or property resulting from any ideas, methods, instructions or products referred to in the content.

Article

The Adsorption Effect of Methane Gas Molecules on Monolayer PbSe with and without Vacancy Defects: A First-Principles Study

Xing Zhou and Yuliang Mao *

Hunan Key Laboratory for Micro–Nano Energy Materials and Devices, School of Physics and Optoelectronic, Xiangtan University, Xiangtan 411105, China
* Correspondence: ylmao@xtu.edu.cn

Abstract: In this paper, the adsorption effect of methane (CH_4) gas molecular on monolayer PbSe with and without vacancy defects is studied based on first-principles calculations. The effects of the adsorption of methane molecular on monolayer PbSe and on the Se vacancy (V_{Se}) and Pb vacancy (V_{Pb}) of monolayer PbSe are also explored. Our results show that methane molecules exhibit a good physical adsorption effect on monolayer PbSe with and without vacancy defects. Moreover, our simulations indicate that the adsorption capacity of CH_4 molecules on monolayer PbSe can be enhanced by applying strain. However, for the monolayer PbSe with Vse, the adsorption capacity of CH_4 molecules on the strained system decreases sharply. This indicates that applying strain can promote the dissociation of CH_4 from V_{Se}. Our results show that the strain can be used as an effective means to regulate the interaction between the substrate material and the methane gas molecules.

Keywords: methane; monolayer PbSe; first principles; strain

Citation: Zhou, X.; Mao, Y. The Adsorption Effect of Methane Gas Molecules on Monolayer PbSe with and without Vacancy Defects: A First-Principles Study. *Nanomaterials* 2023, 13, 1566. https://doi.org/10.3390/nano13091566

Academic Editor: Thomas Dippong

Received: 15 April 2023
Revised: 28 April 2023
Accepted: 29 April 2023
Published: 6 May 2023

Copyright: © 2023 by the authors. Licensee MDPI, Basel, Switzerland. This article is an open access article distributed under the terms and conditions of the Creative Commons Attribution (CC BY) license (https://creativecommons.org/licenses/by/4.0/).

1. Introduction

Methane is a colorless, odorless gas that is the main ingredient in natural gas, biogas, etc. It has wide applications for fuel and as a feedstock for synthesizing other substances, such as carbon monoxide and hydrogen. In environmental science, methane has a more significant impact on the Earth's greenhouse effect than carbon dioxide, and there is a risk of suffocation in environments with high methane concentrations. Therefore, exploring a suitable material for methane gas detection and capture is essential to protect the Earth's environment and personal safety.

As the first two-dimensional (2D) material discovered, graphene has attracted broad interest due to its unique properties. Two-dimensional materials, due to their structural characteristics, have many potential applications, such as photodetectors [1–4], field effect transistors [5], solar cells [6–8], and gas sensors [9–11]. In recent years, more attention has been paid to improving the performance of 2D materials for the applications as gas sensors, such as studies of doping [12–15], defects [16,17], construction of heterojunctions [18,19], application of external electric fields [20,21] and strain [22–25]. These methods can effectively tune the interaction between substrate materials and gas molecules.

Among 2D materials, 2D group IV–VI monochalcogenides have attracted much attention due to their unique orthogonal structures, which are like black phosphorus folds. For example, 2D GeSe has attracted significant attention in the study of its application in sensors due to its smooth surface state, good stability, and anisotropic structure [26–28]. Recent studies have shown that 2D SnSe has good prospects in near-infrared detectors, high-performance supercapacitors, and solar cells [29–35]. PbSe, a member of group IV-VI monochalcogenides, was predicted to be used in diodes, infrared lasers, and sensors [36–38]. There are currently several methods experimentally used to prepare PbSe. For example, it has been reported that a 2D PbSe semiconductor with large transverse size and ultra-thin thickness was successfully synthesized by van der Waals epitaxy technology [39]. Various lead selenide (PbSe) nanostructures were also prepared using aqueous solutions of $Pb(NO_3)_2$ and NaHSe by varying the molar ratio of Pb and Se and the mixing sequence of NH_4OH with $Pb(NO_3)_2$ or NaHSe [40,41]. The above methods can produce different shapes of PbSe nanomaterials. A recent study reported that hexagonal PbSe nanostructures can be observed at the interface between gas and liquid, while 2D PbSe superlattices with orthogonal structures can be further synthesized from the formed hexagonal structures [42]. Although different phases of PbSe crystals were prepared experimentally, according to the existing theoretical analysis and research, the hexagonal and orthogonal structures of the monolayer PbSe exhibit good stability at room temperature. Furthermore, it was reported that the orthogonal structure of PbSe is relatively lower in energy than that of the hexagonal PbSe [43,44]. The adsorption of toxic gases (SO_2 and Cl_2) on hexagonal PbSe with a single layer, multiple layer, and doped cases was reported [45,46]. In this paper, we mainly choose the relatively more stable orthogonal structure of 2D PbSe as the substrate for the adsorption study. As mentioned above, although there are many related studies on the preparation of PbSe and its application potential in terms of various aspects, the research on gas adsorption of PbSe with a 2D orthogonal structure is still lacking.

In this paper, by using first-principles calculations based on density functional theory (DFT), the adsorption effects of methane gas molecules on 2D PbSe (P-PbSe) with and without vacancy defects (Pb defect: VPb, or Se defect: VSe) are explored. The results show that both P-PbSe and PbSe with Se atomic vacancy (VSe) exhibit significant adsorption effects on CH4. In addition, strain control is a valid method for regulating two-dimensional materials, and there are a large number of related studies on the theoretical calculations of gas adsorption on two-dimensional materials [47–49]. Some studies have reported the effect of strain on the electronic and optical properties of PbSe with a 2D orthogonal structure. In this paper, we mainly aim to explore the adsorption of methane molecules on 2D PbSe after the biaxial strain is applied. To ensure that the structure of the calculated material is not distorted under biaxial strain, only the strain in a range from −5% to 5% is considered in our study (where a negative value indicates compressive strain and a positive value indicates tensile strain). In previous reports on gas adsorption research, for substrate materials, only the structure subject to defects or the pristine substrate structure under strain was considered to explore the changes in their adsorption performance. However, research on the application of external strain to the structure under defect conditions in gas adsorption has not yet been explored. Therefore, in this article, the adsorption of methane molecules on the optimal sites of P-PbSe and VSe after applying biaxial strain is systematically studied.

2. Computational Method and Model

The first-principles calculations are performed based on DFT. The geometric structure and electronic properties of the system are simulated using the 5.44 version of Vienna ab initio simulation package (VASP.5.4.4) [50,51]. The exchange-correlation interaction between electrons was described by the Perdew–Burke–Ernzerhof (PBE) function in Generalized Gradient Approximation (GGA) [52,53]. The cutoff energy for the expanding plane wave function was set to 500 eV. The convergence criterion in energy and force was set to 10^{-6} eV and 0.01 eV/Å, respectively. In the integration of the Brillouin zone (BZ), the $7 \times 7 \times 1$ K-grid mesh was used. The DFT-D3 method was applied to describe the van der Waals forces between substrate materials and gas molecules [54,55]. A vacuum layer of 25 Å was set along the z-axis to ensure there is no interaction between the adjacent layers. Strain can affect the bonding between atoms in the material. When strain is applied externally, the lattice constants are changed due to the rearrangement of atoms in the material, and the atomic spacing of the crystal is also changed. Under compressive strain, the lattice constant will decrease. Thus, the distance between the atoms is reduced, and the atoms are arranged more closely. In contrast, under tensile strain, the lattice constant will be increased. Correspondingly, the distance between atoms will become larger, which leads to the weakening of the interaction force between atoms and bonds. In this case, the crystal structure becomes looser. Considering the rationality of the configurations, we calculated the molecular dynamics of the systems under different strains. Ab initio molecular dynamics (AIMD) simulations were performed to verify the thermodynamic stability of the relaxed configurations. The NVT ensemble was chosen in AIMD simulations under 300 K. A 3 × 1 × 1 supercell was used to predict the thermodynamic stability of the system under different strains. The optimized adsorption energy (Ead) of CH_4 molecular adsorbed on the PbSe substrate is defined as:

$$E_{ad} = E_{tol} - E_p - E_{gas}$$

where Etol represents the total energy of the adsorbed configurations, and E_p and E_{gas} represent the total energy of the PbSe monolayer and the gas molecules, respectively.

The biaxial strain was applied to the PbSe substrate, and the strain variability is defined as:

$$\frac{n - n_0}{n_0} \times 100\%$$

where n and n_0 are the lattice constants of the system with and without strain, respectively.

3. Results and Discussion

3.1. Adsorption of Methane Molecules on Pristine 2D PbSe

In order to ensure the rationality of the calculated structures, the structure of the unit cell of PbSe is firstly optimized. In unit cell, the optimized lattice constants are a = b = 4.40 Å. A 3 × 3 × 1 supercell containing 18 Pb atoms and 18 Se atoms was used as the adsorbent substrate material in our lateral calculations with the optimized lattice constant a = b = 13.205 Å. The band structure and density of states (DOS) of the optimized system were calculated. The results show that PbSe is a direct bandgap semiconductor with a bandgap of 1.28 eV, which is consistent with previous reports [56,57]. Methane gas molecules have a regular tetrahedral structure, so only one configuration is considered in the study of the adsorption configuration, i.e., the adsorption configuration of the bonding between the C atom and the H atom in the methane molecule is perpendicular to the PbSe surface. The top view and side view of the optimized structure of PbSe after cell expansion are shown in Figures 1a and 1b, respectively. Four different adsorption sites are proposed, which are named I (the top site of the Se atom, named the TSe site), II (located directly above the bonding between the Pb atom and Se atom, named the Bridge site), III (the top site of the Pb atom, named the T_{Pb} site), and IV (hollow site) corresponding to the different high symmetry adsorption sites of methane molecules on the P-PbSe substrate, as shown in Figure 1c.

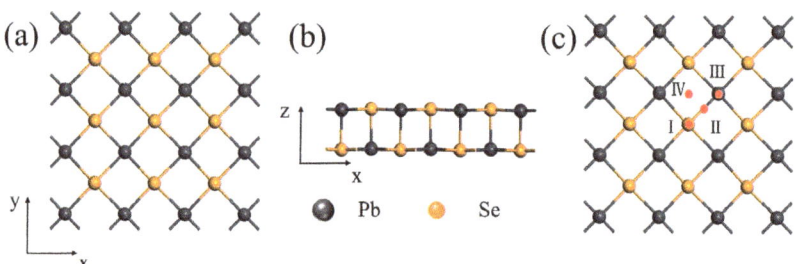

Figure 1. (**a**,**b**) represent the top and side views of the single-layer PbSe, respectively. (**c**) represents four different adsorption sites of P-PbSe (I: TSe site, II: Bridge site, III: TPb site, IV: Hollow site); the gray and green balls represent Pb and Se atoms, respectively.

In order to obtain the optimal adsorption site of methane molecules adsorption on 2D PbSe, the adsorption energies of methane molecules on four different sites of P-PbSe were calculated, and the results are listed in Table 1. Based on the above-mentioned formula of adsorption energy, the larger the absolute value of the adsorption energy E_{ad}, the stronger the adsorption capacity. By comparing the E_{ad} listed in Table 1, the proposed site for methane molecule adsorption on P-PbSe is the T_{Pb} site. As can be found from the data in Table 1, the difference in the adsorption energies of methane molecules on the four sites of P-PbSe or V_{Se} is not very obvious, which means that methane molecules have almost the same adsorption stability on different sites. Therefore, only the adsorption sites with the lowest adsorption energy are considered for further analysis in the subsequent discussion.

Table 1. Adsorption energies of methane molecule adsorption on P-PbSe and PbSe with V_{Se}.

Site	T_{Pb}	T_{Se}	Bridge	Hollow
P-PbSe (eV)	−4.8259	−4.7887	−4.8158	−4.8177
V_{Se}-PbSe (eV)	−4.6776	−4.6506	−4.6467	−4.6885

Therefore, the T_{Pb} site is mainly considered for the adsorption of methane molecules on 2D P-PbSe substrate. The top view and side view of the relaxed structure of methane molecules adsorbed on the T_{Pb} site are shown in Figure 2a. The distance from the carbon atom in the methane molecule to the nearest atom (Pb) on the substrate plane is 2.639 Å. The band structure with and without the adsorption of methane molecules on P-PbSe substrate is shown in Figure 2c, and the electronic density of states after adsorption is shown in Figure 2d. The charge of each atom of a single methane molecule and methane molecules adsorbed on T_{Pb} sites, and the electron transfer of each atom in methane molecules after adsorption, are listed in Table 2. As shown in Figure 2c, the variation in the band gap in the configurations with and without methane molecules is not remarkable. In Figure 2d, it can be seen that methane molecules do not obviously contribute to the states of valence and conduction bands. Combined with the charge transfer in Table 2, it can be concluded that physical adsorption of methane molecules occurs on P-PbSe, which mainly depends on the van der Waals (vdW) interaction between the substrate and methane molecules.

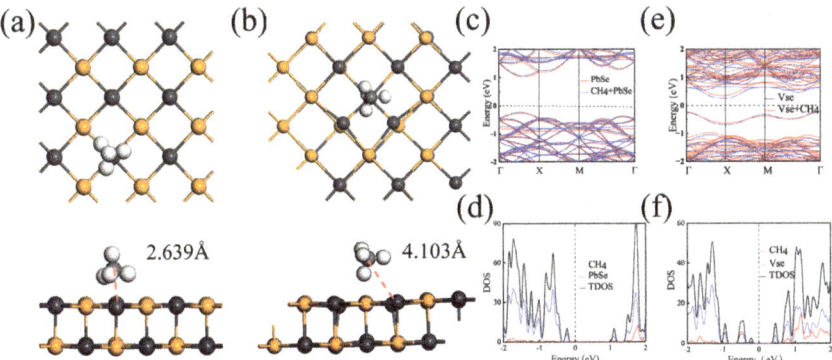

Figure 2. (**a**,**b**) are the configurations of CH$_4$ molecules adsorbed on the sites of P-PbSe and V_{Se}, respectively. (**c**,**d**) and (**e**,**f**) are the corresponding band structure and electronic density of states (DOS), respectively.

Table 2. The Bader charge of each atom in a single methane molecule (CH_4) and methane molecules adsorbed on TPb sites (CH_4-T_{Pb}). The charge transfer of each atom of methane molecules is also listed (positive indicates electron gain, negative indicates electron loss).

Atom	CH_4 (e)	CH_4-T_{Pb} (e)	Transfer (e)
C	4.09581	4.115225	+0.019415
H1	0.967049	0.985145	+0.018096
H2	0.967049	0.955066	−0.011983
H3	0.992434	0.988498	−0.003936
H4	0.977659	0.967289	−0.01037

3.2. Adsorption of Methane Molecules on V_{Se}

When considering the vacancy in monolayer PbSe, two cases are considered. One is Pb atoms vacancy (V_{Pb}) and the other is Se atoms vacancy (V_{Se}). Their configurations are shown in Figures 3a and 3b, respectively.

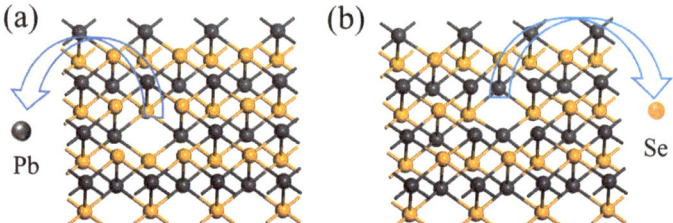

Figure 3. (**a**,**b**) denote vacancy configurations of Pb atoms and Se atoms, respectively.

To gain a further understanding of the two vacancy cases, the energies of the two vacancy configurations are calculated separately. The results show that the energy of the V_{Se} configuration (−139.49 eV) is lower than that of the V_{Pb} configuration (−139.01 eV). From the perspective of structural stability, a more stable V_{Se} configuration is considered for the lateral discussion in this paper.

The top and side views of the V_{Se} optimized configurations are shown in Figures 4a and 4b, respectively. We also consider four different adsorption sites. As shown in Figure 4c, a is the top site of the Pb atom (T_{Pb}), while b is the Bridge site between the Pb atom and Se atom (Bridge). c is the top site of the Se atom (T_{Se}), while h is the site above the Se atom vacancy (Hollow). The adsorption energies of the four sites were calculated, and the results are listed in Table 1. By comparison of the adsorption energies, it is found that the optimal adsorption site of CH_4 on V_{Se} is the Hollow site, and the adsorption energy is −4.6885 eV. It should be noted that the adsorption energies between the four adsorption sites in the V_{Se} system have only minor differences. Usually, only the site with lowest adsorption energy among them is selected as a representative for further calculation and analysis. The top view and side view of the optimized structure are shown in Figure 2b. From our analysis, the distance from the carbon atom in the methane molecule to the nearest atom (Pb) on the substrate plane is 4.103 Å.

In order to further explore the charge transfer of the configurations after adsorption, Bader charge analysis was carried out to analyze the charge of each atom of a single methane molecule adsorbed on the Hollow site of the V_{Se}, and the electron transfer of each atom in methane molecules after adsorption is listed in Table 3. It can be seen that only 0.03 e is transferred from the V_{Se} configuration to CH_4 molecules after adsorption. Combined with the band structure in Figure 2e and the electronic density of states in Figure 2f, it can be concluded that the adsorption of a single methane molecule has a minor influence on the band gap and electronic state of the system, which indicates that the adsorption type between V_{Se} and methane molecules is physical adsorption.

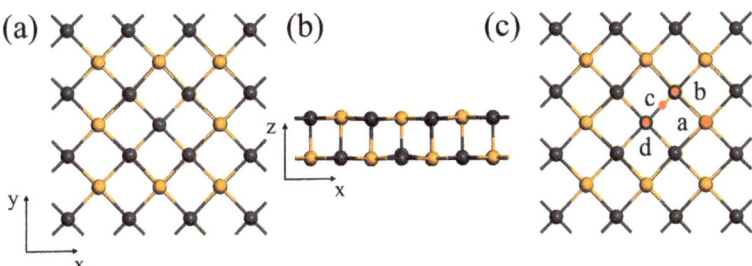

Figure 4. (**a**,**b**) represent the top and side views of optimized Vse configuration. (**c**) Four different sites of Se atomic vacancy (a: T_{Se} site, b: Bridge site, c: T_{Pb} site, d: Hollow site).

Table 3. The charge of each atom of a single methane molecule (CH_4) and methane molecules adsorbed on the Hollow site of the vacancy of the Se atom (CH_4-Hollow). The charge transfer of each atom of the methane molecule (positive indicates electron gain, negative indicates electron loss).

Atom	CH_4 (e)	CH_4-Hollow (e)	Transfer (e)
C	4.09581	4.110206	+0.014396
H1	0.967049	0.980938	+0.013889
H2	0.967049	0.979328	+0.012279
H3	0.992434	0.961884	−0.03055
H4	0.977659	0.999377	+0.021718

3.3. Adsorbtion Methane Molecules on the PbSe Monolayer under Biaxial Strain

When strain is applied to P-PbSe and V_{Se} configurations, the thermodynamic stability of the structure under strain is checked first to ensure the rationality of the explored structure. The variation in free energies in the studied configurations of P-PbSe and V_{Se} under applied strain from −5% compressive strain to 5% tensile strain are shown in Figure 5. It can be seen that the free energies of P-PbSe and V_{Se} only fluctuate in a minimal interval under different strains, which indicates that the structure is stable from −5% compressive strain to 5% tensile strain.

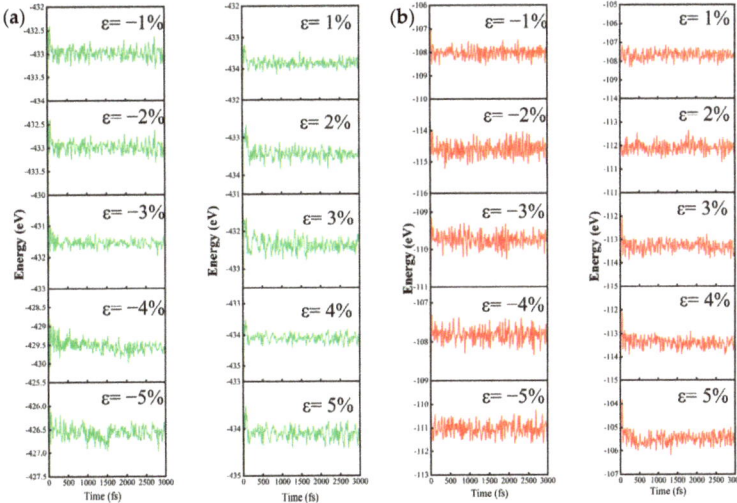

Figure 5. The free energy fluctuations of P-PbSe (**a**) and V_{Se} (**b**) at room temperature after molecular dynamics simulation under different strains.

In order to explore the effect of strain applied to the substrate material on the adsorption energy of gas molecules, the adsorption energy of CH$_4$ molecules on P-PbSe and V$_{Se}$ under different strains is indicated in Figure 6a. It is shown that the strain has an effective control effect on the adsorption capacity of the substrate material. It can be found that when the strain is applied, the adsorption energy of P-PbSe is decreased, but as the strain gradually increases, the adsorption energy of P-PbSe is increased with the increase of the biaxial strain. When the tensile strain of 4% is applied, the adsorption capacity of CH$_4$ molecules is significantly improved when compared with the case without strain. This implies that the adsorption capacity of CH$_4$ molecules on P-PbSe can be enhanced by applying strain, which is beneficial to its application in the capture of CH$_4$ molecules. However, for the V$_{Se}$ system, the adsorption capacity of CH$_4$ molecules on the strained system decreases sharply. This indicates that applying strain can promote the dissociation of CH$_4$ from V$_{Se}$. In other words, the strain can be used as an effective means to regulate the interaction between the substrate material and the methane gas molecules.

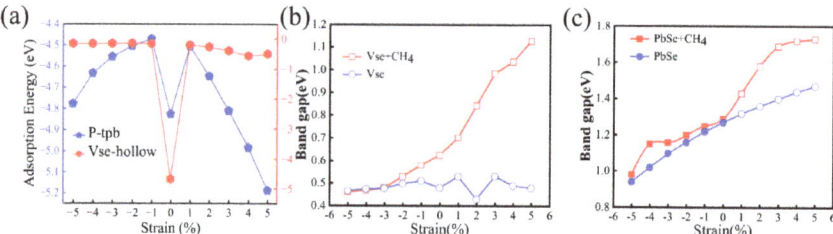

Figure 6. (a) The adsorption energy of P-PbSe (blue part) and V$_{Se}$ (red part) for methane molecules under different strains; (b,c) show the change in the band gap of V$_{Se}$ and P-PbSe with and without adsorbed methane molecules as a function of the response variable strain, respectively.

To further explore the effect of strain on the electronic structure of the adsorbed system, we calculated the band structure of the configurations with and without adsorption of CH$_4$ molecules on P-PbSe and V$_{Se}$ configurations under different strain. The results are shown in Figures 6b and 6c, respectively. It can be seen from Figure 6b that strain does not change the band type of the V$_{Se}$ configuration. However, for the system with adsorbed methane molecules, the band gap is gradually increased with the increase of strain from −5% to 5%. It can be seen that the change in the band gap of the system with and without adsorption under compressive strain is not significant. Especially at strains ranging from −5% to −3%, the effect of strain on its band gap is almost consistent. However, at strains ranging from −2% to 5%, the band gap of the adsorbed system gradually increases, while for the V$_{Se}$ system, the band gap fluctuates within a certain range. However, the band gap of methane molecule adsorption on P-PbSe increases under the applied strain from −5% to 5%. Furthermore, the type of the band structure in these cases changes from direct to indirect during the change from compressive to tensile strain. Although the band gap varies under the strain ranging from −5% to 5%, the types of band gap in the P-PbSe system with and without adsorbing methane molecules are unchanged. Unlike V$_{Se}$, there is a significant difference in the band gap between the system with and without adsorption at strains ranging from −5% to −3%, while at strains ranging from −2% to −1%, the band gap distinction is negligible. However, there is a significant discrepancy in the response of the band gap to strain between the two under 1% to 5% strain. For the V$_{Se}$ configuration and P-PbSe, the band gap of the system after the adsorption of methane molecules has a sensitive response to the strain, and the band gap of the system is changed more significantly, especially under tensile strain.

The band structures of the systems with and without the adsorption of CH$_4$ molecules on P-PbSe and V$_{Se}$ configurations under selected strains are shown in Figure 7. It can be seen from the figure that, whether in the P-PbSe or V$_{Se}$ system, the compressive strain has no noticeable effect on the band structure of the system adsorption of methane molecules. On the contrary, the tensile strain significantly changes the electronic states in the conduction band minimum (CBM) of the adsorbed system, especially for the adsorption of methane molecules on the V_{Se} system.

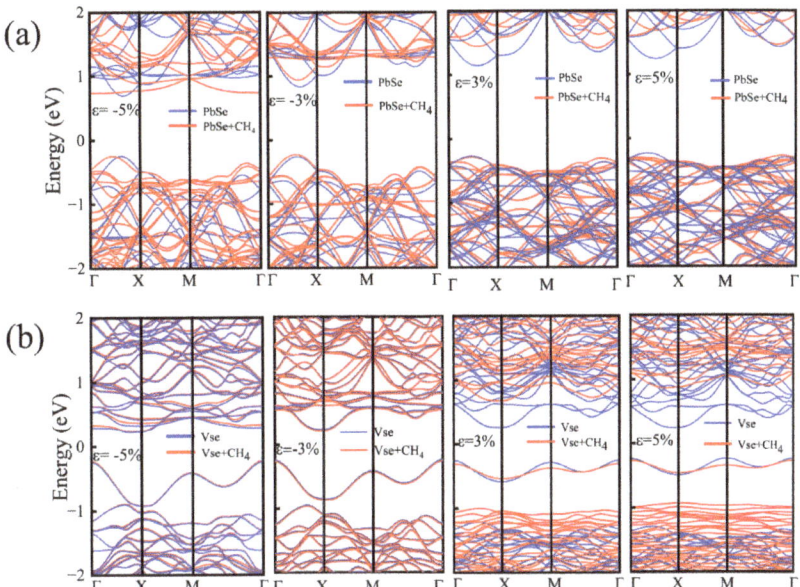

Figure 7. (a,b) are the band structures of P-PbSe and V_{Se} configurations with and without adsorption of methane molecules under 3% and 5% compressive and tensile strains, respectively

The results of the change in adsorption energy, the change in band gap, and the charge transfer between the substrate material and methane molecules of P-PbSe and V_{Se} systems under different strains are listed in Tables 4 and 5, respectively. From the data in the table, it can be seen that strain has a significant regulatory effect on the band gap and adsorption energy of the systems after adsorption of CH_4 molecules, but its role in charge transfer is very limited.

Table 4. Variation in adsorption energy (E_{ad}), band gap (E_{gap}), and charge transfer after adsorption of methane molecules on P-PbSe at different strains. Positive charge transfer means the methane molecule gained electrons.

Strain (%)	E_{ad} (eV)	E_{gap} (eV)	Transfer (e)
−5	−4.767	0.99	0.0175
−4	−4.624	1.15	0.0131
−3	−4.555	1.16	0.0123
−2	−4.504	1.21	0.0107
−1	−4.471	1.26	0.0104
0	−4.826	1.29	0.0112
1	−4.505	1.43	0.0118
2	−4.647	1.58	0.0113
3	−4.81	1.7	0.0108
4	−4.984	1.72	0.0108
5	−5.187	1.73	0.0104

Table 5. Variation in adsorption energy (E_{ad}), band gap (E_{gap}), and charge transfer amount after adsorption of methane molecules on V_{Se} under different strains. Positive charge transfer means the methane molecule gained electrons, respectively.

Strain (%)	E_{ad} (eV)	E_{gap} (eV)	Transfer (e)
−5	−0.161	0.46	0.028
−4	−0.161	0.47	0.028
−3	−0.157	0.48	0.029
−2	−0.15	0.53	0.031
−1	−0.155	0.58	0.031
0	−4.689	0.63	0.03
1	−0.202	0.7	0.03
2	−0.265	0.84	0.029
3	−0.394	0.99	0.029
4	−0.563	1.04	0.228
5	−0.498	1.13	0.027

4. Conclusions

In this paper, based on first-principles calculations, we calculate the adsorption energy, band structure, and Bader charge of CH_4 molecules' adsorption on P-PbSe and V_{Se} substrate materials. The results show that CH_4 molecules exhibit physical adsorption effects on both P-PbSe and V_{Se} substrate. The effect of biaxial strain on the adsorption of CH_4 molecules on P-PbSe and V_{Se} was calculated. Our analysis of band structure reveals that the systems with and without adsorption of CH_4 molecules have a prominent difference. The changes in the band gap of the adsorbed system are more sensitive to strain. However, the effect of strain on charge transfer is very limited. Our study indicates that strain can be used as an effective method to regulate the electronic structure of methane molecules adsorbed on 2D PbSe nanomaterial. Comparing the calculation results of P-PbSe and V_{Se} under strain, it is found that V_{Se} and P-PbSe have significantly different responses to methane molecule adsorption under the same strain. Therefore, applying strain to a substrate material such as 2D PbSe is an effective way to regulate the gas adsorption performance.

Author Contributions: X.Z.: investigation, data curation, writing—original draft preparation. Y.M.: review and editing, supervision, project administration, funding acquisition. All authors have read and agreed to the published version of the manuscript.

Funding: This research is funded by Natural Science Foundation of Hunan Province, China (Grant No. 2021JJ30650) and the research innovation project of postgraduate student in Hunan province (Grant No. QL20210142).

Data Availability Statement: The data used in this study are available from the corresponding author by request.

Conflicts of Interest: The authors declare no conflict of interest.

References

1. Wang, J.; Han, J.; Chen, X.; Wang, X. Design strategies for two-dimensional material photodetectors to enhance device performance. *InfoMat* **2019**, *1*, 33–53. [CrossRef]
2. Kumar, A.; Khan, M.A.; Kumar, M. Recent advances in UV photodetectors based on 2D materials: A review. *J. Phys. D Appl. Phys.* **2021**, *55*, 133002. [CrossRef]
3. Liu, C.; Guo, J.; Yu, L.; Li, J.; Zhang, M.; Li, H.; Shi, Y.; Dai, D. Silicon/2D-material photodetectors: From near-infrared to mid-infrared. *Light Sci. Appl.* **2021**, *10*, 123. [CrossRef] [PubMed]
4. Yan, F.; Wei, Z.; Wei, X.; Lv, Q.; Zhu, W.; Wang, K. Toward High-Performance Photodetectors Based on 2D Materials: Strategy on Methods. *Small Methods* **2018**, *2*, 1700349. [CrossRef]
5. Gao, F.; Yang, H.; Hu, P. Interfacial Engineering for Fabricating High-Performance Field-Effect Transistors Based on 2D Materials. *Small Methods* **2018**, *2*, 1700384. [CrossRef]
6. Ahmad, S.; Fu, P.; Yu, S.; Yang, Q.; Liu, X.; Wang, X.; Wang, X.; Guo, X.; Li, C. Dion-Jacobson Phase 2D Layered Perovskites for Solar Cells with Ultrahigh Stability. *Joule* **2019**, *3*, 794–806. [CrossRef]

7. Shao, M.; Bie, T.; Yang, L.; Gao, Y.; Jin, X.; He, F.; Zheng, N.; Yu, Y.; Zhang, X. Over 21% Efficiency STable 2D Perovskite Solar Cells. *Adv. Mater.* **2022**, *34*, 2107211. [CrossRef]
8. Fu, W.; Wang, J.; Zuo, L.; Gao, K.; Liu, F.; Ginger, D.S.; Jen, A.K.-Y. Two-Dimensional Perovskite Solar Cells with 14.1% Power Conversion Efficiency and 0.68% External Radiative Efficiency. *ACS Energy Lett.* **2018**, *3*, 2086–2093. [CrossRef]
9. Kim, J.-H.; Mirzaei, A.; Kim, H.W.; Kim, S.S. Flexible and low power CO gas sensor with Au-functionalized 2D WS_2 nanoflakes. *Sens. Actuators B Chem.* **2020**, *313*, 128040. [CrossRef]
10. Buckley, D.J.; Black, N.C.G.; Castanon, E.G.; Melios, C.; Hardman, M.; Kazakova, O. Frontiers of graphene and 2D material-based gas sensors for environmental monitoring. *2D Mater.* **2020**, *7*, 32002. [CrossRef]
11. Tang, H.; Sacco, L.N.; Vollebregt, S.; Ye, H.; Fan, X.; Zhang, G. Recent advances in 2D/nanostructured metal sulfide-based gas sensors: Mechanisms, applications, and perspectives. *J. Mater. Chem. A* **2020**, *8*, 24943–24976. [CrossRef]
12. Rad, A.S. First principles study of Al-doped graphene as nanostructure adsorbent for NO_2 and N_2O: DFT calculations. *Appl. Surf. Sci.* **2015**, *357*, 1217–1224. [CrossRef]
13. Jelmy, E.J.; Thomas, N.; Mathew, D.T.; Louis, J.; Padmanabhan, N.T.; Kumaravel, V.; John, H.; Pillai, S.C. Impact of structure, doping and defect-engineering in 2D materials on CO_2 capture and conversion. *React. Chem. Eng.* **2021**, *6*, 1701–1738. [CrossRef]
14. Son, J.; Hashmi, A.; Hong, J. Manipulation of n and p type dope black phosphorene layer: A first principles study. *Curr. Appl. Phys.* **2016**, *16*, 506–514. [CrossRef]
15. Liu, H.; Qu, M.; Du, A.; Sun, Q. N/P-Doped MoS_2 Monolayers as Promising Materials for Controllable CO_2 Capture and Separation under Reduced Electric Fields: A Theoretical Modeling. *J. Phys. Chem. C* **2021**, *126*, 203–211. [CrossRef]
16. Ersan, F.; Gökçe, A.G.; Aktürk, E. Point defects in hexagonal germanium carbide monolayer: A first-principles calculation. *Appl. Surf. Sci.* **2016**, *389*, 1–6. [CrossRef]
17. Lin, L.; Chen, R.; Huang, J.; Wang, P.; Zhu, L.; Yao, L.; Hu, C.; Tao, H.; Zhang, Z. Adsorption of CO, H2S and CH4 molecules on SnS_2 monolayer: A first-principles study. *Mol. Phys.* **2021**, *119*, e1856429. [CrossRef]
18. Zheng, H.; Meng, X.; Chen, J.; Que, M.; Wang, W.; Liu, X.; Yang, L.; Zhao, Y. In Situ phase evolution of TiO_2/Ti_3C_2T heterojunction for enhancing adsorption and photocatalytic degradation. *Appl. Surf. Sci.* **2021**, *545*, 149031. [CrossRef]
19. Zhang, R.; Jian, W.; Yang, Z.-D.; Bai, F.-Q. Insights into the photocatalytic mechanism of the C_4N/MoS_2 heterostructure: A first-principle study. *Chin. Chem. Lett.* **2020**, *31*, 2319–2324. [CrossRef]
20. Jin, Y.; Ding, J.; Chen, H.; Fu, H.; Peng, J. Enhanced adsorption properties of ZnO/GaN heterojunction for CO and H_2S under external electric field. *Comput. Theor. Chem.* **2021**, *1206*, 113495. [CrossRef]
21. Pham, K.D.; Dinh, P.C.; Diep, D.V.; Luong, H.L.; Vu, T.V.; Hoang, D.-Q.; Khyzhun, O.Y.; Ngoc, H.V. Effects of electric field and biaxial strain on the (NO_2, NO, O_2, and SO_2) gas adsorption properties of Sc_2CO_2 monolayer. *Micro Nanostructures* **2022**, *163*, 107135. [CrossRef]
22. Zhao, Z.; Liu, C.; Tsai, H.-S.; Zhou, J.; Zhang, Y.; Wang, T.; Ma, G.; Qi, C.; Huo, M. The strain and transition metal doping effects on monolayer Cr_2O_3 for hydrogen evolution reaction: The first principle calculations. *Int. J. Hydrogen Energy* **2022**, *47*, 37429–37437. [CrossRef]
23. Zhang, H.-P.; Kou, L.; Jiao, Y.; Du, A.; Tang, Y.; Ni, Y. Strain engineering of selective chemical adsorption on monolayer black phosphorous. *Appl. Surf. Sci.* **2020**, *503*, 144033. [CrossRef]
24. Zheng, X.; Ban, S.; Liu, B.; Chen, G. Strain-controlled graphdiyne membrane for CO_2/CH_4 separation: First-principle and molecular dynamic simulation. *Chin. J. Chem. Eng.* **2020**, *28*, 1898–1903. [CrossRef]
25. Yu, X.-F.; Li, Y.-C.; Liu, Z.-B.; Li, Q.-Z.; Li, W.-Z.; Yang, X.; Xiao, B. Monolayer Ti_2CO_2: A promising candidate for NH_3 sensor or capturer with high sensitivity and selectivity. *ACS Publ.* **2015**, *7*, 13707–13713. [CrossRef]
26. Gui, Y.; Liu, Z.; Ji, C.; Xu, L.; Chen, X. Adsorption behavior of metal oxides (CuO, NiO, Ag_2O) modified GeSe monolayer towards dissolved gases (CO, CH_4, C_2H_2, C_2H_4) in transformer oil. *J. Ind. Eng. Chem.* **2022**, *112*, 134–145. [CrossRef]
27. Fan, Q.; Zhang, W.; Qing, H.; Yang, J. Exceptional Thermoelectric Properties of Bilayer GeSe: First Principles Calculation. *Materials* **2022**, *15*, 971. [CrossRef]
28. Yang, Y.; Liu, S.-C.; Wang, Y.; Long, M.; Dai, C.-M.; Chen, S.; Zhang, B.; Sun, Z.; Sun, Z.; Hu, C.; et al. In-Plane Optical Anisotropy of Low-Symmetry 2D GeSe. *Adv. Opt. Mater.* **2019**, *7*, 1801311. [CrossRef]
29. Ma, X.-H.; Cho, K.-H.; Sung, Y.-M. Growth mechanism of vertically aligned SnSe nanosheets via physical vapour deposition. *CrystEngComm* **2014**, *16*, 5080–5086. [CrossRef]
30. Zheng, D.; Fang, H.; Long, M.; Wu, F.; Wang, P.; Gong, F.; Wu, X.; Ho, J.C.; Liao, L.; Hu, W. High-Performance Near-Infrared Photodetectors Based on p-Type SnX (X = S, Se) Nanowires Grown via Chemical Vapor Deposition. *ACS Nano* **2018**, *12*, 7239–7245. [CrossRef]
31. Jagani, H.S.; Gupta, S.U.; Bhoraniya, K.; Navapariya, M.; Pathak, V.M.; Solanki, G.K.; Patel, H. Photosensitive Schottky barrier diodes based on Cu/p-SnSe thin films fabricated by thermal evaporation. *Mater. Adv.* **2022**, *3*, 2425–2433. [CrossRef]
32. Yan, J.; Deng, S.; Zhu, D.; Bai, H.; Zhu, H. Self-powered SnSe photodetectors fabricated by ultrafast laser. *Nano Energy* **2022**, *97*, 107188. [CrossRef]
33. Beltrán-Bobadilla, P.; Beltrán-Bobadilla, P.; Carrillo-Osuna, A.; Rodriguez-Valverde, J.A.; Acevedo-Juárez, B.; De Los Santos, I.M.; Sánchez-Rodriguez, F.J.; Courel, M.; Carrillo-Osuna, A.; Rodriguez-Valverde, J.A.; et al. SnSe Solar Cells: Current Results and Perspectives. *Gen. Chem.* **2021**, *7*, 200012. [CrossRef]

34. Patel, H.; Patel, K.; Patel, A.; Jagani, H.; Patel, K.D.; Solanki, G.K.; Pathak, V.M. Temperature-Dependent I–V Characteristics of In/p-SnSe Schottky Diode. *J. Electron. Mater.* **2021**, *50*, 5217–5225. [CrossRef]
35. Pandit, B.; Jadhav, C.D.; Chavan, P.G.; Tarkas, H.S.; Sali, J.V.; Gupta, R.B.; Sankapal, B.R. Two-Dimensional Hexagonal SnSe Nanosheets as Binder-Free Electrode Material for High-Performance Supercapacitors. *IEEE Trans. Power Electron.* **2020**, *35*, 11344–11351. [CrossRef]
36. Kasiyan, V.; Dashevsky, Z.; Schwarz, C.M.; Shatkhin, M.; Flitsiyan, E.; Chernyak, L.; Khokhlov, D. Infrared detectors based on semiconductor p-n junction of PbSe. *J. Appl. Phys.* **2012**, *112*, 086101. [CrossRef]
37. Sierra, C.; Torquemada, M.C.; Vergara, G.; Rodrigo, M.T.; Gutiérrez, C.; Pérez, G.; Génova, I.; Catalán, I.; Gómez, L.J.; Villamayor, V.; et al. Multicolour PbSe sensors for analytical applications. *Sens. Actuators B Chem.* **2014**, *190*, 464–471. [CrossRef]
38. Li, M.; Luo, J.; Fu, C.; Kan, H.; Huang, Z.; Huang, W.; Yang, S.; Zhang, J.; Tang, J.; Fu, Y.; et al. PbSe quantum dots-based chemiresistors for room-temperature NO_2 detection. *Sens. Actuators B Chem.* **2018**, *256*, 1045–1056. [CrossRef]
39. Jiang, J.; Cheng, R.; Yin, L.; Wen, Y.; Wang, H.; Zhai, B.; Liu, C.; Shan, C.; He, J. Van der Waals epitaxial growth of two-dimensional PbSe and its high-performance heterostructure devices. *Sci. Bull.* **2022**, *67*, 1659–1668. [CrossRef]
40. Díaz-Torres, E.; Flores-Conde, A.; Ávila-García, A.; Ortega-López, M. Electronic transport study of PbSe pellets prepared from self-assembled 2D-PbSe nanostructures. *Curr. Appl. Phys.* **2018**, *18*, 226–230. [CrossRef]
41. Díaz-Torres, E.; Ortega-López, M.; Matsumoto, Y.; Santoyo-Salazar, J. Simple synthesis of PbSe nanocrystals and their self-assembly into 2D 'flakes' and 1D 'ribbons' structures. *Mater. Res. Bull.* **2016**, *80*, 96–101. [CrossRef]
42. Geuchies, J.J.; Van Overbeek, C.; Evers, W.H.; Goris, B.; De Backer, A.; Gantapara, A.P.; Rabouw, F.T.; Hilhorst, J.; Peters, J.L.; Konovalov, O. In Situ study of the formation mechanism of two-dimensional superlattices from PbSe nanocrystals. *Nat. Mater.* **2016**, *15*, 1248–1254. [CrossRef] [PubMed]
43. Singh, A.K.; Hennig, R. Computational prediction of two-dimensional group-IV mono-chalcogenides. *Appl. Phys. Lett.* **2014**, *105*, 042103. [CrossRef]
44. Xiong, F.; Zhang, X.; Lin, Z.; Chen, Y. Ferroelectric engineering of two-dimensional group-IV monochalcogenides: The effects of alloying and strain. *J. Mater.* **2018**, *4*, 139–143. [CrossRef]
45. Zhang, J.; Pang, J.; Chen, H.; Wei, G.; Yan, J. Theoretical Study and Application of Doped 2D PbSe for Toxic Gases. *Phys. Status Solidi* **2023**, *260*, 2200250. [CrossRef]
46. Zhang, J.; Pang, J.; Chen, H.; Wei, G.; Wei, S.; Yan, J.; Jin, S. Study on SO_2 and Cl_2 sensor application of 2D PbSe based on first principles calculations. *RCS Adv.* **2022**, *12*, 8530–8535. [CrossRef]
47. Wang, Y.; Ma, S.; Wang, L.; Jiao, Z. A novel highly selective and sensitive NH_3 gas sensor based on monolayer Hf_2CO_2. *Appl. Surf. Sci.* **2019**, *492*, 116–124. [CrossRef]
48. Li, S.-S.; Li, X.-H.; Cui, X.-H.; Zhang, R.-Z.; Cui, H.-L.J. Effect of the biaxial strain on the electronic structure, quantum capacitance of NH_3 adsorption on pristine Hf_2CO_2 MXene using first-principles calculations. *Appl. Surf. Sci.* **2022**, *575*, 151659. [CrossRef]
49. Ma, S.; Yuan, D.; Wang, Y.; Jiao, Z. Monolayer GeS as a potential candidate for NO_2 gas sensors and capturers. *J. Mater. Chem. C* **2018**, *6*, 8082–8091. [CrossRef]
50. Kresse, G.; Hafner, J. Ab initio molecular dynamics for liquid metals. *Phys. Rev. B* **1993**, *47*, 558. [CrossRef]
51. Kresse, G.; Furthmüller, J. Efficient iterative schemes for ab initio total-energy calculations using a plane-wave basis set. *Phys. Rev. B* **1996**, *54*, 11169. [CrossRef]
52. Wu, Z.; Cohen, R.E. A More Accurate Generalized Gradient Approximation for Solids. *Phys. Rev. B* **2006**, *73*, 235116. [CrossRef]
53. Perdew, J.P.; Burke, K.; Ernzerhof, M. Generalized Gradient Approximation Made Simple. *Phys. Rev. Lett.* **1996**, *77*, 3865–3868. [CrossRef]
54. Grimme, S. Semiempirical GGA-type density functional constructed with a long-range dispersion correction. *J. Comput. Chem.* **2006**, *27*, 1787–1799. [CrossRef]
55. Monkhorst, H.J.; Pack, J.D. Special points for Brillouin-zone integrations. *Phys. Rev. B* **1976**, *13*, 5188–5192. [CrossRef]
56. Mao, Y.; Wu, R.; Ding, D.; He, F. Tunable optoelectronic properties of two-dimensional PbSe by strain: First-principles study. *Comput. Mater. Sci.* **2022**, *202*, 110957. [CrossRef]
57. Li, X.-B.; Guo, P.; Zhang, Y.-N.; Peng, R.-F.; Zhang, H.; Liu, L.-M. High carrier mobility of few-layer PbX (X = S, Se, Te). *J. Mater. Chem. C* **2015**, *3*, 6284–6290. [CrossRef]

Disclaimer/Publisher's Note: The statements, opinions and data contained in all publications are solely those of the individual author(s) and contributor(s) and not of MDPI and/or the editor(s). MDPI and/or the editor(s) disclaim responsibility for any injury to people or property resulting from any ideas, methods, instructions or products referred to in the content.

MDPI
St. Alban-Anlage 66
4052 Basel
Switzerland
www.mdpi.com

Nanomaterials Editorial Office
E-mail: nanomaterials@mdpi.com
www.mdpi.com/journal/nanomaterials

Disclaimer/Publisher's Note: The statements, opinions and data contained in all publications are solely those of the individual author(s) and contributor(s) and not of MDPI and/or the editor(s). MDPI and/or the editor(s) disclaim responsibility for any injury to people or property resulting from any ideas, methods, instructions or products referred to in the content.

www.ingramcontent.com/pod-product-compliance
Lightning Source LLC
LaVergne TN
LVHW070741100526
838202LV00013B/1281